W. H Ballard, A. C McKay, R. A Thompson

The High School Arithmetic

W. H Ballard, A. C McKay, R. A Thompson

The High School Arithmetic

ISBN/EAN: 9783337158873

Printed in Europe, USA, Canada, Australia, Japan

Cover: Foto ©Paul-Georg Meister /pixelio.de

More available books at **www.hansebooks.com**

THE

HIGH SCHOOL ARITHMETIC.

FOR USE IN

High Schools, Collegiate Institutes

AND

SENIOR FORMS OF PUBLIC SCHOOLS.

BY

W. H. BALLARD, M.A.,
Inspector of Schools, Hamilton.

A. C. McKAY, B.A.,
Professor of Mathematics and Physics, McMaster University, Toronto.

AND

R. A. THOMPSON, B.A.,
Mathematical Master, Collegiate Institute, Hamilton.

AUTHORIZED BY THE DEPARTMENT OF EDUCATION.

Toronto:
ROSE PUBLISHING COMPANY (LTD.).
1891.

PREFACE.

THIS Arithmetic has been prepared for the use of pupils in High Schools and Collegiate Institutes, and also for those in the higher forms of Public Schools. It is intended to be not merely a collection of problems, but also a series of graded questions in which the whole theory of arithmetic is developed.

The work is largely a transcript of notes on which we have based our practical teaching for some years. Many of the original problems and methods of solution will be recognized by our former students.

A Glossary of Terms and Tables has been introduced which, though not lengthy, will be found to include all necessary reference for ordinary arithmetical teaching. The table of Powers will be of use in connection with the chapters on Compound Interest and Annuities; it will also apprise pupils of the existence of tables for facilitating the longer computations.

The treatment of Vulgar and Decimal Fractions introduces the pupil to a series of propositions of close mathematical reasoning, with the results of which he is already more or less familiar.

The Roots of Numbers and Mensuration are presented in a novel manner, as far as text books are concerned; the methods here developed have been successfully tested in our classes.

We have thought it advisable to introduce a special collection of simple problems on the Metric System.

The four hundred problems on Fractions and the Elementary Rules appear as a recognition of the necessity felt by teachers of reviewing the earlier work. A number of these have been selected from the Entrance Examination papers.

The Problems Arising from Business Transactions, as well as those on Roots of Numbers and Mensuration, have been arranged in a progressive series which presents a logical development of the principles involved.

Under General Problems we have included one thousand examples arranged in two-page sets of approximate order of difficulty. These will furnish material for review of all previous sections.

We give a few pages of problems illustrating some Properties of Numbers, which will be found a suitable supplement to those given on pages 129-139.

In the Selected University Problems we feel that we have preserved in a more convenient form a number of the best problems that have appeared on Toronto University examination papers for the past thirty years.

The sets of complete Examination Papers, in addition to presenting a number of carefully prepared problems, will serve to acquaint the student with examination standards.

W. H. BALLARD.
A. C. McKAY.
R. A. THOMPSON.

CONTENTS.

ARITHMETIC.

GLOSSARY—Terms and Tables.

AGENT.—A person authorized to transact business for another.

ANNUITY.—A periodical payment made annually or at more frequent intervals, for a fixed term of years or during the life-time of some person.

ASSETS.—The property of a debtor available for the satisfaction of his creditors.

AVERAGE.—The average of several quantities is that quantity which, when repeated as many times as the number of the quantities, will give an aggregate equal to the aggregate of these several quantities.

BANK DISCOUNT.—The amount deducted from the face-value of a note when sold to a bank. It is reckoned as simple interest on the face-value for the time before the note falls due.

BILL OF EXCHANGE.—A written order directing some person (or a bank), to pay to some other person a stated sum of money.

BONDS.—Written or printed contracts, made under seal, promising to pay a certain sum of money at a specified time. They are issued for money borrowed by Governments, Cities, Towns and other Corporations. They usually bear interest which is paid at certain stated times.

BROKER.—A man who buys and sells stocks for others.

BROKERAGE.—Payment made to a broker for buying or selling stocks; it is usually calculated as a percentage on the amount of the stock.

CHORD OF A CIRCLE.—A straight line joining two points on the circumference.

COMMISSION.—An allowance or payment made to a commission merchant or agent for the transaction of business; it is usually calculated as a percentage on the money actually involved in the transaction.

COMPOUND INTEREST.—Interest not only on the principal but on the interest added to the principal when it becomes due.

Banks usually add the interest to the principal every six months.

CONE.—A solid whose base is a circle and whose curved surface tapers uniformly to a point called the vertex.

CONSIGNEE.—The person (or firm) to whom goods are sent.

CONSIGNMENT.—A shipment of goods sent to a commission merchant to be sold.

CONSIGNOR.—The person (or firm) who sends goods to another.

CONSOLS.—An abbreviation for *consolidated annuities*, the funded debt of Great Britain.

CORPORATION.—An association of persons for the transaction of business: its legal rights are in many respects the same as those of a single person.

COUPON.—An interest certificate attached to a bond; to be detached and presented for payment when the interest is due.

COURSE OF EXCHANGE.—The rate connecting the units of value of two countries, according to which bills of exchange for the time being are negotiated.

CREDITOR.—A person (or firm) to whom money is due.

CUBE ROOT.—One of the three equal factors of a number.

CUSTOMS.—Taxes or duties imposed on imported or exported goods.

An *ad valorem* duty is calculated as a certain percentage of the value of the goods as shown in the invoice.

A *specific* duty is assessed upon the number, weight or measure of the goods, without regard to value.

DAYS OF GRACE.—Three *days of grace* are allowed by law after a note is nominally due (that is, after the expiration of the time named in the note for its payment,) before it is legally due.

DEBENTURE.—A deed or contract by which certain property is charged with the repayment of money lent, together with interest at a fixed rate. Debentures usually have interest coupons attached.

DEBTOR.—A person (or firm) owing money to another.

DEFERRED ANNUITY.—An annuity that does not begin until some future time.

DISCOUNT.—An abatement or reduction from a stated price or value.

DIVIDEND.—The sum divided among the stockholders of a company as the profits of the business.

DRAFT.—An order directing some person (or a bank) to pay a specified sum of money to some other person.

DUTY.—See Customs above.

ENDOWMENT POLICY.—An insurance policy that secures to the person insured a certain sum of money at a specified time, or to his heirs if he die before that time.

EQUATED TIME.—The time at which several sums of money due at different times may be paid.

EQUATION OF PAYMENTS.—The process of finding the equated time.

EVOLUTION.—The process of finding roots of numbers.

EXCHANGE.—The system by which persons living in different countries, or persons living in distant parts of the same country, discharge their debts to each other.

FACTOR.—The factors of a number are a set of numbers whose product is the given number.

FRUSTUM.—The portion of a cone or pyramid included between its base and a plane parallel to the base.

INSTALMENT.—A payment made annually or at some other stated period.

INSURANCE.—A guarantee of payment of a specified sum of money in the event of the loss of property by fire, shipwreck, etc., or loss of life.

INTEREST.—The sum paid for the use of money.

LIABILITIES.—Debts.

NET PROCEEDS.—The amount that remains of the money received for property after paying the expenses incurred in disposing of it.

NET PRICE.—The price of goods after the trade discount has been deducted from the marked price.

NOTE.—A written promise to pay a specified sum of money at a stated time and place.

PARALLELOPIPED.—A solid having six faces, each face being a parallelogram. If the faces are all rectangles, the solid is called a rectangular parallelopiped.

PARTIAL PAYMENT.—Part payment of a note, bond or other obligation.

PAR OF EXCHANGE.—The rate connecting the unit of value of one country with the unit of value of another country, the intrinsic value only of the units being considered.

PARTNERSHIP.—The association of two or more persons with joint capital, for the carrying on of some particular business.

PAR VALUE.—The par value of a stock certificate or bond is the original value or the amount stated in the certificate.

PERCENTAGE.—The result obtained by taking a certain number of hundredths of a given quantity.

PERPETUAL ANNUITY, OR PERPETUITY.—An annuity that continues forever.

PLINTH.—A rectangular parallelopiped having two opposite faces square.

POLICY.—The written agreement or contract guaranteeing insurance.

POLL TAX.—A tax levied equally on all taxable persons.

POLYGON.—A plane figure bounded by straight lines.

POLYHEDRON.—A solid bounded by planes.

POWER.—When a product consists of the same factor repeated any number of times it is called a *power* of that factor.

PREFERENCE STOCK.—That part of the stock of a company on which a certain percentage must be paid before any dividend can be declared on the ordinary stock.

PREMIUM.—The sum paid for insurance. Also the excess of market value above par value.

PRESENT WORTH.—The present value of a sum of money due at some future time.

PRINCIPAL.—The sum of money for the use of which interest is paid.

PRISM.—A polyhedron of which two opposite faces are parallel polygons, connected by plane faces at right angles to their parallel faces.

PYRAMID.—A polyhedron of which one face is a polygon, and the other faces are triangles having a common vertex.

RECTANGLE.—A right-angled parallelogram.

RIGHT-ANGLED TRIANGLE.—A triangle, one of the angles of which is a right angle. The side opposite the right angle is called the *hypothenuse*.

ROOT.—A root of a number is one of the equal factors of a number.

SECTOR OF A CIRCLE.—A plane figure bounded by two radii of a circle and the part of the circumference intercepted by these radii.

SHARE.—One of the equal parts into which the capital of a company is divided.

STOCKS.—The shares of companies and the bonds of governments and corporations.

SPHERE.—A solid bounded by a surface, every point of which is equally distant from a certain point within, called the centre.

SQUARE ROOT.—One of the two equal factors of a number is called the square root of the number.

TANGENT.—A straight line which meets a circle, but when produced does not cut it.

TRADE DISCOUNT.—A deduction from the regular or marked price of an article: it is usually expressed as such a "per cent. off."

TRAPEZIUM.—A four-sided plane figure having two sides parallel.

TRUE DISCOUNT.—The difference between the present worth of a debt and the whole debt; hence true discount is equivalent to the interest on the present worth of the debt.

WEDGE.—A triangular prism.

ZONE.—That part of the area of a circle included between two parallel chords.

TABLES OF LENGTH, WEIGHT, VOLUME, Etc.

LENGTH.

The standard unit for the measurement of length is the YARD.

1 yard = 3 feet = 36 inches.
$5\frac{1}{2}$ yards = 1 rod (pole or perch).
1760 yards = 1 mile.
22 yards = 1 chain = 100 links.
220 yards = 1 furlong.
2 yards = 1 fathom.

Also, 4 inches = 1 hand, used in measuring the height of horses.

AREA.

A square foot = 144 square inches.
A square yard = 9 " feet.
A square rod = $30\frac{1}{4}$ " yards.
An acre = 10 square chains.
= 160 " rods.
= 4840 " yards.
A square mile = 640 acres.

Also, 100 square feet = 1 square, used in measuring roofing, flooring, etc.

VOLUME.

A cubic yard = 27 cubic feet.
A cubic foot = 1728 " inches.
A cord (of firewood, etc.) = 128 cubic feet.

CAPACITY.

The standard unit for the measure of capacity is the GALLON, which is the space occupied by ten pounds of distilled water at a temperature of 62° F.

1 gallon = 4 quarts = 8 pints = 32 gills.
2 gallons = 1 peck.
8 gallons = 1 bushel.
$31\frac{1}{2}$ gallons = 1 barrel.
63 gallons = 1 hogshead.

Also, 8 bushels = 1 quarter, used in reporting the British grain market.

For Apothecaries' Fluid Measure the pint is sub-divided as follows:—

1 pint = 20 fluid ounces.
1 fluid ounce = 8 fluid drachms.
1 fluid drachm = 60 minims.

WEIGHT.

The standard unit for the measurement of weight is the POUND AVOIRDUPOIS.

1 pound = 16 ounces = 7000 grains.
100 pounds = 1 cental.
2000 pounds = 1 ton.

196 pounds of flour = 1 barrel.
200 pounds of beef or of pork = 1 barrel.
280 pounds of salt = 1 barrel.
An ounce Troy = 480 grains.
A pound Troy = 12 Troy ounces.
= 5760 grains,

The following table shows the number of pounds re quired to make a LEGAL BUSHEL of each of the substances named ;

14 lbs. of Blue Grass Seed.
34 " " Oats.
36 " " Malt.
40 " " Castor Beans.
44 " " Hemp Seed.
48 " " Barley or Buckwheat.
50 " " Flax Seed.
56 " " Indian Corn, or Rye.
60 " " Beans, Beets, Carrots, Parsnips, Peas, Potatoes, Red Clover Seed, Onions, Turnips, or Wheat.
70 " " Bituminous Coal.

VALUE.

1. Canadian and United States money.

The standard unit of value is the DOLLAR, which is the value of about $23\frac{1}{3}$ grains of pure gold.

1 dollar = 100 cents = 1000 mills.

In the United States there is the eagle (= 10 dollars), and also the dime (= 10 cents); but these denominations are not regarded in business operations, so that the table of U. S. money is practically the same as our own.

2. British money.

4 farthings = 1 penny.
12 pence = 1 shilling.
20 shillings = 1 pound (the unit).
21 shillings = 1 guinea.

The units in British and Canadian money are connected as follows :—

$73 = £15, or, $$4·86\frac{2}{3}$ = £1.

TIME.

The unit for the measurement of time is the MEAN SOLAR DAY.

1 day = 24 hours.
1 hour = 60 minutes.
1 min. = 60 seconds.
7 days = 1 week.
365 days = 1 common year.
366 days = 1 leap year.

The calendar year is divided into 12 months of different lengths, as follows :—

January,	31	days.	July,	31	days.
February,	28	"	August,	31	"
March,	31	"	September,	30	"
April,	30	"	October,	31	"
May,	31	"	November,	30	"
June,	30	"	December,	31	"

In leap year February has 29 days, making 366 days in the whole year.

To ascertain which is leap year, divide the number denoting the year by 4; if there is no remainder the number denotes leap year, while remainders 1, 2, 3, indicate respectively the 1st, 2nd and 3rd year after leap year. Thus, 1891 is the 3rd year after leap year, and 1892 and 1896 will be leap years.

If the number denoting the year ends in two ciphers it must be divisible by 400 in order to indicate leap year; thus the years 2000 and 2400 will be leap years, but not the year 1900.

ANGLES.

The circumference of a circle is divided into 360 equal parts, and each part subtends at the centre an angle called a degree, each degree is divided into 60 equal parts called minutes, and each minute into 60 seconds.

1 degree = 60 minutes = 3600 seconds.
90 degrees = 1 quadrant or right angle.

MISCELLANEOUS.

12 articles = 1 dozen.
12 dozen = 1 gross.
24 sheets of paper = 1 quire.
20 quires = 1 ream.

METRIC SYSTEM OF MEASUREMENT.

LENGTH.

The unit of length is the METRE (= 39·37 inches nearly). To express fractions and multiples of the unit the following significant prefixes are used:

micro	denoting	.... $\frac{1}{1000000}$	abbreviated	μ.
milli	"	.... $\frac{1}{1000}$	"	m.
centi	"	.... $\frac{1}{100}$	"	c.
deci	"	.... $\frac{1}{10}$	"	d.
deka	"	.... 10	"	D.
hecto	"	.... 100	"	H.
kilo	"	.... 1000	"	K.
myria	"	.... 10000	"	M.
mega	"	.... 1000000	"	Mg.

1 metre =	1000000	micrometres, or microns,	μm.
1 metre =	1000	millimetres,	mm.
1 metre =	100	centimetres,	cm.
1 metre =	10	decimetres,	dm.
10 metres =	1	dekametre	Dm.
100 metres =	1	hectometre,	Hm.
1000 metres =	1	kilometre,	Km.
10000 metres =	1	myriametre,	Mm.
1000000 metres =	1	megametre,	Mgm.

AREA

The unit of area is the ARE—equivalent to one square dekametre.

1 are =	1000	milliares,	ma.
1 are =	100	centiares,	ca.
1 are =	10	deciares,	da.
10 ares =	1	dekare,	Da.
100 ares =	1	hectare.	Ha.

VOLUME.

One unit of volume is the STERE—equivalent to one cubic metre ; it is used in measuring wood, excavations, etc.

1 stere =	100 centisteres,	cs.
1 stere =	10 decisteres,	ds.
10 steres =	1 dekastere,	Ds.
100 steres =	1 hectostere,	Hs.
1000 steres =	1 kilostere,	Ks.

Another unit of volume is the LITRE—equivalent to one cubic decimetre ; it is used for fluid measure, or measure of capacity.

1 litre =	100 centilitres,	cl.
1 litre =	10 decilitres,	dl.
10 litres =	1 dekalitre,	Dl.
100 litres =	1 hectolitre,	Hl.
1000 litres =	1 kilolitre,	Kl.
=	1 stere.	

WEIGHT.

The unit of weight (or mass) is the GRAMME—equivalent to the weight of one cubic centimetre of distilled water at its temperature of maximum density.

1 gramme =	1000 milligrammes,	mg.
1 gramme =	100 centigrammes,	cg.
1 gramme =	10 decigrammes,	dg.
10 grammes =	1 dekagramme,	Dg.
100 grammes =	1 hectogramme,	Hg.
1000 grammes =	1 kilogramme,	Kg.
10000 grammes =	1 myriagramme,	Mg.

1 quintal = 100 Kg.
1 millier or tonneau = 1000 Kg.

The following are approximate values of some of the metric units :

1 metre	= 39·37 inches	nearly.
1 centimetre	= $\frac{2}{5}$ inch,	"
1 kilometre	= $\frac{5}{8}$ mile,	"
1 hectare	= $2\frac{1}{2}$ acres,	"
1 litre	= 1 quart,	"
1 stere	= $\frac{1}{4}$ cord,	"
1 kilogramme	= $2\frac{1}{5}$ pounds,	"
1 millier or tonneau	= 2200 pounds,	"

POWERS OF CERTAIN NUMBERS.

No.	1·02	1·025	1·03	1·035	1·04	1·045	No.
1	1·02000	1·02500	1·03000	1·03500	1·04000	1·04500	1
2	1·04040	1·05063	1·06090	1·07123	1·08160	1·09203	2
3	1·06121	1·07689	1·09273	1·10872	1·12486	1·14117	3
4	1·08243	1·10381	1·12551	1·14752	1·16986	1·19252	4
5	1·10408	1·13141	1·15927	1·18769	1·21665	1·24618	5
6	1·12616	1·15969	1·19405	1·22926	1·26532	1·30226	6
7	1·14869	1·18869	1·22987	1·27228	1·31593	1·36086	7
8	1·17166	1·21840	1·26677	1·31681	1·36857	1·42210	8
9	1·19509	1·24886	1·30477	1·36290	1·42331	1·48610	9
10	1·21899	1·28008	1·34392	1·41060	1·48024	1·55297	10
11	1·24337	1·31209	1·38423	1·45997	1·53945	1·62285	11
12	1·26824	1·34489	1·42576	1·51107	1·60103	1·69588	12
13	1·29361	1·37851	1·46853	1·56396	1·66507	1·77220	13
14	1·31948	1·41297	1·51259	1·61869	1·73168	1·85194	14
15	1·34587	1·44830	1·55797	1·67535	1·80094	1·93528	15
16	1·37279	1·48151	1·60471	1·73399	1·87298	2·02237	16
17	1·40024	1·52162	1·65285	1·79468	1·94790	2·11338	17
18	1·42825	1·55966	1·70243	1·85749	2·02582	2·20848	18
19	1·45681	1·59865	1·75351	1·92250	2·10685	2·30786	19
20	1·48595	1·63862	1·80611	1·98979	2·19112	2·41171	20
21	1·51567	1·67958	1·86029	2·05943	2·27877	2·52024	21
22	1·54598	1·72157	1·91610	2·13151	2·36992	2·63365	22
23	1·57690	1·76461	1·97359	2·20611	2·46472	2·75217	23
24	1·60844	1·80873	2·03279	2·28333	2·56330	2·87601	24
25	1·64060	1·85394	2·09378	2·36324	2·66581	3·00543	25
26	1·67342	1·90029	2·15659	2·44596	2·77247	3·14068	26
27	1·70689	1·94780	2·22129	2·53157	2·88337	3·28201	27
28	1·74102	1·99650	2·28793	2·62017	2·99870	3·42970	28
29	1·77584	2·04641	2·35657	2·71188	3·11865	3·58404	19
30	1·81136	2·09757	2·42726	2·80679	3·24340	3·74532	30
31	1·84759	2·15001	2·50008	2·90503	3·37313	3·91386	31
32	1·88454	2·20376	2·57508	3·00671	3·50806	4·08998	32
33	1·92223	2·25885	2·65234	3·11194	3·64838	4·27403	33
34	1·96068	2·31532	2·73191	3·22086	3·79432	4·46636	34

POWERS OF CERTAIN NUMBERS.

No.	1·05	1·06	1·07	1·08	1·09	1·10	No.
1	1·05000	1·06000	1·07000	1·08000	1·09000	1·10000	1
2	1·10250	1·12360	1·14490	1·16640	1·18810	1·21000	2
3	1·15763	1·19102	1·22504	1·25971	1·29503	1·33100	3
4	1·21551	1·26248	1·31080	1·36049	1·41158	1·46410	4
5	1·27628	1·33823	1·40255	1·46933	1·53862	1·61051	5
6	1·34010	1·41852	1·50073	1·58687	1·67710	1·77156	6
7	1·40710	1·50363	1·60578	1·71382	1·82804	1·94872	7
8	1·47746	1·59385	1·71819	1·85093	1·99256	2·14359	8
9	1·55133	1·68948	1·83846	1·99900	2·17189	2·35795	9
10	1·62889	1·79085	1·96715	2·15893	2·36736	2·59374	10
11	1·71034	1·89830	2·10485	2·33164	2·58043	2·85312	11
12	1·79586	2·01220	2·25219	2·51817	2·81266	3·13843	12
13	1·88565	2·13293	2·40985	2·71962	3·06580	3·45227	13
14	1·97993	2·26090	2·57853	2·93719	3·34173	3·79750	14
15	2·07893	2·39656	2·75903	3·17217	3·64248	4·17725	15
16	2·18287	2·54035	2·95216	3·42594	3·97031	4·59497	16
17	2·29202	2·69277	3·15882	3·70002	4·32763	5·05447	17
18	2·40662	2·85434	3·37993	3·99602	4·71712	5·55992	18
19	2·52695	3·02560	3·61653	4·31570	5·14166	6·11591	19
20	2·65330	3·20714	3·86968	4·66096	5·60441	6·72750	20
21	2·78596	3·39956	4·14056	5·03383	6·10881	7·40025	21
22	2·92526	3·60354	4·43040	5·43654	6·65860	8·14027	22
23	3·07152	3·81975	4·74053	5·87146	7·25787	8·95430	23
24	3·22510	4·04893	5·07237	6·34118	7·91108	9·84973	24
25	3·38635	4·29187	5·42743	6·84848	8·62308	10·83471	25
26	3·55567	4·54938	5·80735	7·39635	9·39916	11·91818	26
27	3·73346	4·82235	6·21387	7·98806	10·24508	13·10999	27
28	3·92013	5·11169	6·64884	8·62711	11·16714	14·42099	28
29	4·11614	5·41839	7·11426	9·31727	12·17218	15·86309	29
30	4·32194	5·74349	7·61226	10·06266	13·26768	17.44940	30
31	4·53801	6·08810	8·14511	10·86767	14·46177	19·19434	31
32	4·76494	6·45339	8·71527	11·73708	15·76333	21·11378	32
33	5·00319	6·84059	9·32534	12·67605	17·18203	23·22515	33
34	5·25335	7·25103	9·97811	13·69013	18·72841	25·54767	34

VULGAR FRACTIONS.

The numbers which have so far been considered begin with 1, and go on increasing by 1, each number being greater by 1 than the number just before it. Such numbers are called whole numbers or integers.

Besides these there are other numbers, some of them, such as *one-half*, *three-quarters*, being less than 1, and others such as *two and a half*, *three and a quarter*, being greater than one whole number and less than the next higher whole number.

These numbers are called FRACTIONS, those less than 1 being called PROPER FRACTIONS, and those greater than 1 IMPROPER FRACTIONS.

NOTE.—It will be seen hereafter that whole numbers may also be regarded as fractions, and as such are classed as improper fractions.

PROPER FRACTIONS.

If a unit be divided into two equal parts, each of these parts is called a *half*; if into three equal parts each is called a *third*; if into four equal parts, a *fourth*; and so on, the part in each case taking its name from the number of these parts required to make up the whole unit.

And we *speak* of any number of these parts in the same way that we do of a number of feet, pounds or dollars; but the notation employed when we come to *write* them is very different. Thus for three dollars we write $3; for four pounds, we write 4 lbs.; where the character or symbol denoting the *kind* of thing under consideration is placed to the right or left (and sometimes directly above) that denoting the *number* of things; but for three-fourths we write $\frac{3}{4}$; the 4 in this expression, which tells the kind of thing under consideration, namely, *fourths*, being placed directly below the 3, which tells the number of fourths.

This symbol $\frac{3}{4}$ is called a *Fraction*, and indicates that a unit of some kind has been divided into 4 equal parts and that 3 of these parts are under consideration.

The 4 which tells the kind or *denomination* of the parts (and therefore also the number of parts into which the unit has been divided), is called the *denominator*. The 3, which indicates the number of parts taken (i. e., *enumerates* them) is called the *numerator*. The numerator and denominator are called the *Terms* of a fraction.

If the numerator of a fraction be a smaller number than the denominator it is evident that the number of parts taken is less than the number of parts into which the unit was divided, and therefore that the fraction denotes a number less than the unit or 1. Such fractions are accordingly called *Proper Fractions*.

IMPROPER FRACTIONS.

If we take four fourths or five fifths of any unit it is clear that we take the whole of that unit, and therefore every such fraction as $\frac{4}{4}$, $\frac{5}{5}$ or $\frac{6}{6}$ is equal to 1.

The word fraction (from *fractus*, broken) strictly means a part broken off, and consequently denotes something less than the whole. Such expressions therefore as $\frac{4}{4}$, not indicating a *part* of a unit are called *Improper Fractions*.

If now we suppose two or more units to be divided each into five equal parts, we may consider as many of these parts as we please, and thus we may have such fractions as $\frac{7}{5}$, $\frac{13}{5}$, etc., where the numerator is greater than the denominator. These, for the reason given above, are also called improper fractions. Proper and improper fractions are called *Simple Fractions*.

MIXED NUMBERS.

Let us consider the improper fraction $\frac{13}{5}$; we know that five fifths are equivalent to 1, and that ten fifths are therefore equal to 2, and consequently that the fraction $\frac{13}{5}$ is equivalent to 2 and $\frac{3}{5}$, we thus see that every improper fraction is equivalent to a whole number, or else to a whole number and a proper fraction, and that the whole number can be obtained by finding how often the denominator of the fraction is contained in the numerator, and also that the remainder, after dividing (if any) will be the numerator of the proper fraction.

Such numbers are expressed by writing the whole number first with the fraction close after it; thus, *two and three-fifths* would be written $2\frac{3}{5}$. Such numbers are called *Mixed Numbers.* Thus we see that every improper fraction may be expressed either as a whole or as a mixed number.

EXERCISE.

Express the following improper fractions as whole or mixed numbers:

(1) $\frac{13}{4}$. (2) $\frac{27}{3}$. (3) $\frac{47}{8}$. (4) $\frac{62}{7}$.
(5) $\frac{327}{12}$. (6) $\frac{99}{11}$. (7) $\frac{768}{21}$. (8) $\frac{3482}{43}$.
(9) $\frac{43875}{963}$. (10) $\frac{10000}{909}$. (11) $\frac{10101}{101}$. (12) $\frac{101101}{101}$.
(13) $\frac{834763}{987}$. (14) $\frac{384367}{8976}$. (15) $\frac{763843}{652732}$.

Conversely every mixed number may be expressed as an improper fraction. For if we consider the number $3\frac{4}{5}$, then, since 1 is equal to five fifths, therefore, 3 is equal to fifteen fifths, and therefore $3\frac{4}{5}$ is equivalent to fifteen fifths and four fifths, that is, to nineteen-fifths, which we express thus, $\frac{19}{5}$. We have, then, the following

RULE.

To reduce a mixed number to an improper fraction:

Multiply the whole number by the denominator and to this product add the numerator; this will give the numerator of the improper fraction, while its denominator is the same as that of the fraction in the mixed number.

EXERCISE.

Reduce the following mixed numbers to improper fractions:

(1) $3\frac{1}{2}$. (8) $23\frac{18}{19}$. (15) $10\frac{834760}{916524}$.
(2) $5\frac{1}{4}$. (9) $3476\frac{1}{9}$. (16) $8697\frac{4324}{6547}$.
(3) $6\frac{7}{8}$. (10) $969\frac{19}{23}$. (17) $9009\frac{7007}{8008}$.
(4) $11\frac{11}{12}$. (11) $2\frac{2345}{6789}$. (18) $10001\frac{99}{100}$.
(5) $17\frac{5}{7}$. (12) $576\frac{576}{577}$. (19) $1000\frac{347}{1000}$.
(6) $57\frac{9}{10}$. (13) $1010\frac{101}{102}$. (20) $8736\frac{87}{100}$.
(7) $14\frac{1}{15}$. (14) $1\frac{7613}{8357}$. (21) $9342\frac{856}{1000}$.

COMPOUND FRACTIONS.

It is evident that if 3 fourths be repeated 7 times the result will be 21 fourths ; 6 sevenths repeated 5 times will be 30 sevenths ; and so on. That is,

$$\frac{3}{4} \times 7 = \frac{21}{4}\ ;\ \frac{6}{7} \times 5 = \frac{30}{7}.$$

Hence we have the following

RULE.

To multiply a fraction by a whole number : Multiply the numerator of the fraction by the whole number for the numerator of the product and retain the same denominator.

EXERCISE.

(1) Multiply $\frac{6}{7}$ by 4, 6, 8, 10 and 16.
(2) Multiply $\frac{17}{19}$ by 9, 13, 23 and 47.
(3) Multiply $\frac{100}{137}$ by 49, 93 and 256.

If any unit be divided into 4 equal parts and each of these fourths be divided into 5 equal parts, it is evident that the unit will be divided into 20 equal parts. From this it follows that the fifth part of a fourth is a twentieth ; or, as it is written :

$$\frac{1}{5} \text{ of } \frac{1}{4} = \frac{1}{20}$$

and since 1 fourth is equal to 5 twentieths,
Therefore 3 fourths are equal to 15 twentieths,
And 1 fifth of 3 fourths is equal to 1 fifth of 15 twentieths,
And therefore equal to 3 twentieths : or, as it is written :

$$\frac{1}{5} \text{ of } \frac{3}{4} = \frac{3}{20}.$$

It is clear also that 2 fifths of 3 fourths is twice as much as 1 fifth of 3 fourths, and is therefore equal to 6 twentieths, that is :

$$\frac{2}{5} \text{ of } \frac{3}{4} = \frac{6}{20}.$$

From this it appears that a fraction of a fraction may be reduced to a simple fraction by multiplying the two numerators together for a new numerator, and the two denominators together for a new denominator.

EXERCISE.

Reduce the following to simple fractions :

(1) $\frac{1}{2}$ of $\frac{1}{3}$. (2) $\frac{1}{3}$ of $\frac{2}{3}$. (3) $\frac{1}{8}$ of $\frac{5}{6}$.
(4) $\frac{3}{7}$ of $\frac{6}{31}$. (5) $\frac{11}{12}$ of $\frac{23}{24}$. (6) $\frac{17}{19}$ of $\frac{47}{48}$.

Again, since $\frac{2}{5}$ of $\frac{3}{4} = \frac{6}{20}$

Therefore $\frac{3}{7}$ of $\frac{2}{5}$ of $\frac{3}{4} = \frac{3}{7}$ of $\frac{6}{20} = \frac{18}{140} = \frac{3 \times 2 \times 3}{7 \times 5 \times 4}$.

Hence it appears that if *any* number of fractions are connected by the word 'of' we may reduce the whole expression to a simple fraction by multiplying all the numerators together for a new numerator, and all the denominators together for a new denominator.

Such expressions as $\frac{2}{5}$ of $\frac{3}{4}$, $\frac{3}{7}$ of $\frac{2}{5}$ of $\frac{3}{4}$, are called COMPOUND FRACTIONS.

EXERCISE.

Reduce the following compound fractions to simple ones :

(1) $\frac{1}{2}$ of $\frac{1}{3}$ of $\frac{1}{4}$ (4) $\frac{6}{7}$ of $\frac{8}{11}$ of $\frac{15}{19}$.
(2) $\frac{2}{3}$ of $\frac{4}{5}$ of $\frac{7}{9}$. (5) $\frac{17}{100}$ of $\frac{11}{200}$ of $\frac{9}{1000}$.
(3) $\frac{1}{2}$ of $\frac{1}{3}$ of $\frac{1}{4}$ of $\frac{1}{5}$ of, $\frac{1}{6}$. (6) $\frac{10}{19}$ of $\frac{100}{17}$ of $\frac{16}{3}$.

NOTE.—Mixed numbers must be reduced to improper fractions before applying the rule.

(7) $\frac{2}{3}$ of $5\frac{1}{3}$. (9) $\frac{11}{12}$ of $19\frac{3}{17}$ of $100\frac{9}{10}$.
(8) $\frac{8}{9}$ of $7\frac{1}{3}$ of $4\frac{1}{3}$. (10) $\frac{11}{3}$ of $1\frac{999}{1000}$ of $1\frac{1}{999}$.

Since one-fifth of $\frac{3}{4}$ is $\frac{3}{20}$, and since one-fifth of anything is the result obtained after dividing it by 5, it follows that $\frac{3}{4}$ on being divided by 5 gives $\frac{3}{20}$ for quotient, or

$$\frac{3}{4} \div 5 = \frac{3}{20}.$$

Hence we have the following

RULE.

To divide a fraction by a whole number: Multiply the denominator of the fraction by the whole number for a new denominator and retain the same numerator.

EXERCISE.

1. Divide $\frac{5}{7}$ by 2, 3, 4 and 8.
2. Divide $\frac{7}{18}$ by 8 and by 10.
3. Divide $\frac{144}{375}$ by 125.

If the fraction $\frac{3}{4}$ be multiplied by 5 the result will be $\frac{15}{4}$, and if $\frac{15}{4}$ be divided by 5 the result will be $\frac{15}{20}$; but it is evident that this operation must give us the quantity we started with. It follows, therefore, that $\frac{15}{20}$ and $\frac{3}{4}$ must be equal to one another. Hence, IF BOTH TERMS OF A FRACTION BE MULTIPLIED BY THE SAME NUMBER, THE VALUE OF THE FRACTION REMAINS THE SAME.

Thus, $\frac{3}{7}=\frac{6}{14}=\frac{9}{21}=\frac{21}{49}$, &c.

It will thus appear that we can change a fraction into another whose denominator is any multiple of its denominator by multiplying both terms of the fraction by the number of times the proposed denominator contains the given denominator.

Thus, to reduce $\frac{5}{7}$ to a fraction whose denominator is 56, multiply both terms of the fraction by 8 (the number of times 7 is contained in 56), and the result is $\frac{40}{56}$.

EXERCISE.

1. Change $\frac{3}{4}$ to an equivalent fraction whose denominator is 12, 20, 48, 100.

2. Find the fraction with least denominator which shall be equal to $\frac{5}{7}$, to $\frac{7}{9}$, to $\frac{12}{15}$.

3. Reduce $\frac{5}{7}$ and $\frac{3}{4}$ to fractions which shall have 28 for their denominator.

4. Reduce $\frac{7}{9}$ and $\frac{6}{7}$ to fractions which shall have the same denominator.

5. Reduce $\frac{2}{3}$, $\frac{3}{4}$ and $\frac{4}{5}$ to fractions which shall have 60 for denominator.

6. Reduce $\frac{5}{6}$, $\frac{7}{8}$ to fractions having 24 for denominator.

7. Reduce $\frac{5}{9}$, $\frac{1}{6}$, $\frac{2}{3}$ to fractions having 18 for denominator.

8. Reduce $\frac{5}{6}$, $\frac{7}{9}$, $\frac{11}{12}$, $\frac{1}{2}$ and $\frac{5}{18}$ to equivalent fractions which shall have the same denominator.

Again, if we take $\frac{15}{20}$ and divide both its terms by 5 we obtain $\frac{3}{4}$, which has been shown to be equal to $\frac{15}{20}$; hence, IF BOTH TERMS OF A FRACTION BE DIVIDED BY THE SAME NUMBER, THE VALUE OF THE FRACTION REMAINS THE SAME.

Thus, $\frac{30}{42} = \frac{15}{21} = \frac{5}{7}$.

When both terms of a fraction have been thus divided by all numbers which will divide them both, the fraction is said to be reduced to its lowest terms.

Hence we have the following

RULE.

To reduce a fraction to its lowest terms: Divide both numerator and denominator of the fraction by any number that will divide them both, and continue this operation until a fraction is obtained whose terms have no common factor.

Thus, $\frac{315}{525}$ (dividing by 5) $= \frac{63}{105}$ (dividing by 3) $= \frac{21}{35} = \frac{3}{5}$.

EXERCISE.

Reduce the following fractions to their lowest terms :

1. $\frac{3}{9}$. 2. $\frac{15}{20}$. 3. $\frac{21}{28}$. 4. $\frac{18}{27}$.
5. $\frac{16}{20}$. 6. $\frac{49}{56}$. 7. $\frac{72}{90}$. 8. $\frac{33}{55}$.
9. $\frac{45}{75}$. 10. $\frac{28}{98}$. 11. $\frac{126}{189}$. 12. $\frac{147}{196}$.
13. $\frac{256}{768}$. 14. $\frac{243}{729}$. 15. $\frac{625}{1000}$. 16. $\frac{294}{378}$.

NOTE.—When common divisors can no longer be determined by inspection, the method of finding the G. C. M. of two numbers must be used.

17. $\frac{133}{171}$. 18. $\frac{187}{221}$. 19. $\frac{299}{391}$. 20. $\frac{115}{203}$.
21. $\frac{1547}{1881}$. 22. $\frac{580}{812}$. 23. $\frac{63427}{77121}$. 24. $\frac{47957}{54519}$.

In reducing compound fractions to simple ones the common factors should be struck out before applying the rule. Thus, in the following examples :

Ex. 1. Reduce $\frac{2}{3}$ of $\frac{3}{5}$ to a simple fraction.

$$\tfrac{2}{3} \text{ of } \tfrac{3}{5} = \frac{2 \times 3}{3 \times 5}$$

Now striking out 3, which is a factor of both numerator and denominator, we have the fraction reduced to $\frac{2}{5}$.

Ex. 2. Reduce $\frac{8}{9}$ of $\frac{15}{24}$ of $\frac{3}{10}$.

$$\text{Result} = \frac{8 \times 15 \times 3}{9 \times 24 \times 10}$$

Here we see that 8 occurs in the numerator and is also contained in the denominator as a factor of 24 ; and the same is true of 3 ; striking these out we have left

$$\frac{15}{9 \times 10}$$

Here again we notice that 3 occurs as a factor of 9 and also as a factor of 15, while 5 occurs as a factor of 15 and also as a factor of 10. Striking out these common factors we have left

$$\frac{1}{3 \times 2} = \frac{1}{6}$$

as the final result.

EXERCISE.

Reduce the following compound fraction to simple ones :

1. $\frac{1}{3}$ of $\frac{3}{4}$ of $\frac{4}{5}$. 2. $\frac{2}{3}$ of $\frac{6}{7}$ of $\frac{14}{15}$. 3. $\frac{7}{9}$ of $\frac{81}{147}$.
4. $\frac{12}{13}$ of $\frac{65}{19}$ of $\frac{57}{180}$.
5. $\frac{1}{11}$ of $\frac{3}{5}$ of $\frac{3}{4}$ of $3\frac{2}{3}$.
6. $\frac{5}{2}$ of $\frac{16}{7}$ of $\frac{7}{18}$ of $\frac{11}{10}$ of $\frac{27}{14}$.
7. $\frac{1}{3}$ of $\frac{9}{8}$ of $2\frac{1}{3}$ of $15\frac{2}{7}$.
8. $\frac{5}{7}$ of $\frac{33}{49}$ of $\frac{5}{4}$ of $6\frac{1}{10}$ of $\frac{14}{13}$ of $2\frac{4}{11}$.
9. $\frac{2247}{1017}$ of $\frac{774}{615}$ of $\frac{1017}{903}$ of $\frac{565}{1926}$.
10. $\frac{13}{17}$ of $\frac{19}{23}$ of $\frac{93}{155}$ of $\frac{49}{133}$ of $1\frac{8}{26}$ of $1\frac{10}{105}$ of $1\frac{7}{138}$ of $1\frac{11}{58}$.

To show that $\frac{1}{5}$ of $2 = \frac{2}{5}$.

If each of two units be divided into 5 equal parts each of these parts will be one-fifth, and there will be 10 of them. If now we separate these 10 fifths into groups of 2 each there will be 5 of these groups. We shall have thus divided 2 into 5 equal parts, and we find that each part consists of 2 fifths ; that is

$$\tfrac{1}{5} \text{ of } 2 = \tfrac{2}{5}.$$

Also, since the fifth part of any number is obtained by dividing that number by 5, it follows that

$$\tfrac{2}{5} = \tfrac{1}{5} \text{ of } 2 = 2 \div 5.$$

Therefore EVERY FRACTION EXPRESSES THE QUOTIENT OF THE NUMERATOR BY THE DENOMINATOR.

It will also be seen from this that every whole number may be expressed as a fraction having 1 for its denominator; thus,

$$5 = \frac{5}{1}, \text{ for } \frac{5}{1} = 5 \div 1 = 5.$$

Therefore also a whole number may be expressed as a fraction having any proposed denominator. Thus, to express 7 as a fraction whose denominator is 5, we have

$$7 = \frac{7}{1} = \frac{35}{5}.$$

EXERCISE.

1. Express 9 as a fraction whose denominator is 6, 7, 10, 100.

2. Express 39 as a fraction whose denominator is 14, 19, 23.

TO REDUCE FRACTIONS TO OTHERS HAVING A COMMON DENOMINATOR.

Suppose we have the fractions

$$\frac{2}{3}, \frac{3}{5} \text{ and } \frac{2}{7},$$

and that we are required to reduce them to other fractions which shall have a common denominator. It is evident that this denominator must contain 3, 5 and 7, that is, must be a multiple of 3, 5 and 7, such as 105, 210, 315, etc., and therefore the required fractions would be

$$\frac{70}{105}, \frac{63}{105}, \frac{30}{105},$$

$$\text{or } \frac{140}{210}, \frac{126}{210}, \frac{60}{210},$$

$$\text{or } \frac{210}{315}, \frac{189}{315}, \frac{90}{315},$$

and so on.

In reducing fractions to a common denominator, however, it is most convenient to select the least denominator that will contain all the given denominators, that is, to take the least common multiple of all the denominators. Hence we have the following

RULE.

To reduce fractions to equivalent fractions, which shall have a common denominator:

1. Find the l. c. m. of all the denominators.
2. Multiply both terms of each fraction by the number of times its denominator is contained in this l. c. m.

Ex. To reduce

$$\tfrac{5}{8}, \tfrac{1}{6}, \tfrac{7}{12}, \tfrac{5}{9},$$

to a common denominator.

The l. c. m. of 8, 6, 12, and 9 is 72.

The first denominator 8 is contained in 72 the l. c. m. 9 times, therefore the first fraction $\tfrac{5}{8}$ becomes

$$\frac{5\times 9}{8\times 9}=\frac{45}{72}.$$

Similarly, the second, $\tfrac{1}{6}$ becomes $\tfrac{12}{72}$
the third $\tfrac{7}{12}$ " $\tfrac{42}{72}$
and the fourth " $\tfrac{40}{72}$

The resulting fractions are, therefore,

$$\tfrac{45}{72}, \tfrac{12}{72}, \tfrac{42}{72}, \text{ and } \tfrac{40}{72}$$

EXERCISE.

Reduce the following fractions to equivalent ones with the least common denominator:

NOTE.—If the fractions are not in their lowest terms, they should be reduced to their lowest terms before applying the rule.

1. $\tfrac{7}{9}, \tfrac{8}{15}, \tfrac{1}{6}$.
2. $\tfrac{2}{3}, \tfrac{3}{4}, \tfrac{4}{5}, \tfrac{5}{6}$.
3. $\tfrac{14}{15}, \tfrac{7}{18}, \tfrac{23}{24}, \tfrac{19}{21}$.
4. $\tfrac{1}{10}, \tfrac{1}{100}, \tfrac{1}{1000}, \tfrac{1}{10000}$.
5. $\tfrac{7}{39}, \tfrac{16}{21}, \tfrac{29}{91}$.
6. $\tfrac{5}{7}, \tfrac{30}{119}, \tfrac{1}{51}, \tfrac{2}{3}, \tfrac{1}{21}$.
7. $\tfrac{7}{28}, \tfrac{5}{46}, \tfrac{19}{24}, \tfrac{13}{65}, \tfrac{68}{69}$.
8. $\tfrac{1}{2}$ of $\tfrac{3}{4}$, $6\tfrac{7}{8}$, $\tfrac{5}{9}$ of $7\tfrac{1}{5}$.
9. $\tfrac{7}{13}$ of $\tfrac{4}{5}$, $11\tfrac{1}{13}$, $\tfrac{1}{2}$ of $\tfrac{18}{7}$ of $3\tfrac{8}{9}$.
10. $\tfrac{1}{3}$ of $\tfrac{1}{5}$ of $\tfrac{1}{7}$, $\tfrac{1}{5}$ of $\tfrac{1}{7}$ of $\tfrac{1}{11}$, $\tfrac{1}{3}$ of $\tfrac{1}{11}$ of $\tfrac{1}{13}$.

TO COMPARE FRACTIONS IN MAGNITUDE:

Suppose we wish to know which is the greater of the two fractions, $\tfrac{4}{5}$ and $\tfrac{5}{7}$.

This cannot readily be done so long as they have different denominators, for we cannot at once say whether 4 out of 5 equal parts are greater or less than 5 out of 7 equal parts. But if we change them for their equivalent fractions with a common denominator, they become respectively :

$$\frac{28}{35} \text{ and } \frac{25}{35}.$$

and we can at once say that 28 out of 35 equal parts are greater than 25 of these parts, and consequently we know that $\frac{4}{5}$ is greater than $\frac{5}{7}$.

EXERCISE.

1. Which is the greater $\frac{6}{7}$ or $\frac{7}{8}$?
2. Find the least of the fractions $\frac{5}{6}$, $\frac{7}{9}$ and $\frac{14}{17}$.
3. Arrange $\frac{1}{2}$, $\frac{1}{3}$, $\frac{1}{4}$ in order of magnitude, beginning with the greatest.
4. " $\frac{2}{3}$, $\frac{7}{12}$, $\frac{17}{24}$ in order.
5. " $\frac{1}{8}$ of $\frac{8}{3}$, $\frac{1}{12}$ and $\frac{4}{7}$ of 2.
6. " $\frac{7}{12}$, $\frac{11}{21}$, $\frac{5}{16}$ and $\frac{29}{60}$.
7. " $\frac{16}{7}$, $2\frac{3}{8}$, $\frac{22}{9}$, and $2\frac{5}{11}$.
8. " $\frac{3}{8}$ of $\frac{5}{7}$ of 4, $\frac{2}{7}$ of $\frac{3}{11}$ of 7, $\frac{1}{5}$ of $\frac{1}{3}$ of $5\frac{1}{5}$ and $\frac{13}{27}$.
9. " $\frac{18}{19}$, $\frac{19}{20}$, $\frac{22}{23}$ and $\frac{23}{24}$.

If we take the fraction $\frac{3}{5}$ and form another fraction by adding 2 to each of its terms, namely $\frac{5}{7}$, we shall find that the fraction so formed is greater than the original fraction. By this means we may often determine by inspection which of two fractions is the greater. Thus, to compare $\frac{18}{19}$ and $\frac{19}{20}$, we see that $\frac{19}{20}$ may be formed by adding 1 to each of the terms of $\frac{18}{19}$ and therefore $\frac{19}{20}$ is greater than $\frac{18}{19}$.

So also in the two fractions $\frac{16}{19}$ and $\frac{21}{24}$, the second fraction is formed by adding 5 to each term of the first, and therefore the second fraction is the greater.

Again, in comparing the fractions

$$\frac{5}{7} \text{ and } \frac{9}{10}$$

we observe that the second fraction may be formed from the first by adding 4 to its numerator and 3 to its denominator. If 3 be added to each term of $\frac{5}{7}$ the fraction $\frac{8}{10}$ will be obtained.

$\therefore$ $\frac{8}{10}$ is greater than $\frac{5}{7}$

but $\frac{9}{10}$ " " $\frac{8}{10}$

$\therefore$ $\frac{9}{10}$ " " $\frac{5}{7}$.

From these examples we may infer,

1. That if one fraction can be formed by adding the same number to both terms of another fraction, then the fraction so formed is greater than the other.

2. That if one fraction can be formed by subtracting the same number from both terms of another fraction, then the fraction so formed is less than the other.

3. That if a fraction can be formed from another by adding a greater number to its numerator than to its denominator, then the fraction so formed is greater than the other.

4. That if a fraction can be formed from another by subtracting a smaller number from its numerator than from its denominator, then the fraction so formed is less than the other.

EXERCISE.

Determine by inspection the greatest fraction in each of the following cases :

1. $\frac{6}{7}$, $\frac{7}{8}$.
2. $\frac{16}{17}$, $\frac{18}{19}$.
3. $\frac{11}{12}$, $\frac{12}{13}$.
4. $\frac{17}{35}$, $\frac{27}{45}$.
5. $\frac{13}{36}$, $\frac{7}{30}$.
6. $\frac{7}{17}$, $\frac{16}{25}$.
7. $\frac{6}{7}$, $\frac{11}{13}$ (change $\frac{6}{7}$ to $\frac{12}{14}$ then compare $\frac{12}{14}$ and $\frac{11}{13}$).
8. $\frac{6}{5}$, $\frac{17}{23}$.
9. $\frac{7}{9}$, $\frac{29}{37}$.
10. $\frac{1}{2}$, $\frac{2}{3}$, $\frac{3}{4}$, $\frac{4}{5}$.
11. $\frac{4}{5}$, $\frac{5}{6}$, $\frac{6}{7}$, $\frac{7}{8}$, $\frac{8}{9}$.
12. $\frac{8}{9}$, $\frac{9}{10}$, $\frac{10}{11}$, $\frac{11}{12}$, $\frac{12}{13}$, $\frac{13}{14}$.

Fractions may sometimes be advantageously compared as follows : $\frac{1}{6}$ is evidently greater than $\frac{1}{7}$; $\frac{2}{5}$ is greater than $\frac{2}{7}$; $\frac{3}{8}$ greater than $\frac{3}{11}$; etc. That is, where two fractions have the same numerator it is evident that the fraction having the greater denominator is less than the other.

Also, where fractions not having the same numerator may readily be exchanged for equivalent fractions which have the same numerator, these fractions may thus be compared by comparing their denominators than : $\frac{3}{7}$, $\frac{4}{11}$ and $\frac{6}{19}$ readily reduce to $\frac{12}{28}$, $\frac{12}{33}$ and $\frac{12}{38}$, showing at once that $\frac{12}{28}$ or $\frac{3}{7}$ is the greatest of the three fractions.

This method is preferable in the case of fractions having large denominators.

EXERCISE.

Arrange the following fractions in order of magnitude, beginning with the least:

1. $\frac{3}{19}$, $\frac{5}{31}$, $\frac{6}{37}$.

2. $\frac{8\frac{1}{2}}{111}$, $\frac{7}{92}$, $\frac{8}{105}$.

3. $\frac{3}{97}$, $\frac{5}{161}$, $\frac{11}{353}$, $\frac{7}{225}$.

In this last example the fraction may be written:

$$\frac{1}{32\frac{1}{3}},\ \frac{1}{32\frac{1}{5}},\ \frac{1}{32\frac{1}{11}},\ \frac{1}{32\frac{1}{7}},$$

in which form the comparison can be readily made.

ADDITION OF FRACTIONS.

If 1 dollar, 2 dollars and 3 dollars be added together their sum will be 6 dollars, and so if 1 seventh, 2 sevenths and 3 sevenths be added together their sum will be 6 sevenths; or as it is written,

$$\tfrac{1}{7} + \tfrac{2}{7} + \tfrac{3}{7} = \tfrac{6}{7}.$$

Thus it will be seen that if fractions have the same denominator their sum is a fraction whose numerator is the sum of their numerators, and whose denominator is the same as their denominator.

Hence we have the following

RULE.

To add fractions which have the same denominator:

Add their numerators together for a new numerator, and under this place the common denominator.

EXERCISE.

Add together the following fractions:

1. $\frac{3}{5}$, $\frac{2}{5}$, $\frac{4}{5}$.
2. $\frac{3}{13}$, $\frac{7}{13}$, $\frac{2}{13}$.
3. $\frac{100}{333}$, $\frac{10}{333}$, $\frac{1}{333}$.
4. $\frac{5}{9}$, $\frac{1}{18}$ of 2, $\frac{4}{36}$.
5. $\frac{6}{15}$, $\frac{10}{25}$, $\frac{6}{30}$.
6. $\frac{2}{30}$, $\frac{1}{3}$ of $\frac{5}{6}$, $\frac{1}{2}$ of $\frac{7}{9}$.

If the fractions to be added have not the same denominator, as

$$\frac{2}{3}, \frac{3}{4}, \frac{4}{5},$$

they may be reduced to equivalent fractions having their least common denominator, as

$$\frac{40}{60}, \frac{45}{60}, \frac{48}{60},$$

and their sum can then be found by the method already given, thus:

$$\frac{40}{60} + \frac{45}{60} + \frac{48}{60} = \frac{133}{60} = 2\frac{13}{60}.$$

Hence we have the following

RULE.

To add fractions which have not the same denominator: Reduce the given fractions to equivalent fractions having their least common denominator, then add the numerators and under their sum place the common denominator.

EXERCISE.

Find the sum of,

1. $\frac{3}{5}$ and $\frac{5}{7}$.
2 $\frac{4}{5}$ and $\frac{6}{7}$.
3. $\frac{5}{6}$ and $\frac{7}{8}$.
4. $\frac{3}{7}$ and $\frac{19}{21}$.
5. $\frac{17}{18}$ and $\frac{19}{24}$.
6. $\frac{13}{39}$ and $\frac{28}{42}$.
7. $\frac{2}{3}$, $\frac{3}{4}$ and $\frac{4}{5}$.
8. $\frac{3}{4}$, $\frac{4}{5}$ and $\frac{5}{6}$.
9. $\frac{4}{5}$, $\frac{5}{6}$, $\frac{6}{7}$, $\frac{7}{8}$.
10. $\frac{6}{7}$, $\frac{9}{14}$, $\frac{7}{9}$, $\frac{12}{27}$.
11. $\frac{7}{11}$, $\frac{1}{35}$, $\frac{1}{15}$, $\frac{10}{39}$.
12. $\frac{1}{7}$, $\frac{1}{17}$, $\frac{1}{27}$, $\frac{1}{37}$.
13. $\frac{1}{9}$, $\frac{1}{19}$, $\frac{1}{29}$, $\frac{1}{39}$.
14. $3\frac{1}{2}$, $4\frac{1}{3}$, $5\frac{1}{4}$.

$$\begin{aligned} \text{Sum} &= 3\tfrac{1}{2} + 4\tfrac{1}{3} + 5\tfrac{1}{4}. \\ &= 3 + \tfrac{1}{2} + 4 + \tfrac{1}{3} + 5 + \tfrac{1}{4} \\ &= 3 + 4 + 5 + \tfrac{1}{2} + \tfrac{1}{3} + \tfrac{1}{4} \\ &= 12 + \tfrac{1}{2} + \tfrac{1}{3} + \tfrac{1}{4} \\ &= 12 + \tfrac{13}{12} \\ &= 12 + 1\tfrac{1}{12} \\ &= 13\tfrac{1}{12}. \end{aligned}$$

In such cases, therefore, we first add the whole numbers then the fractions and then add the two results together.

15. $6\frac{1}{2}$ and $7\frac{1}{4}$.
16. $5\frac{6}{7}$ and $4\frac{1}{5}$.
17. $9\frac{3}{11}$ and $11\frac{3}{17}$.
18. $4\frac{3}{7}$, $6\frac{7}{9}$, $7\frac{8}{5}$ and $\frac{3}{4}$.
19. $17\frac{3}{8}$, $\frac{47}{15}$, $\frac{13}{7}$ and $\frac{5}{9}$.

(Reduce the improper fractions to mixed numbers, then proceed as before.)

EXERCISE—(*Continued*).

20. $16\frac{1}{4}$, $7\frac{1}{5}$, $4\frac{6}{7}$, $\frac{18}{5}$, $\frac{19}{6}$, $\frac{27}{8}$.

21. Find the value of

$$\frac{6}{7} + \frac{3}{11} \text{ of } \frac{7}{9} \text{ of } 10\frac{2}{7} + \frac{5}{7} \text{ of } \frac{3}{8} \text{ of } \frac{28}{30}.$$

Here we must first reduce the compound fractions to simple ones, thus,

$\frac{3}{11}$ of $\frac{7}{9}$ of $10\frac{2}{7} = \frac{3}{11}$ of $\frac{7}{9}$ of $\frac{72}{7} = \frac{24}{11} = 2\frac{2}{11}$,

$\frac{5}{7}$ of $\frac{3}{8}$ of $\frac{28}{30} = \frac{1}{4}$,

$$\therefore \frac{6}{7} + \frac{3}{11} \text{ of } \frac{7}{9} \text{ of } 10\frac{2}{7} + \frac{5}{7} \text{ of } \frac{3}{8} \text{ of } \frac{28}{30}$$
$$= \frac{6}{7} + 2\frac{2}{11} + \frac{1}{4}$$
$$= 2 + \frac{6}{7} + \frac{2}{11} + \frac{1}{4}$$
$$= 2 + 1\frac{89}{308}$$
$$= 3\frac{89}{308}.$$

22. Find the value of

$$\frac{3}{4} \text{ of } \frac{7}{8} + \frac{1}{6} \text{ of } 4\frac{1}{2} + \frac{5}{9} \text{ of } \left(\frac{1}{2} + \frac{3}{4}\right) + \frac{19}{20} \text{ of } \left(\frac{2}{5} + \frac{4}{7}\right).$$

Here we must first add the fractions within the brackets, thus,

$$\left(\frac{1}{2} + \frac{3}{4}\right) = 1\frac{1}{4};$$
$$\text{and } \left(\frac{2}{5} + \frac{4}{7}\right) = \frac{34}{35}.$$

Thus the expression becomes reduced to

$$\frac{3}{4} \text{ of } \frac{7}{8} + \frac{1}{6} \text{ of } 4\frac{1}{2} + \frac{5}{9} \text{ of } 1\frac{1}{4} + \frac{19}{20} \text{ of } \frac{34}{35}.$$

Next reduce compound fractions to simple ones, thus,

$$\frac{3}{4} \text{ of } \frac{7}{8} = \frac{21}{32};$$
$$\frac{1}{6} \text{ of } 4\frac{1}{2} = \frac{3}{4};$$
$$\frac{5}{9} \text{ of } 1\frac{1}{4} = \frac{25}{36};$$
$$\frac{19}{20} \text{ of } \frac{34}{35} = \frac{323}{350}.$$

Thus the expression becomes reduced to

$$\frac{21}{32} + \frac{3}{4} + \frac{25}{36} + \frac{323}{350},$$
$$\text{which } = 3\frac{1187}{50400}.$$

Find the value of

23. $\frac{2}{7} + 5\frac{2}{11} + 1\frac{1}{3}$ of $2\frac{1}{2} + 6\frac{1}{4}$.

24. $\frac{2}{7}$ of $\frac{5}{14} + \frac{4}{5}$ of $\frac{1}{10} + \frac{3}{5}$ of $\left(\frac{1}{2} + \frac{1}{14}\right) + \frac{3}{70}$ of $\left(\frac{2}{7} + \frac{4}{5}.\right)$

SUBTRACTION OF FRACTIONS.

If from a group of 7 apples 4 apples be taken there will be 3 apples left, and so if from 7 ninths 4 ninths be taken there will be 3 ninths left, or as it is written,

$$\frac{7}{9} - \frac{4}{9} = \frac{3}{9}$$

Thus it is seen that if two fractions have the same denominator their difference is a fraction whose numerator is the difference of their numerators, and whose denominator is the same as their denominator. Hence, we have the following

RULE.

To find the difference between two fractions which have the same denominator, take the difference between their numerators for a new numerator, and under this place the common denominator.

EXERCISE.

Find the difference between

1. $\frac{1}{5}$ and $\frac{4}{5}$. 2. $\frac{3}{8}$ and $\frac{5}{8}$. 3. $\frac{13}{17}$ and $\frac{11}{17}$.
4. $\frac{17}{29}$ and $\frac{1}{29}$. 5. $\frac{37}{38}$ and $\frac{18}{38}$. 6. $\frac{13}{117}$ and $\frac{39}{117}$.

Find the value of

7. $\frac{11}{29} - \frac{5}{29}$. 8. $\frac{57}{113} - \frac{19}{113}$. 9. $3\frac{3}{4} - \frac{1}{4}$.
10. $7\frac{8}{9} - \frac{5}{9}$. 11. $\frac{7}{8} - \frac{3}{4}$. 12. $\frac{7}{8} - \frac{5}{6}$.

Suppose we require the difference between

$$\frac{5}{7} \text{ and } \frac{3}{8},$$

We cannot find it in their present form, but if we reduce them to their equivalent fractions

$$\frac{40}{56} \text{ and } \frac{21}{56}$$

we see at once that their difference is

$$\frac{19}{56}.$$

Hence we have the following

RULE

To find the difference between two fractions which have not the same denominator:

Reduce them to equivalent fractions having their least common denominator; take the difference between these new numerators, and under this difference place the common denominator.

EXERCISE.

Find the difference between

1. $\frac{1}{2}$ and $\frac{1}{3}$. 2. $\frac{1}{3}$ and $\frac{1}{4}$. 3. $\frac{1}{11}$ and $\frac{1}{13}$.
4. $\frac{9}{13}$ and $\frac{12}{17}$, 5. $\frac{11}{24}$ and $\frac{15}{32}$ 6. $\frac{17}{5}$ and $\frac{10}{25}$.
7. $\frac{98}{99}$ and $\frac{65}{66}$. 8. $\frac{6}{73}$ and $\frac{6}{71}$. 9. $\frac{19}{20}$ and $\frac{346}{347}$.
10. $37\frac{8}{9}$ and $14\frac{2}{7}$. 11. $\frac{1}{4}$ of $\frac{1}{5}$ and $\frac{3}{7}$ of $\frac{2}{9}$.
12. $\frac{1}{8}$ of $\frac{3}{7}$ of 6 and $\frac{4}{5}$ of $\frac{7}{8}$ of $2\frac{1}{2}$.
13. The sum of $\frac{1}{2}$, $\frac{1}{3}$ and $\frac{1}{4}$ and the sum of $\frac{6}{7}$ and $\frac{4}{5}$.
14. The sum of $\frac{1}{18}$ and $\frac{1}{19}$ and their difference.
15. $\frac{19}{20}+\frac{4}{5}$ and $\frac{7}{8}+\frac{3}{4}+\frac{11}{12}$.
16. $\frac{3}{8}+\frac{4}{11}$ of $3\frac{2}{3}$ and $1\frac{1}{3}+\frac{5}{9}$ of $\frac{27}{50}$ of $1\frac{1}{4}$.

Find the value of

17. $\frac{7}{8}-\frac{6}{7}$. 18. $\frac{18}{19}-\frac{17}{18}$. 19. $\frac{28}{29}-\frac{27}{28}$.
20. $\frac{3}{7}$ of $\frac{14}{21}-\frac{1}{7}$. 21. $\frac{6}{7}+\frac{3}{10}-\frac{19}{20}$. 22. $5\frac{1}{3}-3\frac{1}{5}$.

In examples like (22), where the difference between two mixed numbers is required, it is not necessary to reduce the mixed numbers to improper fractions, for, since 3 and $\frac{1}{5}$ are both to be taken from $5\frac{1}{3}$, we may first take away 3 and then take $\frac{1}{5}$ from what is left. When 3 is subtracted the remainder is $2\frac{1}{3}$; we have next, therefore, to find the value of

$$2\frac{1}{3}-\frac{1}{5}.$$

Now, it does not matter whether $\frac{1}{5}$ is taken from 2 or from $\frac{1}{3}$ so long as the whole quantity $2\frac{1}{3}$ is diminished by $\frac{1}{5}$, we may, therefore, take $\frac{1}{5}$ from 2 (and this part of the operation can always be done mentally), leaving $1\frac{4}{5}$, and also the $\frac{1}{3}$ which must now be added to $1\frac{4}{5}$. This is done in the way already explained, and therefore the result required is

$$1\frac{4}{5}+\frac{1}{3} \text{ which} = 2\frac{2}{15}.$$

Find the value of

23. $17\frac{3}{11}-15\frac{5}{9}$. 24. $3\frac{113}{117}-2\frac{12}{13}$.
25. $136\frac{1}{247}-120\frac{253}{280}$. 26. $\frac{19}{7}-\frac{12}{5}$.

(Change the improper fractions $\frac{19}{7}$ and $\frac{12}{5}$ to mixed numbers, then subtract).

27. Find the value of

$$\frac{7}{8}-\frac{2}{5}+\frac{15}{16}-\frac{1}{4}-\frac{1}{2}.$$

This value may be found in several ways :

EXERCISE—*(Continued)*

1. We may proceed thus

find the value of $\frac{7}{8} - \frac{2}{5}$, which is $\frac{19}{40}$,
to this add $\frac{15}{16}$ which gives $1\frac{33}{80}$
from this take $\frac{1}{4}$ which leaves $1\frac{13}{80}$
and from this take $\frac{1}{2}$ and we have $\frac{53}{80}$

as the final result.

2. Since additions and subtractions may be performed in any order, the fractions may be arranged thus:

$$\frac{7}{8} + \frac{15}{16} - \frac{2}{5} - \frac{1}{4} - \frac{1}{2}.$$

Then $\frac{7}{8} + \frac{15}{16} = 1\frac{13}{16}$;

$$1\frac{13}{16} - \frac{2}{5} = 1\frac{33}{80};$$
$$1\frac{33}{80} - \frac{1}{4} = 1\frac{13}{80};$$
$$1\frac{13}{80} - \frac{1}{2} = \frac{53}{80}.$$

3. Since $\frac{2}{5}$, $\frac{1}{4}$ and $\frac{1}{2}$ are all to be taken from the sum of $\frac{7}{8}$ and $\frac{15}{16}$, we may, instead of subtracting these fractions one at a time, subtract them all at once, that is, find their sum and then subtract it from the sum of $\frac{7}{8}$ and $\frac{15}{16}$, thus:

$$\frac{7}{8} + \frac{15}{16} = 1\frac{13}{16};$$
$$\frac{2}{5} + \frac{1}{4} + \frac{1}{2} = 1\frac{3}{20};$$
$$1\frac{13}{16} - 1\frac{3}{20} = \frac{53}{80}.$$

This last method, in most cases, though not always, gives a readier solution than either of the others.

When fractions are enclosed in brackets their value must first be found and then the result must be added or subtracted according as the bracket has the sign + or — before it.

28. Simplify $8\frac{7}{15}$ of $2\frac{7}{19} - (3\frac{4}{7} + 7\frac{5}{9} - 3\frac{4}{5})$.

$$8\frac{7}{15} \text{ of } 2\frac{7}{19} = \frac{381}{19} = 20\frac{1}{19};$$
$$3\frac{4}{7} + 7\frac{5}{9} - 3\frac{4}{5} = 7\frac{103}{315};$$
$$\text{and } 20\frac{1}{19} - 7\frac{103}{315} = 12\frac{4343}{5985}.$$

29. Simplify $\frac{1}{2}$ of $\frac{3}{7} - \frac{2}{11}$ of $3\frac{1}{7} + \frac{3}{7}$ of $2\frac{5}{8}$.

30. Simplify $4\frac{4}{7}$ of $3\frac{1}{16} - 2\frac{2}{3}$ of $1\frac{5}{16} + \frac{8}{9} - \frac{3}{14}$ of $\frac{7}{4}$ of $4\frac{4}{7}$.

31. Find the value of

$$3\frac{1}{2} + 4\frac{1}{3} + 5\frac{1}{4} + \frac{3}{4} \text{ of } \frac{2}{9} + \frac{1}{2} \text{ of } \frac{2}{3} \text{ of } \frac{5}{8}.$$

32. Find the difference between

$$1 - (\tfrac{1}{2} + \tfrac{1}{3} + \tfrac{1}{24}) \text{ and } 1 - (\tfrac{1}{2} \text{ of } \tfrac{1}{3} \text{ of } \tfrac{1}{24}).$$

EXERCISE—(*Continued*).

33. Find the sum of the greatest and least of

$$\frac{27}{28}, \frac{28}{29}, \frac{29}{30}, \frac{30}{31};$$

also the sum of the other two, and then find the difference between these sums.

MULTIPLICATION OF FRACTIONS.

If 8 be multipled by 12 the result will be 96; if 8 be multiplied by 6, the half of 12, the result is 48, which is half of 96; if 8 be multiplied by one-third of 12 the product is one-third of 96; if by one-fourth of 12 the product is one-fourth of 96. Thus we see that in multiplying by whole numbers, whatever part one multiplier is of another, its product is the same part of the product of the other. In the example cited, all the parts taken give whole numbers, that is, one-half, one-third, and one-fourth of 12 are all whole numbers. We assume the same law to hold where the parts taken do not give whole numbers; so that if any number is multiplied by one-fourth of 3 the product will be one-fourth of that obtained by multiplying the number by 3. But one-fourth of 3 is $\frac{3}{4}$, therefore, to multiply any number by $\frac{3}{4}$ we must first multiply the number by 3 and then divide the result by 4. Thus, to multiply $\frac{5}{7}$ by $\frac{3}{4}$, we first multiply $\frac{5}{7}$ by 3, which gives $\frac{15}{7}$; we next divide $\frac{15}{7}$ by 4, which gives $\frac{15}{28}$. Therefore $\frac{5}{7}$ when multiplied by $\frac{3}{4}$ gives $\frac{15}{28}$ as the product; or, as it is written,

$$\frac{5}{7} \times \frac{3}{4} = \frac{15}{28}.$$

Since $\frac{15}{28} = \frac{5 \times 3}{7 \times 4}$, therefore $\frac{5}{7} \times \frac{3}{4} = \frac{5 \times 3}{7 \times 4}$

And thus we see that the method of finding the product when one fraction is to be multiplied by another is precisely the same as that for reducing a compound fraction to a simple one.

If we multiply $\frac{3}{4}$ by $\frac{5}{7}$ we get $\frac{15}{28}$ for the product; from this we infer that when two fractions are to be multiplied together either of them may be taken as the multiplier. Now suppose we require the product of three fractions,

$$\frac{2}{3},\ \frac{4}{5},\ \text{and}\ \frac{2}{7}.$$

We may first take the product of $\frac{2}{3}$ and $\frac{4}{5}$, which is $\frac{8}{15}$; then take the product of $\frac{8}{15}$ and $\frac{2}{7}$, which is $\frac{16}{105}$; therefore

$$\frac{2}{3} \times \frac{4}{5} \times \frac{2}{7} = \frac{16}{105} = \frac{2 \times 4 \times 2}{3 \times 5 \times 7}$$

Hence, to multiply any number of fractions together, we have the following

RULE.

Multiply all the numerators together for the numerator of the product, and all the denominators together for its denominator.

NOTE.—Before applying the rule all factors common to a numerator and a denominator should be struck out.

EXERCISE.

Find the product of

1. $\frac{2}{3}$ and $\frac{4}{3}$.
2. $\frac{5}{7}$ and $\frac{2}{3}$.
3. $\frac{6}{7}$ and $\frac{5}{9}$.
4. $\frac{2}{5}$, $\frac{1}{3}$ and $\frac{4}{7}$.
5. $\frac{7}{11}$, $\frac{3}{5}$ and $\frac{3}{4}$.
6. $\frac{7}{10}$, $\frac{7}{9}$ and $\frac{7}{8}$.
7. $\frac{3}{4}$, $\frac{4}{5}$, $\frac{5}{6}$ and $\frac{6}{7}$.
8. $\frac{7}{9}$, $\frac{3}{11}$, $\frac{3}{8}$ and $\frac{16}{17}$.
9. $4\frac{1}{3}$ and $\frac{3}{13}$.
10. $\frac{3}{4}$ of $\frac{15}{17}$ and $\frac{5}{9}$ of 6.
11. $19\frac{7}{9}$, $\frac{2}{3}$ of $\frac{6}{11}$ of $8\frac{1}{4}$ and $\frac{3}{89}$.
12. $\frac{2}{3}$ of $\frac{4}{7}$ and $\frac{17}{19}$ of $\frac{9}{34}$ of $2\frac{5}{8}$.
13. $2\frac{1}{2}$, $3\frac{1}{3}$, $3\frac{3}{4}$ and $5\frac{1}{5}$.
14. $\frac{1}{2}+\frac{1}{3}$ and $\frac{1}{3}+\frac{1}{4}$.
15. $\frac{1}{4}+\frac{1}{5}$ and $\frac{1}{4}-\frac{1}{5}$.
16. $\frac{3}{4}$ of $(\frac{5}{6}+\frac{3}{8}-\frac{2}{3})$ and $(\frac{1}{3}+\frac{1}{5}) \times 7\frac{1}{2}$.
17. $3\frac{1}{3}$, $\frac{99}{100}$, $\frac{17}{66}$, $\frac{16}{136}$, $17\frac{6}{7}-11\frac{11}{14}$, $\frac{7}{17}$ and $\frac{3}{4}$ of $\frac{8}{15}$.

Find the value of

18. $\frac{2}{3} \times \frac{3}{4} \times 4\frac{1}{2} \times \frac{8}{27}$.
19. $1\frac{3}{4} \times 4\frac{4}{7} \times \frac{3}{14}$.
20. $\frac{3}{4} \times \frac{6}{7} - \frac{1}{2}$.
21. $\frac{3}{4} \times (\frac{6}{7} - \frac{1}{2})$.
22. $\frac{7}{8} - \frac{2}{3} \times \frac{3}{4} + \frac{2}{9}$.

Proceed thus:

$$\frac{2}{3} \times \frac{3}{4} = \frac{1}{2}.$$
$$\frac{7}{8} - \frac{1}{2} = \frac{3}{8}.$$
$$\frac{3}{8} + \frac{2}{9} = \frac{43}{72}.$$

EXERCISE—(*Continued*).

23. $\frac{5}{6} \times \frac{9}{10} - \frac{3}{7} \times \frac{7}{9} + \frac{6}{7} \times (\frac{2}{3} - \frac{1}{2}.)$

In such cases as this the order of reduction is as follows:

1. Find the value of the fractions in brackets.
2. Perform the multiplications indicated.
3. Perform the additions and subtractions indicated.

Thus—

$$\frac{2}{3} - \frac{1}{2} = \frac{1}{6}.$$
$$\frac{6}{7} \times \frac{1}{6} = \frac{1}{7}.$$
$$\frac{3}{7} \times \frac{7}{9} = \frac{1}{3}.$$
$$\frac{5}{6} \times \frac{9}{10} = \frac{3}{4}.$$
$$\frac{3}{4} - \frac{1}{3} + \frac{1}{7} = \frac{47}{84}.$$

24. $\frac{7}{5} \times 3\frac{1}{3} + \frac{5}{7} \times 1\frac{10}{11} + \frac{341}{363}.$

25. $4\frac{4}{7} \times 3\frac{1}{16} - 2\frac{2}{3} \times 1\frac{5}{16} + \frac{8}{9} - 1\frac{3}{4} \times 4\frac{4}{7} \times \frac{3}{14}.$

26. $(\frac{3}{5}$ of $7\frac{1}{2} - \frac{8}{17}) \times \frac{9}{11}.$

27. $\frac{18}{17} \times (1 - \frac{64}{81}) + \frac{8}{11} \times \frac{1}{6} \times (\frac{1}{2} + \frac{5}{12}).$

28. $\frac{5}{16} + \frac{7}{12} \times 3\frac{1}{4} - (\frac{7}{8} \times \frac{37}{21} - \frac{1}{3}).$

29. Multiply the excess of $\frac{19}{20}$ over the sum of $\frac{1}{5}$ and $\frac{1}{8}$, by the difference between $\frac{7}{8}$ of $6\frac{4}{7}$ and $\frac{5}{9}$ of $8\frac{1}{5}$.

30. What number divided by $7\frac{1}{4}$ will give $13\frac{4}{5}$ for quotient?

31. What number on division by $\frac{2}{3} + \frac{7}{8}$ of $6\frac{1}{2}$ will give $\frac{5}{7}$ of $(\frac{7}{8} - \frac{4}{5}) \times 7$ for quotient?

DIVISION OF FRACTIONS.

If 60 be divided by 12 the quotient is 5.
" " one-half of 12 the quotient is twice 5;
" " one-third " " 3 times 5;
" " one-fourth " " 4 times 5.

Similarly, if any number be divided by one-fourth of 3 the quotient will be 4 times as great as that obtained by dividing the same number by 3, but one-fourth of 3 is $\frac{3}{4}$; therefore, to divide any number by $\frac{3}{4}$ we must first divide the number by 3 and then multiply the result by 4.

Thus, to divide $\frac{5}{7}$ by $\frac{3}{4}$,

We first divide $\frac{5}{7}$ by 3, which gives $\frac{5}{21}$;

We next multiply $\frac{5}{21}$ by 4, " $\frac{20}{21}$.

Therefore, $\frac{5}{7}$ when divided by $\frac{3}{4}$ gives $\frac{20}{21}$ as the quotient; or, as it is written,

$$\frac{5}{7} \div \frac{3}{4} = \frac{20}{21}.$$

Since $\dfrac{20}{21} = \dfrac{5 \times 4}{7 \times 3} = \dfrac{5}{7} \times \dfrac{4}{3}$;

therefore $\dfrac{5}{7} \div \dfrac{3}{4} = \dfrac{5}{7} \times \dfrac{4}{3}$.

Thus we see that to divide by $\frac{3}{4}$ is the same as to multiply by $\frac{4}{3}$, and consequently to divide by a fraction we have simply to multiply by that fraction inverted.

Hence we have the following

RULE

To divide a fraction (or a whole number) by a fraction: Invert the divisor and then proceed as in multiplication.

EXERCISE.

Divide

1. $\frac{2}{5}$ by $\frac{1}{2}$.
2. $\frac{3}{7}$ by $\frac{3}{5}$.
3. $\frac{5}{9}$ by $\frac{5}{18}$.
4. $\frac{17}{19}$ by $\frac{1}{38}$.
5. $\frac{48}{49}$ by $\frac{13}{14}$.
6. $\frac{99}{100}$ by $\frac{999}{1000}$.
7. $6\frac{7}{9}$ by $7\frac{6}{7}$.
8. $19\frac{4}{17}$ by $6\frac{7}{17}$,
9. $\frac{187}{209}$ by $\frac{204}{216}$.
10. $\frac{5}{9}$ of $17\frac{1}{4}$ by $\frac{37}{40}$.
11. $\frac{19}{20}$ by $\frac{4}{5}$ of $6\frac{1}{4}$.
12. $\frac{15}{16}$ of $\frac{64}{100}$ by $\frac{17}{18}$ of $1\frac{2}{3}$.
13. $\frac{9}{10} \times \frac{5}{7}$ of 14 by 99.
14. $(\frac{6}{7} + \frac{4}{5} - \frac{5}{7})$ by $\frac{6}{7}$ of $(\frac{9}{10} - \frac{8}{9})$.
15. $\frac{3}{4}$ of $(6\frac{1}{2} + 3\frac{1}{3}) \times (\frac{5}{7}$ of $5\frac{3}{5})$ by $5\frac{1}{3} - 2\frac{1}{5}$.
16. The product of $\frac{32}{51}$ and $\frac{189}{207}$ by the product of $1\frac{27}{85}$ and $\frac{36}{23}$.

Find the value of,

17. $\frac{5}{8} \div \frac{8}{10}$.
18. $\frac{11}{19} \div \frac{22}{38}$.
19. $\frac{999}{1000} \div \frac{99}{100}$.
20. $\frac{7}{8}$ of $4\frac{1}{2} \div \frac{7}{16}$.
21. $\frac{5}{9}$ of $\frac{37}{40} \div \frac{2}{3}$ of $\frac{9}{16}$.

EXERCISE—(*Continued*).

22. $\frac{6}{7} \div \frac{4}{5} \times \frac{7}{15} \div \frac{5}{9}$ of $\frac{7}{10} \times (\frac{8}{9} + 2\frac{1}{4}) \div (4\frac{1}{3} - 3\frac{1}{2}) \div 4\frac{59}{70}$.

Proceed as follows:

1. Find the value of the fractions in brackets.

$$\frac{8}{9} + 2\frac{1}{4} = 3\frac{5}{36} = \frac{113}{36}; \text{ and } 4\frac{1}{3} - 3\frac{1}{2} = \frac{5}{6}.$$

This reduces the expression to

$$\frac{6}{7} \div \frac{4}{5} \times \frac{7}{15} \div \frac{5}{9} \text{ of } \frac{7}{10} \times \frac{113}{36} \div \frac{5}{6} \div 4\frac{59}{70}.$$

2. Simplify compound fractions

$$\frac{5}{9} \text{ of } \frac{7}{10} = \frac{7}{18};$$

this reduces the expression to

$$\frac{6}{7} \div \frac{4}{5} \times \frac{7}{15} \div \frac{7}{18} \times \frac{113}{36} \div \frac{5}{6} \div 4\frac{59}{70}.$$

3. Invert every fraction before which the sign ÷ occurs; this reduces the expression to

$$\frac{6}{7} \times \frac{5}{4} \times \frac{7}{15} \times \frac{18}{7} \times \frac{113}{36} \times \frac{6}{5} \times \frac{70}{339}.$$

4. Multiply the fractions together, and the expression becomes $= 1$.

It may be noticed here that there is an important difference between "of" and "×" when placed between two fractions. The fractions between which "of" is placed, form parts of the same compound fraction, whereas, those between which "×" is placed, are two distinct fractions. Thus:

$$\frac{6}{7} \div \frac{2}{3} \text{ of } \frac{4}{5}$$

means that $\frac{6}{7}$ is to be divided by the fraction $\frac{2}{3}$ of $\frac{4}{5}$, that is by $\frac{8}{15}$, and the quotient is, therefore, $\frac{45}{28}$; whereas

$$\frac{6}{7} \div \frac{2}{3} \times \frac{4}{5}$$

means that $\frac{6}{7}$ is to be divided by $\frac{2}{3}$, and that the quotient so obtained is to be multiplied by $\frac{4}{5}$: that is, we are to invert $\frac{2}{3}$ and then multiply together the fractions $\frac{6}{7}$, $\frac{3}{2}$ and $\frac{4}{5}$, the result in this case being $\frac{36}{35}$.

In the first case the sign ÷ occurs before the fraction $\frac{2}{3}$ of $\frac{4}{5}$; in the second case before the fraction $\frac{2}{3}$, so that in both cases we have followed the rule and inverted the fraction before which the sign of division is placed.

23. Simplify $\frac{2}{3} \div \frac{4}{5} + \frac{3}{7} \times \frac{5}{6} - \frac{7}{8} \div \frac{3}{4}$ of 6.

Here we first reduce the compound fraction

$$\frac{3}{4} \text{ of } 6 = \frac{9}{2}.$$

Next invert the fractions before which ÷ occurs. This reduces the expression to

$$\frac{2}{3} \times \frac{5}{4} + \frac{3}{7} \times \frac{5}{6} - \frac{7}{8} \times \frac{2}{9}$$

EXERCISE—(*Continued*).

Next perform the multiplications indicated. The expression thus becomes

$$\frac{5}{6}+\frac{5}{14}-\frac{7}{36}.$$

Lastly, perform the addition and subtraction indicated, and we obtain $\frac{251}{252}$ as the required result.

Simplify

24. $\frac{5}{7}\times 1\frac{6}{13}\div 2\frac{5}{7}\times 3\frac{1}{4}$.
25. $(3\frac{4}{5}+5\frac{1}{9}-\frac{1}{45})\times(4\frac{1}{5}-3\frac{1}{4})\div(1\frac{5}{11}+2\frac{1}{8}-2\frac{9}{16}+\frac{1}{8}+\frac{1}{22})$.
26. $\frac{1}{23}\times 6\frac{13}{17}$ of $24\frac{11}{13}-4\frac{13}{18}\times 3\frac{33}{34}\div 3\frac{37}{96}$.
27. $8\frac{17}{19}\times 5\frac{14}{39}\div 4\frac{15}{32}-7\frac{19}{20}\times 5\frac{11}{65}\div 14\frac{21}{25}$.
28. $1\frac{7}{9}$ of $\frac{27}{64}\div\frac{11}{12}$ of $9\frac{9}{11}\div(4\frac{4}{7}$ of $\frac{21}{160}\div 2\frac{5}{6}$ of $\frac{15}{34})$.
29. $\left\{2\frac{3}{4}+\frac{5}{7}\text{ of }(7\div 3\frac{4}{5})-1\frac{2}{3}\div 2\frac{1}{2}\right\}\div 1\frac{77}{128}$.

COMPLEX FRACTIONS.

Since every fraction expresses the quotient of the numerator by the denominator, the division of one whole number by another may always be expressed in the form of a fraction. Thus if we wish to divide 376 by 195 we may indicate the operation in the form of a fraction, thus:

$$\frac{376}{195}.$$

The same method is employed to indicate the division of one fraction by another; $7\frac{1}{2}\div\frac{4}{5}$ being expressed in the form of a fraction thus:

$$\frac{7\frac{1}{2}}{\frac{4}{5}}$$

Also $(\frac{3}{4}+\frac{2}{5})\div\frac{2}{3}$ of $6\frac{3}{5}$ may be written

$$\frac{\frac{3}{4}+\frac{2}{5}}{\frac{2}{3}\text{ of }6\frac{3}{5}}$$

When the division of one fraction by another is indicated in this way, the resulting expression is called a COMPLEX FRACTION, and the two fractions so used are called respectively the numerator and denominator of the complex fraction. Thus in the above fraction $\frac{3}{4}+\frac{2}{5}$ is called the numerator, and $\frac{2}{3}$ of $6\frac{3}{5}$ the denominator.

Complex fractions may be reduced to simple ones by performing the division indicated thus,

$$\frac{\frac{3}{4}+\frac{2}{5}}{\frac{2}{3}\text{ of }6\frac{3}{5}}=(\tfrac{3}{4}+\tfrac{2}{5})\div\tfrac{2}{3}\text{ of }6\tfrac{3}{5}$$
$$=\tfrac{23}{20}\div\tfrac{22}{5}$$
$$=\tfrac{23}{20}\times\tfrac{5}{22}$$
$$=\tfrac{23}{88}.$$

or by first reducing the numerator and denominator to simple fractions, and then performing the division thus,

$$\frac{\frac{3}{4}+\frac{2}{5}}{\frac{2}{3}\text{ of }6\frac{3}{5}}=\frac{\frac{23}{20}}{\frac{22}{5}}$$
$$=\tfrac{23}{20}\div\tfrac{22}{5}$$
$$=\tfrac{23}{20}\times\tfrac{5}{22}$$
$$=\tfrac{23}{88}.$$

Every complex fraction whose numerator and denominator are simple fractions can at once be reduced to a simple fraction, for

$$\frac{\frac{5}{7}}{\frac{2}{3}}=\tfrac{5}{7}\div\tfrac{2}{3},$$
$$=\tfrac{5}{7}\times\tfrac{3}{2}$$
$$=\frac{5\times 3}{7\times 2}$$

$$=\frac{\text{product of highest and lowest numbers.}}{\text{product of the other two numbers.}}$$

EXERCISE.

Simplify

1. $\dfrac{\frac{8}{9}}{\frac{5}{7}}$ 2. $\dfrac{3\frac{4}{5}}{\frac{2}{3}}$ 3. $\dfrac{7}{\frac{5}{6}}$

4. $\dfrac{\frac{7}{5}}{6}$ 5. $\dfrac{\frac{1}{2}-\frac{1}{3}}{\frac{1}{2}+\frac{1}{3}}$ 6. $\dfrac{\frac{2}{3}\text{ of }\frac{1}{5}\times 2}{\frac{4}{5}\text{ of }3\frac{1}{2}}$

7. $\dfrac{\frac{3}{7}\times 1\frac{4}{7}}{\frac{5}{9}\div 2\frac{6}{7}}$ 8. $\dfrac{\frac{6}{7}\times\frac{5}{6}\div 7}{\frac{1}{9}+\frac{2}{3}\times\frac{6}{7}}$ 9. $\dfrac{\frac{2}{3}}{\frac{4}{5}}+\dfrac{\frac{3}{5}}{\frac{7}{8}}$

10. $\dfrac{7\frac{1}{2}}{3\frac{1}{5}}+\dfrac{\frac{2}{3}\text{ of }6}{\frac{1}{5}\text{ of }4}$ 11. $\dfrac{\frac{6}{7}+\frac{2}{9}}{\frac{2}{3}-\frac{1}{6}}+\dfrac{\frac{2}{3}+\frac{1}{6}}{\frac{6}{7}-\frac{2}{9}}$

EXERCISE—(*Continued*).

12. $\dfrac{\frac{2}{7}+\frac{4}{5}\text{ of }\frac{10}{16}}{(\frac{2}{7}+\frac{4}{5})\times\frac{10}{16}}-\dfrac{\frac{49}{56}}{\frac{56}{64}}$ 13. $\dfrac{7\frac{1}{9}}{9\frac{1}{7}}+\dfrac{8\frac{1}{11}}{11\frac{1}{8}}-\dfrac{3\frac{1}{33}}{33\frac{1}{3}}$

14. $6\times\dfrac{7\frac{2}{9}\text{ of }12\frac{2}{3}}{2\frac{8}{9}\text{ of }15\frac{5}{6}}-3\times\dfrac{2\frac{1}{7}\text{ of }4\frac{1}{3}}{2\frac{1}{14}\text{ of }2\frac{1}{2}}$

15. $\dfrac{1\frac{7}{9}+8\frac{1}{4}}{7\frac{5}{6}-4\frac{2}{9}}\div\dfrac{2\frac{1}{19}+\frac{7}{38}}{\frac{7}{38}-\frac{7}{76}}\times\dfrac{45\frac{1}{8}.}{21\frac{1}{4}.}$

16. $\left(\dfrac{7\frac{1}{4}}{14\frac{3}{10}}-\dfrac{2\frac{1}{2}+\frac{26}{100}}{4\frac{1}{4}+\frac{21}{100}}\right)\div\dfrac{4\frac{1}{4}}{\frac{23}{28}-\frac{18}{91}-\frac{19}{26}}$

17. $\frac{1}{5}\ (3\frac{1}{3}+1\frac{1}{4})+\frac{1}{4}\text{ of }\dfrac{1\frac{1}{8}-\frac{1}{3}\times1\frac{5}{6}}{\frac{1}{10}\times3\frac{1}{3}+\frac{13}{72}}\times\frac{95}{100}\text{ of }\frac{1}{4}.$

18. $\dfrac{1}{3+\dfrac{1}{7\frac{1}{16}}}$ 19. $\dfrac{1}{\frac{1}{2}+\frac{1}{3}+\dfrac{\frac{1}{6}}{\frac{3}{4}-\frac{1}{3}.}}$

20. $\dfrac{6+\dfrac{1}{6-\frac{1}{6}}}{6}$ 21. $6+\dfrac{1}{\dfrac{6+\dfrac{1}{6-\frac{1}{6}}}{6}}$

22. $\dfrac{1}{4-\dfrac{1}{2-\dfrac{1}{1-\frac{5}{13}}}}$ 23. $\dfrac{6+\dfrac{1}{6-\frac{1}{6}}}{4-\dfrac{1}{4-\frac{1}{4}.}}\times10\frac{8}{9}.$

24. $\dfrac{\frac{1}{23}\times6\frac{13}{17}\times24\frac{1}{13}-4\frac{13}{18}\times3\frac{33}{34}\div3\frac{37}{96}}{8\frac{17}{19}\times5\frac{14}{39}\div4\frac{15}{32}-7\frac{19}{20}\times5\frac{11}{65}\div14\frac{21}{25}}\times4\frac{8}{23}.$

GREATEST COMMON MEASURE AND LEAST COMMON MULTIPLE OF FRACTIONS.

The G. C. M. of 9, 12 and 15 is 3, and the G. C. M. of 9 quarts, 12 quarts and 15 quarts is 3 quarts,

In the same way the G. C. M. of 9 seventeenths, 12 seventeenths and 15 seventeenths is 3 seventeenths. That is, the G. C. M. of

$\frac{9}{17}$, $\frac{12}{17}$ and $\frac{15}{17}$ is $\frac{3}{17}$.

So that if any number of fractions have the same denominator their G. C. M. is found by taking the L. C. M. of their numerators and placing under it the common denominator.

If the fractions have not the same denominator they can be reduced to equivalent fractions which have a common denominator, and then their G. C, M. can be found in the manner already stated.

If any number of fractions in their lowest terms, having different denominators be reduced to equivalent fractions having the same denominator, the G. C. M. of the numerators of the new fractions is the same as that of the old.

First, take two fractions,

$$\frac{14}{15} \text{ and } \frac{21}{40}.$$

The equivalent fractions are

$$\frac{112}{120} \text{ and } \frac{63}{120}$$

That is

$$\frac{14 \times 8}{15 \times 8} \text{ and } \frac{21 \times 3}{40 \times 3}.$$

Where we see that the first numerator is multiplied by a factor of the second denominator, and this factor cannot occur in the second numerator, since the fractions are in their lowest terms, and therefore the factor introduced into the first numerator is not one which already occurs in the second numerator, and therefore the new numerator of the first (112) and the old numerator of the second fraction (20), have no new common factor; neither, by the same reasoning, will the new numerator of the second and the old numerator of the first.

Lastly, the factors 8 and 3 introduced into the numerators can have no common factor; for if they had, 120 would not be the L. C. M, of 15 and 40, but 120 divided by the common factor of 8 and 3.

Hence, we conclude that the process of converting two fractions to others having a common denominator leaves the G. C. M. of their numerators the same, and this will be true of every pair of fractions if there are more than two.

From this it appears that in finding the G. C. M. of fractions in their lowest terms with different denominators we need only find the common denominator and then we can find the G. C. M. of the old numerators instead of the new.

Hence we have the following

RULE.

For finding the G. C. M. of any number of fractions, in their lowest terms:

Find the G. C. M. of their numerators and under this place the L. C. M. of their denominators.

TO FIND L. C. M. OF ANY FRACTIONS.

We shall first show that the quotient of one fraction divided by another is the same as the quotient of the reciprocal of the second fraction divided by the reciprocal of the first.

Thus:—

$$\begin{aligned} &\tfrac{5}{7} \div \tfrac{2}{3} \\ &= \tfrac{5}{7} \times \tfrac{3}{2} \\ &= \tfrac{3}{2} \times \tfrac{5}{7} \\ &= \tfrac{3}{2} \div \tfrac{7}{5} \end{aligned}$$

From this it follows that the quotients arising from dividing fractions into their L. C. M. are the same as those arising from the division of the reciprocals of these fractions by the reciprocal of their L. C. M., and since no fraction less than the L. C. M. will contain the fractions, it follows that no fraction greater than its reciprocal will divide the reciprocal of the fractions, and hence it follows that the L. C. M. of any number of fractions is the reciprocal of the G. C. M. of their reciprocals.

Thus, to find the L. C. M. of

$$\tfrac{2}{3}, \tfrac{5}{6} \text{ and } \tfrac{4}{9},$$

the reciprocals of these fractions are

$$\tfrac{3}{2}, \tfrac{6}{5}, \tfrac{9}{4}.$$

The G. C. M. of these last fractions, is

$$\frac{\text{G. C. M. of } 3, 6 \text{ and } 9}{\text{L. C. M. of } 2, 5 \text{ and } 4}$$

and the L. C. M. is the reciprocal of this, or

$$\frac{\text{L. C. M. of 2, 5 and 4}}{\text{G. C. M. of 3, 6 and 9}}$$

∴ the L. C. M. of

$$\tfrac{2}{3},\ \tfrac{5}{6} \text{ and } \tfrac{4}{9}$$

is formed by taking the L. C. M. of their numerators and placing under it the G. C. M. of their denominators.

Hence we have the following

RULE.

For finding the L. C. M. of any fractions in their lowest terms:

Find the L. C. M. of their numerators and under it place the G. C. M. of their denominators.

DECIMALS AND DECIMAL FRACTIONS.

Let us consider the number

7648.

Here the first digit to the right (8) denotes so many simple units or *ones*. The second digit (4) denotes so many *tens*; the third, *hundreds*; the fourth, *thousands*, etc. We thus see that each digit denotes so many units, each of which is ten times as large as that denoted by the digit to its right.

Commencing now at the left hand digit: each unit denoted by 7 is one thousand, each unit denoted by 6 is one hundred, which is one-tenth of one thousand, each unit denoted by 4 is one-tenth of one hundred, each unit denoted by 8 is one-tenth of each unit denoted by 4.

If now we write down

8935,

and suppose 8 to denote units, each of which is unity or *one*, then the 9 immediately to its right will denote units, each of which is one-tenth of 1 or $\frac{1}{10}$: these units are accordingly called *tenths*.

The next digit 3 denotes units, each of which is one-tenth of each unit denoted by 9, therefore each unit denoted by 3 is one-tenth of one-tenth, but $\frac{1}{10}$ of $\frac{1}{10} = \frac{1}{100}$; these units are therefore called *hundredths*.

In the same way each unit denoted by 5 is one-tenth of one-hundredth

$$\text{but } \tfrac{1}{10} \text{ of } \tfrac{1}{100} = \tfrac{1}{1000}.$$

These units are accordingly called *thousandths*.

If more digits were placed to the right of the 5, they would denote *ten-thousandths, hundred-thousandths, millionths, ten-millionths,* etc., respectively.

It will appear from what has been said that in any group of digits we must know which is the units' digit before we can say what number the group of digits represents.

Thus in

3426,

if 4 be taken as the units' digit, the group will represent 3 tens, 4 units, 2 tenths, and 6 hundredths; but if 2 be taken as the units' digit, the same group will represent three hundred and forty-two and 6 tenths.

The units' digit is distinguished from the others by placing a dot after it, thus in 34·26, 4 is the units' digit; while in 342·6 the units' digit is 2.

If a number does not contain any units, or any higher denomination, it may be written thus:

0·376,

where the cipher takes the units' place, but it is more frequently written

·376,

which denotes 3 tenths, 7 hundredths, and 6 thousandths.

If there are no tenths in a number, a 0 takes its place, as

·076.

In the same way we may have

·003, ·00004, and so on,

which last denotes 4 hundred-thousandths.

Such numbers are called Decimals, and the dot so used is called the Decimal Point.

The expression

$$\cdot 67$$

denotes 6 tenths and 7 hundredths ; that is,

$$\frac{6}{10} \text{ and } \frac{7}{100}$$

$$\text{but } \frac{6}{10} \text{ and } \frac{7}{100} = \frac{6}{10} + \frac{7}{100} = \frac{67}{100}.$$

Therefore ·67 is read *sixty-seven hundredths.*

In the same way ·347 is read *three hundred and forty-seven thousandths*, and so on. That is, every decimal is read as a whole number of the denomination indicated by the last digit to the right.

The expression

$$396 \cdot 89$$

is read three hundred and ninety-six, and eighty-nine hundredths.

EXERCISE.

Write the following numbers in words :

1. 7·6. 2. 39·3. 3. 4·89. 4. 762.
5. ·762. 6. 762·762. 7. 1234·5678. 8. 123·45678.
9. 2400·0036. 10. ·2136. 11. ·0006.
12. ·000006. 13. ·000000006.

Express in digits :

14. Seventy-six and eighty-nine hundredths.
15. Fourteen and three thousandths.
16. One hundred, and three ten-thousandths.
17. One hundred and three ten-thousandths.
18. Thirty thousand and seventy, and one thousand and eighty-three millionths.

In practice we do not read decimals in the manner explained above, but simply name the decimal point and then the digits in order as they occur. 37·34056 is read "37, decimal 3, 4, 0, 5, 6 ;" ·7854 is read "decimal 7, 8, 5, 4."

$$\text{Since } \cdot 6 \text{ is six-tenths } = \frac{6}{10}$$

$$\text{and } \cdot 69 \text{ is sixty-nine hundredths} = \frac{69}{100}$$

$$\text{and similarly } \cdot 347 = \frac{347}{1000} \text{ and } \cdot 03 = \frac{3}{100}$$

it appears that a decimal may be expressed in the form of a fraction by taking the decimal, after removing the point, for numerator and for denominator, 1 followed by as many 0's as there are digits to the right of the decimal point.

And conversely any fraction whose denominator is 1 follow by 0's may be expressed as a decimal by omitting the denominator and placing the decimal point in the numerator in such a way that there will be as many digits to the right of it as there were 0's in the denominator.

$$\text{Thus } \frac{347}{1000} = \cdot347$$

$$\frac{347}{10000} = \cdot0347$$

In this last example, since there are 4 0's in the denominator and only 3 digits in the numerator, a cipher is placed between the point and the first digit of the numerator.

In the same way,

$$\frac{7}{1000} = \cdot007$$

ADDITION OF DECIMALS.

It will be evident from what has been said about decimals that the method to be pursued in adding them is almost identical with that adopted in adding whole numbers. There is this apparent difference, however, that in adding whole numbers we begin by placing units under units, tens under tens, etc., proceeding from right to left in arranging the different numbers under one another, whereas in adding decimals we commence by placing the decimal points under one another, the tenths under tenths, hundredths under hundredths, and so on going from left to right. It will be seen that both these methods amount to this, that we must begin with that part of the number which is certain to exist, for, as every whole number must have a digit in the units' place, though not necessarily in that of tens, hundreds, etc., so every decimal must have a decimal point and also a digit in the tenths' place, though not necessarily in that of hundredths, thousandths, etc.

We might in fact go a step further and assume that, as every whole number must have a digit in the units' place, it must also have a decimal point expressed or understood, since the place of the decimal point is always to the right of the units' figure; we should then have the same method in all cases, viz., to place the decimal points under one another and then arrange the digits of each number in the proper order to the right or left of their respective points.

Thus, to add ·3487, 16·396, 3·04324, and 746·28, arrange thus:

```
   ·3487
 16·396
  3·04324
746·28
766·06794
```

Here we find that in the fifth place from the decimal point there is but one digit, 4. We therefore set it down below the line as 4 hundred-thousandths. In the fourth place we find 7 and 2 and their sum 9 is also set down as 9 ten-thousandths. In the third or thousandths' place we find 8, 6 and 3; their sum is 17, that is 17 thousandths or 1 hundredth and 7 thousandths. We therefore place the 7 down under thousandths and reserve the 1 to add to the column of hundredths, which thus becomes equal to 26 hundredths or 2 tenths and 6 hundredths; we set down the 6 under hundredths and add the 2 to the column of tenths which thus amounts to 10 tenths, but as 10 tenths are equal to unity we place a 0 under the column of tenths and reserve the 1 to be added to the column of units which will thus amount to 16, and so on, until the addition is completed.

EXERCISE.

Add together:

1. 36·4158, 3426, ·3246, 59·637 and 3·6.
2. 57·29577, ·01, ·7854, 4·732 and 1·4142.
3. 3·14159265, 2509, ·1771213 and ·00004848.
4. ·30103, 10, ·5236, 14641 and 2·7182818.
5. ·1, ·901, ·0001, ·000001, ·0000091.

SUBTRACTION OF DECIMALS.

Subtraction of Decimals is performed in the same manner as in whole numbers. Suppose we have to subtract 3·14159265 from 57·29577, we proceed thus :

$$\begin{array}{r} 57\cdot29577 \\ 3\cdot14159265 \\ \hline \end{array}$$

arranging the numbers with regard to the decimal point as in addition.

It will simplify the process at first if ciphers be added to the right of the minuend, thus :

$$\begin{array}{r} 57\cdot29577000 \\ 3\cdot14159265 \\ \hline 54\cdot15417735 \end{array}$$

as these ciphers do not alter the denomination expressed by any of the digits, they can have no effect on the value of the number to which they are affixed.

MULTIPLICATION OF DECIMALS.

If the decimal point in a number be moved one place to the right the number will be multiplied by 10.

Take the number 67·9428 and remove the decimal point one place to the right and we obtain

679·428.

Comparing this with the original number we observe that

8 ten-thousandths has become 8 thousandths.			
2 thousandths	"	"	2 hundredths.
4 hundredths	"	"	4 tenths.
9 tenths	"	"	9 units.
7 units	"	"	7 tens.
6 tens	"	"	6 hundreds,

so that each digit represents ten times as much in the new number as in the old, and therefore the new number is ten times the old one.

If the point be again moved one place to the right it is clear that we shall have 100 times the original number; 3 places, 1000 times, and so on.

It will be evident, too, that if we move the point one place to the left we shall divide a number by 10; if two places, by 100; if three places, by 1000, etc.

In the case of a pure decimal the decimal point is moved to the left by placing ciphers between the point and the nearest figure, thus: ·67 when divided by 10 becomes ·067; and this when divided by 10 becomes ·0067, and so on.

Suppose we wish to multiply 37·643 by 3·86.

We have $$37{\cdot}643 = 37\frac{643}{1000} = \frac{37643}{1000};$$

and $$3{\cdot}86 = 3\frac{86}{100} = \frac{386}{100}.$$

Therefore $$37{\cdot}643 \times 3{\cdot}86$$
$$= \frac{37643}{1000} \times \frac{386}{100}$$
$$= \frac{37643 \times 386}{100000}$$
$$= \frac{14530198}{100000}$$
$$= 145{\cdot}30198$$

Now in this result we observe that if the decimal point be omitted we shall have the product of 37643 and 386, which are the multiplicand and multiplier with their decimal points left out, and that the number of digits after the decimal point in the product is equal to the sum of those in the multiplicand and multiplier, there being 3 in the multiplicand, 2 in the multiplier, and 5 in the product.

Hence, we have the following

RULE.

To multiply together two numbers containing decimals.

Disregard the decimal points and multiply as though the numbers were whole numbers.

Then point off as many decimal places in the product as there are in both the multiplicand and multiplier.

It will sometimes be necessary to place ciphers to the

left of the figures in the product in order to obtain the requisite number of decimal figures, thus,

$$\cdot 3 \times \cdot 2 = \frac{3}{10} \times \frac{2}{10} = \frac{6}{100} = \cdot 06.$$

Where the product of 3 and 2 consists of one digit only, but the product requires two decimal digits, which we obtain by the introduction of the cipher.

EXERCISE.

Find the product of:

1. 1·732 and 1·4142.
2. 2·236 and 2·4495.
3. 147·4771 and ·30103.
4. 57·29577 and 1·01.
5. 20403·1416 and ·000378.
6. ·00004848 and 5463.
7. 101.0101, 37·203 and 3·2.
8. ·04, ·06, ·003 and ·1.
9. ·1, .01, ·001, ·0001 and 1000.
10. ·1, ·1, ·1, ·01, ·01, ·01 and 1000000000.

DIVISION OF DECIMALS.

Suppose we require to divide 57·20575 by 6·79, we proceed thus:

$$57 \cdot 20575 \div 6 \cdot 79.$$

$$= 57\frac{20575}{100000} \div 6\frac{79}{100}$$

$$= \frac{5720575}{100000} \div \frac{679}{100}$$

$$= \frac{5720575}{100000} \times \frac{100}{679}$$

$$= \frac{5720575 \times 100}{100000 \times 679}$$

$$= \frac{5720575 \times 100}{679 \times 100000}$$

$$= \frac{5720575}{679} \times \frac{100}{100000}$$

$$= \frac{5720575}{679} \times \frac{1}{1000}$$

$$= \frac{5720575 \div 679}{1000}$$

$$= \frac{8425}{1000}$$

$$= 8{\cdot}425$$

Here we observe:

1. That the digits in the result, 8425, are obtained by dividing 5720575 by 679, that is, by disregarding the decimal points and dividing as though the given numbers were whole numbers.

2. That the number of decimal places in the quotient is the number in the dividend diminished by the number in the divisor.

Therefore, when the number of decimal places in the dividend exceeds the number in the divisor:

Divide as though both were whole numbers, and point off a number of decimal places in the quotient equal to the difference between that in the dividend and that in the divisor.

It is evident from this that if the number in the dividend is equal to the number in the divisor, there will be no decimal places in quotient.

EXERCISE.

Divide

1. 1·728 by 1·44, by 14·4 and by 144.

2. 4·096 by 10·24, by 102·4, by 25·6, by ·32, by ·064.

3. 53·1441 by 6·561, by 65·61, by 59·049, by 19·683, by ·243 and by 177147.

4. ·390625 by 25, by ·005, by ·000625, by 15625, by ·625.

5. 8·23543 by 117·649, by 16807, by 2401, by ·00343.

6. 16·807 by 3430.

```
3430)16·807(4
     13720        ·004.
      3087
```

Here we find that on dividing as in whole numbers there is a remainder, 3087.

If, however, we had taken 16·8070 for the dividend we should get:

```
3430)16·8070(49
     13720          ·0049
     ─────
      30870
      30870
```

Now, since a cipher placed to the right of a decimal does not alter its value, 16·8070 is the same as 16·807, and therefore ·0049 is the quotient of 16·807 by 3430.

It follows, therefore, that if in any case the division does not terminate, ciphers may be added to the right of the dividend until the division does terminate or until a sufficient number of digits in the quotient has been obtained.

Divide

7. 177·1561 by 146410, 1100, 12100, 13310.
8. 34·2 by 25, 625, 12·5, 250, 500.
9. 1771·561 by 1·4641.

In examples like this when the number of decimal places in the divisor exceeds the number in the dividend, we may add ciphers to the right of the dividend until the number of places equals the numbers in the divisor, and then proceed as before; thus:

```
1·4641)1771·5611(1210
       14641
       ─────
        30746
        29282
        ─────
         14641
         14641
```

and the quotient is therefore 1210.

Divide

10. 16796·16 by ·006, ·216, ·1296, ·07776.
11. 2799·36 by 7·776, ·01296, 4·6656.
12. 1 by ·1, ·01, ·001, ·0001.
13. 1 by 2, 5, 8, 25, 64, 125, 256, 625.

In all the examples given thus far the division terminates if carried far enough. It does not generally happen, however, that the division may terminate, but a stated number of decimal places may be required.

Ex. Divide 3·14159 by 2·718 to four decimal places.

Since the divisor contains 3 decimal places and the quotient is to contain 4, the dividend must contain 7, and as it contains but 5, two ciphers must be added, so that the process of dividing will be as follows:

```
2·718)3·1415900(1.1558
        2718
        ----
         4235
         2718
         -----
         15179
         13590
         -----
          15890
          13590
          -----
           23000
           21744
           -----
```

Divide to 3 places of decimals:

14. 3·1416 by 57·29.
15. 57·2957 by 1·732.
16. 2·4495 by 1·4142.

And to 5 places;

17. 1·09861 by 2·302585.
18. 1 by 2·31258.
19. ·30103 by ·43429.
20. 113 by 355.

TO CONVERT A DECIMAL INTO A VULGAR FRACTION.

Since ·9375 may be expressed as

$$\frac{9375}{10000}$$

We may reduce this fraction to its lowest terms in the manner already indicated. Thus, if we divide successively by 5 we obtain $\frac{15}{16}$.

Similarly $\cdot 5104 = \frac{5104}{10000} = \frac{319}{625}$.

EXERCISE.

Reduce to equivalent vulgar fractions :

1. ·96875, ·8125, ·0064, ·00032.

It will be observed in reducing these decimals that no divisor but 2 and 5 or some powers of them are used, the reason being that 10 and any power of 10, as 100, 1000, etc., have no factors but 2 and 5, or powers of 2 and 5.

TO CONVERT A VULGAR FRACTION TO A DECIMAL.

We may divide 11 by 16 thus :

```
16)11·0000(6875
   9 6
   ----
   1 40
   1 28
   ----
     120
     112
     ---
      80
      80
     ---
```

and since we have placed in the dividend 4 ciphers to the right of the decimal point, there will be 4 decimal places in the quotient, which is therefore ·6875.

But since the quotient obtained by dividing 11 by 16 is expressed by the fraction $\frac{11}{16}$ it follows that

$$\frac{11}{16} = \cdot 6875.$$

This exemplifies the method to be pursued in all cases.

EXERCISE.

Reduce to equivalent decimals :

1. $\frac{1}{4}$, $\frac{1}{2}$, $\frac{3}{4}$, $\frac{1}{8}$, $\frac{3}{8}$, $\frac{5}{8}$, $\frac{7}{8}$.

2. $\frac{3}{16}$, $\frac{5}{16}$, $\frac{7}{16}$, $\frac{9}{16}$, $\frac{11}{16}$, $\frac{13}{16}$, $\frac{15}{16}$.

3. $\frac{1}{5}, \frac{2}{5}, \frac{3}{5}, \frac{4}{5}$.

4. $\frac{24}{25}, \frac{124}{125}, \frac{624}{625}$.

5. $\frac{31}{32}, \frac{63}{64}, \frac{127}{128}, \frac{1}{256}$.

Express as a decimal with 6 decimal places:

6. $\frac{1}{7}, \frac{2}{7}, \frac{3}{7}, \frac{4}{7}, \frac{5}{7}, \frac{6}{7}$.

7. $\frac{31}{37}, \frac{25}{37}, \frac{59}{111}, \frac{146}{333}, \frac{386}{999}$.

8. $\frac{1}{3}, \frac{2}{3}, \frac{1}{6}, \frac{5}{6}, \frac{7}{9}, \frac{8}{11}, \frac{9}{13}$.

CIRCULATING DECIMALS.

If we reduce the fractions $\frac{7}{8}$ and $\frac{5}{6}$ to decimals, we get for the former ·875 and for the latter ·8333 with remainder 2, which will give another 3 in the quotient and another remainder 2, and so on forever. Such decimals as ·8333, etc., are called *Repeating* or *Circulating* decimals. The repeating digits may occur either singly or in groups and may either commence immediately after the decimal point or at some distance from it.

Thus: $\frac{1}{3}$ = ·333, etc., which is written $\cdot\dot{3}$

$\frac{1}{6}$ = ·16666, etc., " " $\cdot1\dot{6}$.

$\frac{11}{12}$ = ·916666, etc., " " $\cdot91\dot{6}$.

$\frac{8}{11}$ = ·72727272, etc. " " $\cdot\dot{7}\dot{2}$.

$\frac{19}{37}$ = ·513513513, etc. " " $\cdot\dot{5}1\dot{3}$.

$\frac{17}{44}$ = ·38636363, etc. " " $\cdot38\dot{6}\dot{3}$.

We have now to examine into the nature of the vulgar fractions which produce these various kinds of decimals.

Since 2 is a factor of 10

Therefore 2×2 is a factor of 10×10

Therefore $2 \times 2 \times 2$ " " $10 \times 10 \times 10$.

That is, any power of 2 is a factor of the same power of 10.

Therefore, by placing ciphers to the right of any number we may make it divisible by any power of 2 without remainder.

Therefore, no fraction which has a power of 2 for its denominator can produce a circulating decimal.

And the same reasoning shows that no fraction whose denominator is a power of 5 can produce a circulating decimal.

And the same is true if powers of 2 and 5 occur in the same denominator, for the numerator may first be divided by the power of 2, and the quotient so obtained divided by the power of 5.

If, therefore, the denominator of a fraction contains no factors but 2 and 5 (either singly or repeated), that fraction cannot produce a circulating decimal.

Again, if we proceed to reduce $\frac{3}{7}$ to a decimal we find that since 10 contains only the factors 2 and 5, and therefore powers of 10 contain only powers of 2 and 5, therefore we cannot by successive multiplications by 10 introduce the factor 7, and therefore we can never make the dividend contain 7 without a remainder. Therefore the decimal obtained from a fraction, whose denominator is 7, will never terminate.

From this it is clear that if the denominator of a fraction (in its lowest terms) contain factors other than 2 and 5, the corresponding decimal can never terminate.

We have next to show that these non-terminating decimals must repeat.

$$7)3\cdot000000000000$$
$$\cdot428571\overline{428571}$$

Reducing the Fraction $\frac{3}{7}$ to a decimal, we find that the remainders in order are

2, 6, 4, 5, 1, 3,

and the dividends must therefore be (after the first)

20, 60, 40, 50, 10, 30.

Now since the last of these is the same as the dividend with which we started,

Therefore the quotient obtained from it will be the same as the first quotient.

Therefore the remainder obtained from it will be the same as the first remainder.

Therefore the next dividend will be the same as the one next to the first, and so on.

Therefore the second set of 6 digits in the quotient must be the same as the first set of 6 digits ; and for the same reason the third set will be the same as the second.

Thus we see that the digits in the quotient must recur in the same order.

And since in dividing by seven there can be only 6 different remainders, it follows that there can be only 6 digits in the repeating part of the decimal. Similarly, if the denominator of the fraction is 17, there can be only 16 digits in the repeating part ; if the denominator is 19, only 18 digits, and so on. So that the number of digits in the repeating part of a decimal cannot be greater than the number of units in the divisor diminished by 1.

The number of digits in the repeating part will not always be so great as this, for if the denominator be 3, 6, or 9 only one digit will repeat ; if 11, two digits ; 13, six digits ; 31, fifteen digits ; and 37 gives but three digits ; 41, five digits.

To determine where the decimal will begin to repeat :

We find that $\frac{1}{3} = \cdot\dot{3}$.

" " $\frac{1}{6} = \cdot 1\dot{6}$.

" " $\frac{7}{12} = \cdot 58\dot{3}$.

" " $\frac{19}{24} = \cdot 791\dot{6}$.

In the first of these the factor 2 does not occur in the denominator, and the decimal begins to repeat at the decimal point.

In the second case the factor 2 occurs *once* in the denominator and one digit occurs between the decimal point and the repeating digit.

In the third case two 2's occur in the denominator and two digits between the decimal point and the first repeating digit.

In the fourth, three, and so on, every 2 which occurs in the denominator having the effect of moving the first repeating digit one place further to the right, and the same will be found to be true if 5 be put for 2 in what has just been stated.

If 2's and 5's both occur in the denominator, the same thing will be true of the one that occurs oftenest, and the other will not affect the position of the first repeating digit. Thus if 2 occurs four times, and 5 three times, there will be four intervening digits; if 5 occur three times and 2 once, there will be 3 intervening digits.

The following examples will shew why this should be the case:

$$1.\quad \frac{19}{28}=\frac{1}{4}+\frac{3}{7}$$
$$=\cdot 25+\cdot\dot{4}2857\dot{1}$$
$$=\cdot 25+\cdot 428571428571\ldots\ldots$$
$$=\cdot 67857142857142\ldots\ldots$$
$$=\cdot 67\dot{8}5714\dot{2}.$$

From this it will be seen that the decimal would repeat from the decimal point were it not for the presence of the ·25, which changes the first two digits and therefore moves the repeating period two places to the right.

$$2.\quad \frac{7}{24}=\frac{2}{3}-\frac{3}{8}$$
$$=\cdot\dot{6}-\cdot 375$$
$$=\cdot 6666666\ldots\ldots-\cdot 375$$
$$=\cdot 291666\ldots\ldots$$
$$=\cdot 291\dot{6}$$

$$3.\quad \frac{13}{15}=\frac{1}{5}+\frac{2}{3}$$
$$=\cdot 2+\cdot\dot{6}$$
$$=\cdot 8\dot{6}$$

4. $\frac{274}{275} = \frac{9}{25} + \frac{7}{11}$

$= \cdot 36 + \cdot\dot{6}\dot{3}$

$= \cdot 36 + \cdot 6363 \ldots\ldots$

$= \cdot 99\dot{6}\dot{3}$

NOTE.—If either of the factors 2 or 5 is common to both terms of any fraction it is obvious that this factor must be struck out before applying the test.

The reduction of a vulgar to a decimal fraction may generally be much shortened, as in the following examples:

1. Reduce $\frac{7}{19}$ to a decimal fraction.

$$19)70(\cdot\dot{3}68421052631578947\dot{3}6$$
```
19)70(
   57
   ---
   130
   114
   ---
    160
    152
    ---
     80.
```

It is evident that the result obtained by dividing 19 into 80 will be half that obtained by dividing 19 into 160. We may therefore begin at the figure obtained from 160 as dividend and divide by 2, and continue this operation. Thus we shall divide 2 successively into 8, 4, 2, 1, 10, 5, 12, 6, etc., and obtain the quotients 4, 2, 1, 0, 5, 2, 6, etc.

2. Reduce $\frac{12}{13}$ to a decimal.

$$13)120(\cdot\dot{9}2307\dot{6}9$$
```
13)120(
   117
   ---
     3.
```

Here, since 3 is $\frac{1}{4}$ of 12, we may commence at once and divide 4 into 9, 12, 3, 30, 27, 6.

Thus, whenever in the operation of reducing a vulgar fraction to a decimal, we reach a dividend which is an exact divisor of any previous dividend, we may commence dividing at the digit obtained as quotient to that previous dividend.

EXERCISE.

Reduce to decimal fractions :

1. $\frac{18}{19}$, $\frac{12}{19}$, $\frac{10}{17}$, $\frac{6}{17}$, $\frac{8}{13}$, $\frac{9}{23}$, $\frac{8}{21}$.

2. $\frac{14}{23}$, $\frac{13}{29}$, $\frac{30}{31}$, $\frac{36}{37}$, $\frac{1}{41}$, $\frac{19}{53}$.

3. $\frac{80}{91}$, $\frac{17}{63}$, $\frac{76}{77}$, $\frac{100}{259}$, $\frac{98}{189}$, $\frac{1}{2849}$, $\frac{1000}{1001}$.

If we reduce $\frac{1}{7}$, $\frac{2}{7}$, $\frac{3}{7}$, $\frac{4}{7}$, $\frac{5}{7}$, $\frac{6}{7}$ to decimals, we have

$$\frac{1}{7} = \cdot\dot{1}4285\dot{7}, \qquad \frac{3}{7} = \cdot\dot{4}2857\dot{1},$$

$$\frac{2}{7} = \cdot\dot{2}8571\dot{4}, \qquad \frac{6}{7} = \cdot\dot{8}5714\dot{2},$$

$$\frac{4}{7} = \cdot\dot{5}7142\dot{8}, \qquad \frac{5}{7} = \cdot\dot{7}1428\dot{5}.$$

In these cases the same digits occur in all the decimals, and in the same order. Thus, 1 is always followed by 4 ; 4, by 2 ; 2, by 8, and so on.

The following examples furnish illustrations of the application of this result :

1. Knowing that $\frac{1}{7} = \cdot\dot{1}4285\dot{7}$, find the decimal equivalent to $\frac{5}{7}$.

By division we get 7 as the first digit in the quotient ; then, knowing that 7 is followed by 1, 1 by 4, etc., we have at once 714285, and therefore

$$\frac{5}{7} = \cdot\dot{7}1428\dot{5}.$$

2. Reduce $\frac{27}{28}$ to a circulating decimal.

```
28)270(·964
   252
   ---
    180
    168
    ---
     120,
```

In this case, since there are two 2's as factors of the denominator, there will be two digits between the decimal point and the first of the repeating digits. Therefore 4 is the first repeating digit, and the others are known at once to be 2, 8, 5, 7, 1, and therefore the required result is $\cdot 96\dot{4}2857\dot{1}$.

Reduce $\frac{1}{175}$ to a circulating decimal.

Here the factors of the denominator are 5, 5, 7; therefore there will be two digits between the decimal point and the repetend, and the repetend will contain six digits. Hence the same method as in Ex. 2, gives the result $\cdot 00\dot{5}7142\dot{8}$.

From the value of $\frac{7}{19}$ given on page 58, we can find the value of any other proper fraction having 19 for denominator. Take for example $\frac{3}{19}$:

On dividing, the first digit in the quotient is 1, but as 1 occurs twice in the repetend, we must obtain by division another digit in the quotient; this is 5. Then we are at once enabled to write down the remaining digits of the repetend, namely, 789, etc., and therefore the required result is $\cdot\dot{1}5789473684210526\dot{3}$.

TO REDUCE CIRCULATING DECIMALS TO VULGAR FRACTIONS.

We know that

$$\frac{1}{9} = .\dot{1}$$

$$\frac{2}{9} = \cdot\dot{2}$$

$$\frac{3}{9} = \cdot\dot{3}$$

Therefore, a pure decimal in which only one digit repeats, is reduced to a vulgar fraction by placing 9 under the repeating digit.

Again,

$$\frac{1}{99} = \cdot\dot{0}\dot{1}$$

$$\frac{2}{99} = \cdot\dot{0}\dot{2}$$

Now suppose we wish to reduce a decimal in which two digits repeat, to a vulgar fraction, for example, $\cdot\dot{3}\dot{7}$.

$$
\begin{aligned}
\cdot\dot{3}\dot{7} &= \cdot 37373737 \text{ - - - -} \\
&= \cdot 30303030 \text{ - - - -} \\
&\quad + \cdot 07070707 \text{ - - - -} \\
&= 10 \times \cdot 03030303 \text{ - - - -} \\
&\quad + \cdot 07070707 \text{ - - - -} \\
&= 10 \times \cdot 0\dot{3} + \cdot\dot{0}\dot{7} \\
&= 10 \times \frac{3}{99} + \frac{7}{99} \\
&= \frac{30}{99} + \frac{7}{99} \\
&= \frac{37}{99}
\end{aligned}
$$

Therefore, when the repeater consists of two digits the corresponding vulgar fraction will consist of these two digits for numerator and two 9's for denominator.

Again $\frac{1}{999} = \cdot\dot{0}0\dot{1}$, $\frac{2}{999} = \cdot\dot{0}0\dot{2}$, &c.

$$
\begin{aligned}
\text{Therefore } \cdot\dot{6}7\dot{3} &= \cdot 673673673 \text{ - - - - -} \\
&= \cdot 600600600 \text{ - - - - -} \\
&\quad + \cdot 070070070 \text{ - - - - -} \\
&\quad + \cdot 003003003 \text{ - - - - -} \\
&= 100 \times \cdot 006006006 \text{ - - - - -} \\
&\quad + 10 \times \cdot 007007007 \text{ - - - - -} \\
&\quad + \cdot 003003003 \text{ - - - - -} \\
&= 100 \times \cdot\dot{0}0\dot{6} + 10 \times \cdot\dot{0}0\dot{7} + \cdot\dot{0}0\dot{3} \\
&= 100 \times \tfrac{6}{999} + 10 \times \tfrac{7}{999} + \tfrac{3}{999} \\
&= \tfrac{600}{999} + \tfrac{70}{999} + \tfrac{3}{999} \\
&= \tfrac{673}{999}.
\end{aligned}
$$

And we may proceed in a similar manner with a decimal having any number of digits in the repeating part.

Therefore, in every case we put the repeating digits for the numerator and as many 9's as there are digits in the repeating part for the denominator of the vulgar fraction.

Thus, $\cdot\dot{3}04\dot{7} = \frac{3047}{9999}$,

and $\cdot\dot{0}09\dot{9} = \frac{99}{9999} = \frac{1}{101}$.

EXERCISE.

Reduce to vulgar fractions in their lowest terms :

1. $\cdot\dot{7}\dot{2}$, $\cdot\dot{5}\dot{4}$, $\cdot\dot{8}\dot{1}$, $\cdot\dot{1}\dot{0}$, $\cdot\dot{0}\dot{1}$, $\cdot\dot{9}$.

2. $\cdot\dot{3}8\dot{7}$, $\cdot\dot{6}02\dot{1}$, $\cdot\dot{7}1428\dot{5}$, $\cdot\dot{0}7692\dot{3}$, $\cdot\dot{0}243\dot{9}$.

The method of reducing a mixed circulating decimal to a vulgar fraction may be shown by an example thus :

$$\cdot 38\dot{7}2\dot{5} = \frac{1}{100} \text{ of } 38\cdot\dot{7}2\dot{5}$$

$$= \frac{1}{100} \text{ of } 38\tfrac{725}{999}.$$

$$= \frac{1}{100} \times \frac{38 \times 999 + 725}{999}$$

$$= \frac{1}{100} \times \frac{38 \times (1000 - 1) + 725}{999}$$

$$= \frac{38 \times (1000 - 1) + 725}{99900}$$

$$= \frac{38000 - 38 + 725}{99900}$$

$$= \frac{38725 - 38}{99900}$$

From this we deduce the following

RULE.

To reduce a mixed circulating decimal to a vulgar fraction :

Write down the digits in the decimal as far as the end of the first period.

From this subtract the part that does not repeat. The remainder will be the numerator of the vulgar fraction, and its denominator will consist of as many 9's as there are digits in the repeating part, followed by as many ciphers as there are digits in the non-repeating part.

Example 1. To reduce $\cdot 387\dot{4}\dot{6}$ to a vulgar fraction;

$$\frac{\begin{array}{r}38746\\387\\\hline 38359\end{array}}{99000}$$

EXERCISE.

Reduce to vulgar fractions in their lowest terms:

(1) $\cdot\dot{2}87\dot{1}$. (2) $\cdot\dot{0}0099\dot{9}$. (3) $\cdot\dot{4}6153\dot{8}$.

(4) $\cdot 1\dot{9}4\dot{5}$. (5) $\cdot 990\dot{3}8461\dot{5}$. (6) $\cdot 982\dot{1}4285\dot{7}$.

The last example may be worked thus:

$$\begin{aligned}\cdot 982\dot{1}4285\dot{7} &= \cdot 857\dot{1}4285\dot{7} + \cdot 125\\ &= \cdot\dot{8}5714\dot{2} + \cdot 125\\ &= \tfrac{6}{7} + \tfrac{1}{8} = \tfrac{55}{56}.\end{aligned}$$

If whole numbers and decimals are combined, the same method is to be pursued in their reduction, thus:

To reduce $27\cdot 3\dot{4}\dot{6}$:

$$\begin{aligned}27\cdot 3\dot{4}\dot{6} &= \tfrac{1}{10} \text{ of } 273\cdot\dot{4}\dot{6}\\ &= \tfrac{1}{10} \text{ of } 273\tfrac{46}{99}\\ &= \frac{1}{10} \text{ of } \frac{273 \times 99 + 46}{99}\\ &= \frac{273 \times 99 + 46}{990}\\ &= \frac{273(100 - 1) + 46}{990}\\ &= \frac{27346 - 273}{990}\end{aligned}$$

That is, we must subtract 273 (the digits which do not repeat) from 27346, and under this result place as many 9's as there are digits in the repeater, followed by as many ciphers as there are digits between the decimal point and the first of the repeating digits.

Ex. Reduce $264\cdot 38\dot{7}42\dot{6}$ to a vulgar fraction.

$$\frac{\begin{array}{r}264387426\\26438\\\hline 264360988\end{array}}{999900}.$$

EXERCISE.

Reduce to vulgar fractions in their lowest terms.

(1) $3{\cdot}4\dot{1}\dot{6}$. (2) $19.3\dot{8}4\dot{2}$. (3) $217{\cdot}0\dot{9}$. (4) $31416{\cdot}00\dot{3}$. (5) $100{\cdot}\dot{1}4285\dot{7}$.

ADDITION AND SUBTRACTION OF CIRCULATING DECIMALS.

1. When absolute accuracy is not required.

A result sufficiently correct for most purposes may be obtained in the manner exemplified in the following example :

Add together the following numbers so as to obtain their sum correct to 4 decimal places :

$$\cdot\dot{3}\dot{7},\ \cdot2\dot{3}\dot{5},\ \cdot35\dot{1}4285\dot{7},\ \cdot\dot{7},\ \cdot0\dot{3}\dot{6}.$$

Extend each decimal to 6 places—2 more than the required number, thus :

$$\begin{array}{r} \cdot373737 \\ \cdot235353 \\ \cdot351428 \\ \cdot777777 \\ \cdot036363 \\ \hline 1{\cdot}774658 \end{array}$$

The difference between this result and the true result is less than one ten-thousandth. That is, it differs from the true result by less than 1 in the fourth decimal place.

EXERCISE.

Add

1. $\cdot\dot{7}$, $\cdot\dot{7}\dot{3}$, $\cdot\dot{8}1\dot{6}$, $\cdot\dot{9}31\dot{2}$, $\cdot\dot{6}320\dot{5}$ to 4 places.

2. $\cdot\dot{6}\dot{3}$, $\cdot\dot{0}6\dot{4}$, $\cdot00\dot{6}\dot{7}$, $\cdot937\dot{6}$, $\cdot04\dot{0}\dot{6}$ to 4 places.

3. $\cdot734\dot{2}$, $\cdot94\dot{7}6\dot{5}$, $\cdot3764\dot{1}\dot{0}$, $\cdot100\dot{1}00\dot{0}$, $38.\dot{7}\dot{4}$ to 5 places.

4. $106{\cdot}\dot{3}\dot{0}$, $937{\cdot}\dot{1}$, $89{\cdot}001$, $387{\cdot}159\dot{1}$, $101{\cdot}1\dot{0}\dot{1}$, $1000{\cdot}\dot{1}000\dot{1}$ to 6 places.

2. When absolute accuracy is required.

We may either reduce the decimals to their equivalent

vulgar fractions and then add—which is apt to be a tedious operation—or we may proceed as follows :—

Required to add $\cdot\dot{3}\dot{6}$, $\cdot 5\dot{3}\dot{9}$, $\cdot 032\dot{4}3\dot{7}$.

First make the repeaters all *begin* at the same distance from the decimal point.

Now, since $\cdot\dot{3}\dot{6} = \cdot 36363636$ - - - - -, it may be written in any of the following ways :

$$\cdot 3\dot{6}\dot{3}$$

$$\cdot\dot{3}63\dot{6}$$

$$\cdot 36\dot{3}\dot{6}$$

$$\cdot 363\dot{6}36\dot{3}$$

$$\cdot 363\dot{6}3636\dot{3}, \text{ etc.}$$

For all of these, when the points are removed, give the same result, namely,

$\cdot 3636363636$ - - - - -

Thus we see that the place where the repeater begins may be moved to the right, but not to the left. We must therefore select that decimal which has its first repeating digit farthest to the right, as our guide in arranging the others.

This is $\cdot 032\dot{4}3\dot{7}$,

therefore the others become $\cdot 363\dot{6}\dot{3}$

and $\cdot 539\dot{3}\dot{9}$.

Secondly, we must make the repeaters all *end* at the same distance from the decimal point.

We see that the first of these must take some such form as,

$$\cdot 032\dot{4}3\dot{7}$$

$$\cdot 032\dot{4}3743\dot{7}$$

$$\cdot 032\dot{4}3743743\dot{7}$$

That is, the number of digits in the repeater must be a multiple of 3.

Similarly in the other two the number of digits in the repeater must be a multiple of 2, and since 6 is a multiple of both 3 and 2, we may have 6 digits in the repeater in each case. Thus these numbers become

$$\begin{array}{r} \cdot 032\dot{4}3743\dot{7} \\ \cdot 363\dot{6}3636\dot{3} \\ \cdot 539\dot{3}9393\dot{9} \\ \hline \cdot 935\dot{4}6774\dot{0} \end{array}$$

We then proceed to add them together, remembering to increase the right-hand column by the number to be carried as the result of adding the digits which would be to the right of this if the decimals were extended.

EXERCISE.

Add

1. $\cdot\dot{3}$, $\cdot\dot{2}6\dot{5}$, $\cdot0\dot{3}\dot{7}$.
2. $\cdot\dot{3}\dot{6}$, $\cdot\dot{3}6\dot{4}$, $\cdot\dot{9}87\dot{3}$.
3. $\cdot000\dot{3}\dot{7}$, $\cdot\dot{4}03\dot{8}$, $\cdot\dot{9}$, $\cdot\dot{2}\dot{7}$.
4. $\cdot\dot{3}9815\dot{7}$, $\cdot00\dot{3}7\dot{5}$, $\cdot000\dot{1}\dot{8}$.

The method to be pursued in subtraction needs no further explanation.

MULTIPLICATION AND DIVISION OF CIRCULATING DECIMALS.

1. When absolute accuracy is required.

Reduce the decimals to their equivalent vulgar fractions:

Perform the required operation:

Reduce the resulting fraction to a decimal.

2. When absolute accuracy is not required.

The requisite degree of accuracy may be secured by the methods of contracted multiplication and division of decimals exemplified below.

CONTRACTED MULTIPLICATION OF DECIMALS.

1. If ·679 be multiplied by ·345, what two digits multiplied together will give tenths in the product? Hundredths? Thousandths? Millionths? If the product of 9 by 5 be omitted, how will the whole product be affected? How far will the new product coincide with the old?

2. Obtain the sixteen partial products whose sum forms the complete product of ·6789 by ·5432; set them down in order of magnitude, beginning with ·30 and add them together.

What digits multiplied together affect only the first four places in the product? Only the last four? Find the sum of the remaining partial products. What decimal places does this sum occupy? What products could be omitted without affecting the first four decimal places of the result?

3. Multiply ·5678 by ·9243, obtaining only those partial products which give digits in the first four decimal places of the product. What digits have you multiplied by 9? By 2? By 4? By 3? What part of the multiplicand have you multiplied by 9? Set it down and multiply it by 9 in the usual way. What part have you multiplied by 2? Multiply it by 2 in the usual way. What part by 4? By 3? Multiply these also in the usual way. What is the denomination of each of these four products? Add them together. How far does this sum coincide with the complete product of ·5678 by ·9243? Explain why this result is not correct to four decimal places when none of the omitted products give digits in the first four places.

4. Explain the following operation:—

```
 5678
 3429
-----
51102
 1134
  224
   15
------
·52475
```

Why is 2 placed under 7? 4 under 6? To how many decimal places is this result correct?

5. Multiply when the numbers are arranged thus;

```
 5678
3429
```

NOTE.—In this case there is no digit in the multiplier by which 8 can be multiplied, nor any digit in the multiplicand which can be multiplied by 3. The products 8 by

9, and 5 by 3 which occurred in the previous result will therefore not appear in this. What other products have also been omitted? To how many decimal places is this result correct.

6. If the multiplier had been 1·9243 where would the 1 have been placed in the process of multiplying?

If the multiplication of 38·34265 by 7·846 be arranged as follows, how many decimal places will there be in the product?

```
38·34265
   6487
-------
2683982
 306736
  15336
   2298
-------
```

Under which decimal place in the multiplicand is the units' digit of the multiplier placed? Where should it be placed to get three decimal places in the result? To get five? Obtain the results to three, four and five decimal places. Find, by comparing it with the complete product, how far each of these results is correct.

Generally, unless the multiplier and multiplicand contain many digits, the product will be correct in all but the last two digits, so that if the result is required to be correct to, say, three decimal places, it will be necessary to multiply so as to obtain five decimal places in the product. That is, to place the units' digit in the multiplier under the fifth decimal place in the multiplicand.

EXERCISE.

1. Find the product of 47·3846 by 3·14159 correct to three decimal places.
2. Find $(1·06)^5$ correct to two places.
3. Find $186·75 × $(1·07)^4$ correct to cents.
4. Find the product of 1·07, 1·035, 1·05, 1·025 correct to four places.
5. Find $597·67 × $(1·05)^5$ ÷ 1·025 correct to cents.
6. Find 37·25 × 37·25 × 3·1416 correct to three places.
7. Find 4 × 3·14159 × $(2·37)^2$ to four places.

8. Find $\frac{4}{3} \times 3{\cdot}1416 \times (238{\cdot}5)^3$ to four places.

9. Find $57{\cdot}2958 \times 3{\cdot}1416$ to three places.

10. Multiply $13{\cdot}\dot{1}4285\dot{7}$ by $9{\cdot}\dot{9}$; $17{\cdot}\dot{3}7$ by $385{\cdot}04\dot{3}9\dot{7}$; and $14{\cdot}285\dot{7}$ by $\cdot0139$, each correct to four places.

CONTRACTED DIVISION OF DECIMALS.

Divide 234398766 by 23456789 to six digits in the quotient.

Omit the 9 in the divisor and divide to six digits. How far does this quotient coincide with the first?

Omit the 8 also and divide to six digits. How far is the quotient correct in this case?

If you divide by 23456 instead of the complete divisor, how far will the quotient be correct?

What relation do you notice between the number of digits in the divisor and the number of digits to which the quotient is correct in each case?

NOTE.—If in any case we divide by, say, the first seven digits of the divisor, the quotient will generally be correct to at least five digits, and may be correct to six or seven. Thus in every case we have only to take as much of the divisor as will contain two digits more than the number required in the quotient.

Divide 2·718281828 by 3·14159265 correct to five digits in the quotient. Take the first seven digits for divisor, thus :—

3141592)2718281828(8
2513273б
2050082

As we have now to obtain four digits in the quotient, we divide by six digits of the divisor, thus :—

314159)2050082(6
1884954
165128

Similarly, we next divide by five digits, thus :—

31415)165128(5
157075
8053

Next dividing similarly by four and three digits of the divisor we get the other two digits of the quotient, which is therefore ·86525, the position of the decimal point being determined in the same way as in ordinary division of decimals.

The complete operation may be arranged thus:

```
3·14159265)2·718281828(
     52568    25132736
              --------
               2050082
               1884954
               -------
                165128
                157075
                ------
                  8053
                  6282
                  ----
                  1771
                  1570
                  ----
                   201
```

A convenient arrangement is to place each digit in the quotient under the right-hand digit of the divisor used in obtaining it; thus, 8 was obtained by dividing by 3141592, and is accordingly placed under 2; 6 is placed under 9 of its divisor 314159; and so on. This enables the multiplications by these digits to be readily effected, and also serves to indicate the next divisor to be used. The quotient, when obtained, may be arranged in proper order in its usual place, and the position of the decimal point determined.

EXERCISE.

Find each of the following to three decimal places:

(1) $\frac{30}{1·00017}$; (2) $\frac{7000}{8·16834}$; (3) $\frac{83·365}{58·995}$;

(4) $\frac{29·995}{1·001294}$; (5) $\frac{13631·361}{1002·766}$;

(6) $\frac{110356·81}{9·815 \times 7990}$; (7) $\frac{1·01315}{1·0012} \times 252·286$;

(8) $19·9568 \times \frac{100}{61·02705} \times \frac{30}{29·997} \times \frac{1}{1·057}$;

(9) $\dfrac{\cdot 6931471}{2\cdot 30258509}$ to six places;

(10) $\dfrac{1\cdot 0986122}{2\cdot 30258509}$ to six places;

(11) $1 \div 2\cdot 30258509$ to eight places.

POWERS OF NUMBERS.

I.

Write the following products in the exponential form, and find their values :—

1. 2×2.
2. 3×3.
3. 10×10.
4. 17×17
5. 20×20.
6. 100×100.
7. 300×300.
8. $\cdot 1 \times \cdot 1$.
9. $\cdot 4 \times \cdot 4$.
10. $\cdot 06 \times \cdot 06$.
11. $\cdot 235 \times \cdot 235$.
12. $\frac{1}{2} \times \frac{1}{2}$.
13. $\frac{1}{7} \times \frac{1}{7}$.
14. $\frac{3}{11} \times \frac{3}{11}$.
15. $5 \times 5 \times 5$.
16. $10 \times 10 \times 10$.
17. $6 \times 6 \times 6 \times 6$.
18. $\cdot 1 \times \cdot 1 \times \cdot 1$.
19. $\cdot 02 \times \cdot 02 \times \cdot 02 \times \cdot 02$.
20. $\frac{3}{2} \times \frac{3}{2} \times \frac{3}{2} \times \frac{3}{2} \times \frac{3}{2}$.
21. $1\frac{2}{3} \times 1\frac{2}{3} \times 1\frac{2}{3} \times 1\frac{2}{3}$.

II.

Find the value of :—

1. 4^2, 5^2, 23^2.
2. $\cdot 3^2$, $\cdot 04^2$, $\cdot 016^2$, $\cdot 518^2$.
3. $(\frac{1}{2})^2$, $(\frac{3}{4})^2$, $(\frac{11}{14})^2$.
4. $(\cdot 21689)^2$, $(\cdot 04635)^2$, $(4\cdot 623)^2$, correct to four places of decimals.
5. $(10)^2$, $(100)^2$, $(1000)^2$.
6. 7^3, 14^3, 163^3.
7. $\cdot 1^3$, $\cdot 02^3$, $\cdot 0001^3$.
8. $(\frac{2}{3})^3$, $(\frac{9}{13})^3$, $(1\frac{3}{10})^3$.
9. $(\cdot 6932)^3$, $(\cdot 02475)^3$, $(16\cdot 4325)^3$, correct to five places of decimals.
10. $4^2 \times 3^3$, $17^2 \times 5^3$, $12^2 \times 10^3$.
11. 7^4, 3^5, 2^7.
12. $(1\cdot 046)^9$, $(2\cdot 0482)^7$, $(1\cdot 06)^{15}$, correct to four decimal places.

ROOTS OF NUMBERS.

1. Obtain the square roots of the following :—1, 4, 9, 16, 25, 36, 49, 64, 81, 100, 144, 196, 256, 324, 64×81, 3600, 8100, $16 \times 25 \times 49$, $64 \times 36 \times 100$, 640000, 81000000.

2. Obtain the cube roots of the following :—1, 8, 27, 64, 125, 216, 343, 512, 729, 1000, 8×343, 27×512, 125×729, 8×1000, 125000, 512000, $27 \times 216 \times 343$, $27 \times 729 \times 1000$, 8000000, 729000000.

3. By resolving into prime factors, obtain the square roots of the following :—36, 64, 100, 576, 729, 1024, 1296, 1764, 6400, 1089, 57600, 640000, 313600, 78400, 11025.

4. Obtain the cube roots of the following in a similar manner :—8, 27, 512, 1000, 216000, 27000, 1728, 9261, 1331, 1331000, 27000000, 42875.

5. Obtain the fifth root of the following in a similar manner :—32, 243, 3125, 1024, 7776, 3200000, 4084101.

6. Obtain the square roots of the following in a similar manner :—$\frac{1}{4}$, $\frac{1}{9}$, $\frac{9}{64}$, $\frac{49}{100}$, $\frac{81}{10000}$, $\frac{64}{729}$, $\frac{100}{729}$, $\frac{64}{11025}$, ·01, ·04, ·09, ·64, 5·76, 10·24, 17·64, 110·25, 1·1025, $\frac{448}{5103}$.

7. Obtain the cube roots of the following in a similar manner :—$\frac{1}{8}$, $\frac{1}{27}$, $\frac{64}{125}$, $\frac{343}{1000}$, $\frac{1000}{729}$, $\frac{64}{9261}$, $\frac{42875}{4096}$, ·001, ·008, ·064, 1·728, 1·331, 9·261, 42·875, ·000001, ·001728, ·042875.

SQUARE ROOT.

I.

1. Multiply 35 by 35.

2. In this example, how many *units* are there in the product of 5 by 5?

3. How many *units* in the product of 3 by 5, and how often is this product found in the whole operation?

4. How many *units* in the product of 3 by 3?

5. What do we obtain if we add the results of examples 2, 3, and 4 together?

6. What is the sum?

7. If the product of 3 by 3 is subtracted from the *proper part* of this sum, what is the remainder?

8. Could any square integer greater than 3^2 or 9 be subtracted in a similar way from 1225?

9. How then can the tens digit of the square root of 1225 be found?

10. How are the digits 3 and 5 combined to form the remainder in example 7?

11. If twice 3 tens is divided into this remainder (since twice 3 tens × 5 forms the greater part of it); what is the quotient?

12. What should be added to the divisor twice 3 tens or 60, to make the division exact?

13. Divide 65 into the remainder of example 7, and find the remainder.

14. As this remainder is 0, what has been subtracted altogether from 1225?

15. What is the tens digit of the square root? The units digit? The whole square root?

16. How was the tens digit of the square root obtained?

17. How was the units digit of the square root *first* obtained?

18. What was added to the trial divisor of example 11 to give a divisor which would divide exactly?

19. If the number had been 1230, after going through the same operation to find its square root, what would the remainder be?

20. When we obtain 5 for a remainder, what has been subtracted from 1230?

21. What is the number subtracted from 1230 the square of?

22. Extract the square roots of 625, 631, 961, 970, and give the remainder in each case.

23. In an example in *long* division, what is used as a trial divisor to obtain the different digits of the quotient?

24. Does the trial divisor always indicate accurately the number of times the whole divisor will be contained in the dividend? Why?

25. In a case where it does not give the right digit, what is your next step?

26. Should the trial divisor in example 11 always give a quotient which will hold when the complete divisor is used?

27. When it does not, how should you proceed?

II.

1. What is the square of 3?

2. What must a number be multiplied by to increase it ten fold?

3. If 3 is increased ten fold, how many fold must the square of 3 be increased to obtain the square of 30?

4. If 30 is increased ten fold, how many fold must the square of 30 be increased to obtain the square of 300?

5. If any number is increased ten fold, how many fold must the square of the number be increased to obtain the square of the new number?

6. How many digits in the product of a units digit by a units digit? of a tens by a tens? of a hundreds by a hundreds?

7. In obtaining any product, what is the digit of *lowest denomination* which is affected by the product of a units digit by a units digit? of a tens by a tens? of a hundreds by a hundreds?

8. What part of 568516 can be omitted, if we wish to obtain only the hundreds digit in the square root?

9. What part can be omitted if we wish to obtain both the hundreds and tens digits?

10. Find the hundreds and tens digits in the square root of this number. (Use only the digits of the number which are required.)

11. How many tens in 7 hundreds+5 tens?

12. When 75 tens squared is subtracted from 5685 hundreds, what is the remainder?

13. When it is subtracted from 568516, what is the remainder?

14. When we have obtained this remainder and the total number of tens in the square root, what is the trial divisor for finding the units digit?

15. What is the complete divisor?

16. What is the remainder after this divison?

17. What is the square root of 568516?

18. Every time we found a digit in the square root, how many digits of the original number did we use?

19. Draw vertical lines between the digits of the num-

ber 840889, so as to indicate the digits which enter together into the operation of finding the different digits of its square root; and find its square root.

20. Find the square roots of 687241, 151321, 45369, 87025.

III.

1. What is the square of 1, ·1, ·01, ·001?

2. What is the square root of 1, ·01, ·0001, ·000001?

3. How many digits must we have in a decimal to give 1 decimal place in the square root? to give 2 places? to give 3 places?

4. For every digit in the square root, how many must we have in the decimal?

5. From what place should we begin to mark off the decimal in periods containing two digits each?

6. Extract the square roots of ·09, ·0049, ·000081, ·000004, 1·21, 1·44, 50·41, 26·01, 6772·41, 4·5369, 297·6423, 354·5, 2649·5, ·4, to 4 places when possible.

CUBE ROOT.

I.

1. How do the digits 7 and 4 enter into combination to obtain the product of 74 by 74?

2. If this product be multiplied by 74, how often will 7 tens cubed be found in the product?

3. How often will 7 tens squared multiplied by 4 be found in the product? How often will 7 tens multiplied by 4 squared be found? How often 4 cubed?

To obtain the answers to the foregoing examples the work will be simplified by proceeding as follows:—

$$70 + 4 = 74$$
$$70 + 4 = 74$$
$$70 \times 4 + 4^2 = 296$$
$$70^2 + 70 \times 4 = 5180$$
$$70^2 + 2(70 \times 4) + 4^2 = 5476 = 74^2$$
$$70 + 4 = 74$$
$$70^2 \times 4 + 2(70 \times 4^2) + 4^3 = 21904$$
$$70^3 + 2(70^2 \times 4) + (70 \times 4^2) = 383320$$
$$70^3 + 3(70^2 \times 4) + 3(70 \times 4^2) + 4^3 = 405224 = 74^3.$$

4. Under what digits in the number 405224 must the cube of 7 be placed to obtain the same remainder, as when 70 cubed is subtracted from it?

5. Could any cube greater than that of 7 be subtracted from 405224 in a similar manner?

6. How then can the tens digit in the cube root of 405224 be found?

7. How are the digits 7 and 4 combined to form the remainder in example 4?

8. If 3×70^2 is divided into this remainder [since $3(70^2 \times 4)$ forms the greater part of it], what is the quotient?

9. What must be added to 3×70^2 to make the division exact?

10. Divide 15556 into the remainder in example 4, and find the new remainder.

11. As this remainder is 0, what has been subtracted altogether from 405224?

12. What is the tens digit of the cube root? the units? the whole cube root?

13. How was the tens digit of the cube root first obtained? The units digit?

14. What was added to the trial divisor in example 8 to make the division exact?

15. If the number had been 405250, after going through the same operation to find its cube root, what would the last remainder be?

16. When the remainder is 26, what has been subtracted from 405250?

17. What is the number subtracted the cube of?

18. Extract the cube roots of 13261, 13269, 830584, 830597, and give the remainder in each case.

19. Should the trial divisor in example 8 always give a quotient which will hold when the complete divisor is used?

20. When it does not, how should you proceed?

II.

1. Every time a number is increased ten fold, how many fold must the cube of that number be increased to give the cube of the new number?

2. How many digits are there in the cube of a units digit?

3. How many digits are there in the cube of a tens digit?

4. How many digits are there in the cube of a hundreds digit?

5. In the cube of any number, what is the digit of *lowest denomination* which is affected by the cube of the units digit? of the tens digit? of the hundreds digit?

6. What part of 594823321 can be omitted, if we wish to obtain only the hundreds digit in the cube root?

7. What part can be omitted if we wish to obtain both the hundreds and tens digits?

8. Find the hundreds and tens digits in the cube root of this number. (Use only the digits of the number which are required.)

9. How many tens in 8 hundreds + 4 tens?

10. When 84 tens cubed is subtracted from 594823 thousands, what is the remainder?

11. When 84 tens cubed is subtracted from 594823321, what is the remainder?

12. When we have obtained the remainder and the total number of tens in the cube root, what is the trial divisor for finding the units digit?

13. What is the complete divisor?

14. What is the remainder after this division?

15. What is the cube root of 594823321?

16. Every time we found a digit in the cube root, how many digits of the original number did we introduce into this operation?

17. Draw vertical lines between the digits of this number, so as to indicate the digits which enter together into the operation of finding the different digits of its cube root.

18. Find the cube roots and the remainders of 771095213, 771095689, 15813251, 15814697.

III.

1. What is the cube of 1, ·1, ·01?

2. What is the cube root of 1, ·001, ·000001?

3. How many digits must there be in a decimal to give

1 decimal place in the cube root? to give 2 places? to give 3 places?

4. For every digit in the cube root, how many must there be in the decimal?

5. From what place should we begin to mark off a decimal in periods of 3 digits each?

6. Extract the cube roots of ·008, ·027, ·000008, ·000512, 1·331, 9·261, 389·017, 14·613, 29·4, ·1, ·08, ·00027, to 3 places of decimals when possible.

MENSURATION.

RECTANGLE.

I.

1. Place 3 units (square feet) in the form of a rectangle.

2. What is the length of this rectangle?

3. What is the width of this rectangle?

4. Place a second rectangle, equal in all respects to the former, with a side contiguous throughout to a side of the former, and in the same plane.

5. What figure do the two rectangles form?

6. What is the length of this rectangle?

7. What is the width of this rectangle?

8. What is its area in terms of the area of the first rectangle?

9. What is its area in square feet?

10. Place a third rectangle, equal to either of the others, alongside the second, in a manner similar to that in example 4.

11. What is the length of this figure?

12. What is the width of this figure?

13. What is its area in terms of the area of the first rectangle?

14. What is its area in square feet?

15. What is its length in yards?

16. What is its width in yards?

17. What name do you give the figure?

18. How many square feet in a square yard?

II.

1. Place 4 square units in the form of a square.
2. What is the length of a side of this square?
3. If this square unit is one-half foot long and one-half foot wide, what is the length of a side of the whole square?
4. How many of these units in a square foot?
5. Make a rectangle whose length is 8 and width 5 of the linear units corresponding to the square unit in example 3.
6. What is the area of this rectangle in those square units?
7. What is the area of this rectangle in square feet?
8. What is the area of a rectangle 4 feet in length and $2\frac{1}{2}$ feet in width?

Find the areas of the following rectangles:

9.	7 rods	in length	and	$5\frac{1}{2}$ rods	in width.
10.	$3\frac{1}{4}$ yds.	"	"	$2\frac{3}{4}$ yds.	"
11.	7 yds. 2 ft. 4 in.	"	"	2 ft. 7 in.	"
12.	6·947 ft.	"	"	3·46 ft.	"
13.	2 mi. 7 per.	"	"	$46\frac{3}{4}$ ft.	"

14. Prove that $30\frac{1}{4}$ square yards are equal to 1 square rod, being granted that $5\frac{1}{2}$ linear yards are equal to 1 linear rod.

TRIANGLE.

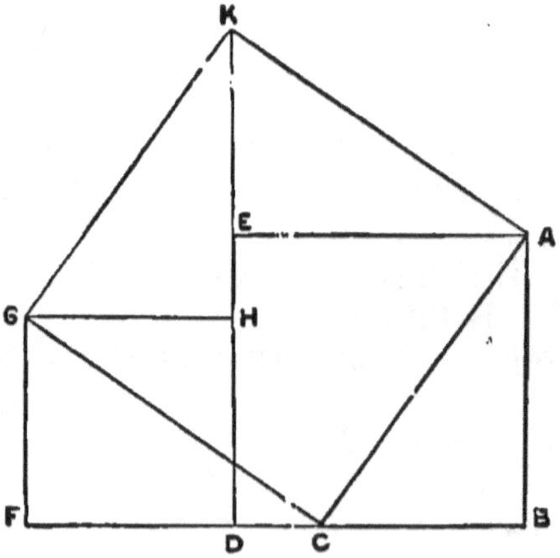

Let ABC be a right angled triangle, with the angle ABC a right angle.

Let ABDE be a square described on AB, and DFGH a square equal to one described on BC, and let BD and FD be coterminous at D, and in the same straight line.

Join GC.

Remove the triangle GFC and place it in the position of GHK.

Also remove the triangle ABC and place it as AEK.

The figure ACGK, which is the square on CA, is composed of the same parts as the two squares ABDE and DFGH.

∴ The squares described on AB and BC are equal to the square described on CA.

NOTE.—If a piece of paper shaped like ABFGHE be cut along the lines AC and GC, then the parts can be put together in the shape of the figure ACGK.

ANOTHER PROOF.

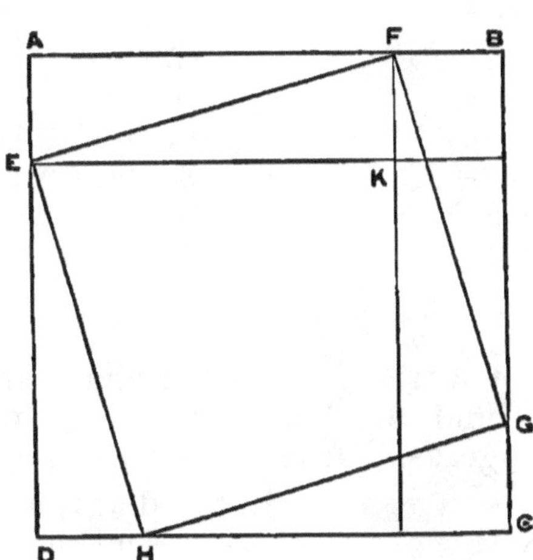

CG, DH, AE, FB are all equal.

The triangles AEF, FBG, GHC, and DEH are all equal.

The triangle AEF $= \frac{1}{2}$ the rectangle AK.

∴ The 4 triangles mentioned are equal to the 2 rectangles AK, and KC.

HEFG is the square on FG.
BK " " FB.
DK is equal to the square on BG.

HEFG + 4 triangles, AEF, FBG, HGC and EDH, are equal to the square AC.

BK + DK + AK + KC are equal to the square AC.

But the 4 triangles were shewn equal to AK + KC.

∴ HEFG is equal to BK + DK.

i.e., The square on FG is equal to the sum of the squares on BF and BG.

Find the length of the hypotenuse in each of the following right angled triangles, given:—

1. The sides 5 ft. and 12 ft. respectively.
2. " 9 ft. " 40 ft. "
3. " 13 ft. " 84 ft. "
4. " 8 yds. " 15 yds. "

5.	The sides 14 in.	and 2 ft.		respectively.
6.	" 2 ft. 2 in.	"	4 ft. 6 in.	"
7.	" 3 yds. 2 ft.	"	5 yds. 4 in.	"
8.	" $7\cdot\dot{6}4$ ft.	"	$2\cdot\dot{9}$ ft.	"
9.	" $7\frac{3}{11}$ yds.	"	$4\frac{4}{13}$ yds.	"
10.	" $17\cdot\dot{5}\dot{6}$ in.	"	$9\cdot\dot{4}$ in.	"

Find the third side in each of the following right angled triangles, given:

11.	The hypotenuse 10 ft.	one side	8 ft.
12.	" " 41 ft.	"	9 ft.
13.	" " 113 ft.	"	15 ft.
14.	" " 14 yds.	"	6 yds.
15.	" " $18\frac{3}{4}$ in.	"	14 in.
16.	" " 4 yds. 2ft. 7 in.	"	3 yds. 1 ft. 9 in.
17.	" " 9·72 ft.	"	8·6 ft.
18.	" " 11·345 ft.	"	107·69 in.
19.	" " $4\cdot6\dot{3}$ yds.	"	$3\cdot\dot{7}$ yds.

EXAMPLE.—The sides of a triangle are 3, 4 and 5 units in length respectively; find the length of the perpendicular dropped on the longest side from the opposite angle.

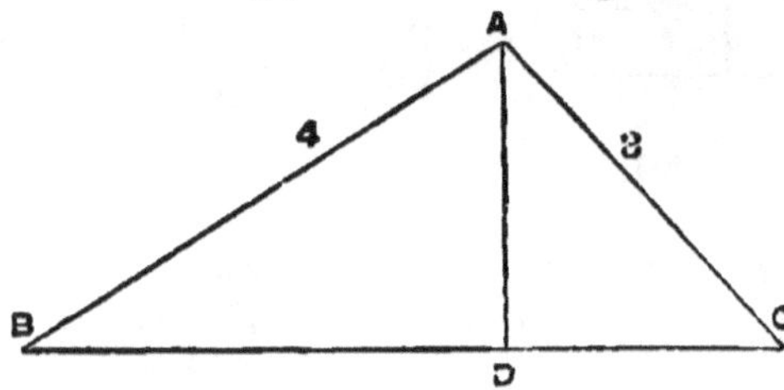

In the diagram let AD represent the perpendicular, and let it contain p units. Also, let DC contain x units.

∴ DB contains 5 – x units.

In the triangle ADC, $p^2 + x^2 = 3^2$ (1)

And in the triangle ADB, $p^2 + (5 - x)^2 = 4^2$ (2)

Subtract (1) from (2) and we get $(5 - x)^2 - x^2 = 4^2 - 3^2$

Therefore $25 - 10x = 7.$

" $10x = 18.$

" $x = \frac{18}{10} = 1\frac{4}{5}.$

Substitute this value for x in (1) and $p^2 + (1\frac{4}{5})^2 = 3^2.$

Therefore $p^2 = \frac{144}{25}.$

" $p = \sqrt{\frac{144}{25}} = \frac{12}{5} = 2\frac{2}{5}$ units.

Find the segments of the base of a triangle into which the perpendicular dropped from the opposite angle divides it; also the length of the perpendicular, the base being given last in each case.

1. Sides are 5, 12, 13 units in length respectively.
2. " 9, 40, 41 ft. " "
3. " 26, 168, 170 yds. " "
4. " 4, 9, 8 ft. " "
5. " 6ft., 7 ft., 19 in. " "
6. " $3\frac{1}{2}$ yds., $9\frac{1}{2}$ ft., $23\frac{1}{3}$ in. " "
7. " $5{\cdot}2$, $7{\cdot}9$, $4{\cdot}6\dot{3}$ yds. " "
8. " $6{\cdot}\dot{4}$, $5{\cdot}\dot{9}3$, $8{\cdot}\dot{4}\dot{5}$ in. " "

---

1. Draw an acute angled triangle having an altitude 10 units, and a base 7 units in length.

2. On the base describe a rectangle having the same altitude and towards the same parts as the triangle.

3. Into how many parts does the altitude divide the triangle?

4. Into how many parts does the altitude divide the rectangle?

5. What is the area of either part of the rectangle in terms of the corresponding part of the triangle?

6. What is the area of the whole rectangle in terms of the whole triangle?

7. What is the area of the whole triangle in terms of the whole rectangle?

8. What is the area of the rectangle in square units? And what the area of the triangle?

Find the areas of the following triangles:—

9. Base 4 ft. and altitude 3 ft.
10. " 2·79 yds. " 8 "
11. " 2·59 in. " 4·68 ft.
12. The sides being 3, 4, 5 ft. respectively.
13. " " " 7, 9, 10 yds. "
14. " " " 11, 15, 17 in. "
15. " " " 2 yds. 5 ft. 15 in. "

CIRCLE.

The circumference of a circle bears a constant ratio to its diameter; this ratio is $3\frac{1}{7}$ nearly, but closer approximations are 3·1416 and $\frac{355}{113}$, and is always denoted by the letter π.

EXERCISE.

Given $\pi = 3\frac{1}{7}$, find the circumference:

1. If the diameter is 4 ft. in length.
2. " " " $5\frac{3}{4}$ yds. "
3. " " " 4 yds. 2 ft. 3 in. in length.
4. " " " 7·468 ft "

Find the diameter:

5. If the circumference is 9 ft. in length.
6. " " " 7 ft. 3 in. "
7. " " " 9·46 yds. "
8. " " " $17{\cdot}4\dot{3}$ yds. "
9. " " " $\cdot 04\dot{6}\dot{9}$ per. "
10. " " " $27\frac{84}{93}$ in. "

Find the circumference:

11. If the radius is 12 ft. in length.
12. " " $4\frac{1}{3}$ yds. "
13. " " $2{\cdot}7\dot{9}\dot{4}$ in "
14. " " 4 yds. 2 ft. 7 in. in length.
15. " " $\cdot 04\dot{6}\dot{7}$ rods "

POLYGON AND CIRCLE.

Take a number of equal isosceles triangles.

1. Express the sum of the areas of two of the triangles in terms of the area of one of them.
2. Express the sum of two bases in terms of one base.
3. What relation is there between the measure of the sum of the areas in example 1, and the measure of the sum of the bases in example 2?
4. Express the sum of the areas of 5 of the triangles in terms of the area of one of them.
5. Express the sum of 5 bases in terms of one base.
6. Find the relation which exists between the measures in examples 4 and 5.

7. If the triangles are of such a shape that when placed in a plane, with a common vertex, their bases will form the perimeter of a complete figure; what is this figure called?

8. If there are 20 triangles in this polygon, express the area of the polygon in terms of one triangle.

9. Express the perimeter of the polygon in terms of one base.

10. Express the measure of the area of one of these triangles in terms of the measures of its altitude and base.

11. Express the measure of the area of the polygon in terms of the measures of the altitude and the base of one triangle.

NOTE.—The base and the altitude of one triangle are respectively equal to the base and the altitude of any other.

12. Express the measure of the area of the polygon in terms of the measures of its perimeter and the altitude of one triangle.

13. What is to be noticed regarding the number of triangles required to make a complete polygon, if the bases of the triangles become very short, while the sides remain the same length?

14. What other figure does this one resemble?

15. If we round off the corners of this figure, so as just to destroy the angles, how much of the figure is removed?

16. What is the difference between the areas of the figures in examples 14 and 15.

17. What is the name of the figure in example 15?

18. What is the difference in length between the side of one of the triangles and its altitude in example 13?

19. In example 15, what is a side of a triangle called?

20. In example 15, what is the perimeter called?

21. Find the measure of the area of the figure in example 15 in terms of the measures of its radius and circumference.

22. Find the area of a circle whose radius is $3\frac{1}{2}$ feet and circumference 22 feet.

23. Find the area of a circle whose radius is 4 feet.

24. Find the area of a circle whose radius is 5 yds. 2 ft.
25. " " " " 6·42 ft.
26. " " " diameter is $2\frac{3}{4}$ in.
27. " " circumference is 11 ft.
28. " " " " 9 ft. 4 in.

SECTOR OF A CIRCLE.

1. In a circle whose circumference is 144 in., what part is an arc of 12 in. of the whole circumference?

2. In the same circle, what part is a sector of a circle, which stands on this arc, of the whole area?

3. If the circumference of a circle is 140 ft., find the length of the arc on which a sector, which is one-fifth of the whole area, stands.

4. In a circle which is 72 ft. in circumference, find the area of the sector which stands on an arc 9 ft. in length?

5. Find the area of a sector of a circle which stands on an arc 10 ft. in length, if the radius of the circle is 7 ft.

6. Find the area of the sector of a circle which stands on an arc equal in length to the radius of the circle; given the radius 3 ft.

CYLINDER.

1. Take a sheet of foolscap paper, bring the two ends into contact, and have each side in the form of a circumference of a circle.

2. If the space inclosed were a solid mass, what would it be called?

3. What does an end of the paper represent on this mass?

4. What does a side represent?

5. Find the area of the surface of the paper required to enclose this solid, if the end of the paper is 8 inches and the side 14 inches in length.

6. Find the area of the curved surface of a cylinder the height of which is 7 ft. and the circumference of the base 12 ft.

7. The radius of the circular base of a cylinder is $3\frac{1}{2}$ ft. and the height 3 ft.; find the area of the curved surface.

8. The curved surface of a cylinder contains 24 sq. ft. and the height is 4 ft.; find the circumference of the base.

9. In the preceding example, find the radius of the base.

10. If the radius of the base of a cylinder is 3 ft. and the curved surface contains 49 sq. ft.: find the height.

CONE.

1. Take a piece of paper in the form of a sector of a circle, and bring the bounding radii into contact, so that the arc will form the circumference of a circle.

2. If the space inclosed were a solid mass, what would it be called?

3. What does a bounding radius represent on this mass?

4. What does the arc represent?

5. What does the centre of the sector represent?

6. If the arc is 7 ft. and the radius 6 ft. in length, find the area of the surface of the paper required to enclose this solid.

7. Find the area of the curved surface of a cone whose slant side is 6 ft. and the circumference of the base 7 ft.

8. Given the slant side of a cone 9 ft. and the circumference of the base 12 ft., find the area of the curved surface.

9. Given the slant side of a cone 11 in. and the radius of the circular base 7 in., find the area of the curved surface.

10. Given the area of the curved surface of a cone 49 sq. in. and the slant side 7 in., find the circumference of the base.

11. In the preceding example, find the radius of the base.

12. Given the area of the curved surface 77 sq. in., and the radius of the base 2 in., find the slant side.

13. Given the slant side of a cone 5 ft., and the radius of the base 3 ft., find the perpendicular height.

14. Given the slant side 7 ft., and the radius of the base 4 ft., find the perpendicular height.

15. Given the slant side 9 in., and the circumference of the base 14 in., find the perpendicular height.

16. Given the perpendicular height 4 ft., and the radius of the base 3 ft., find the slant side.

17. Given the perpendicular height 11 in., and the circumference of the base 21 in., find the slant side.

18. Given the perpendicular height 12 in., and the radius of the base 5 in., find the area of curved surface.

19. Given the perpendicular height 13 in., and the radius of the base 6 in.. find the area of the curved surface.

20. Given the perpendicular height 8 ft., and the circumference of the base 28 ft., find the area of the curved surface.

21. Given the perpendicular height 12 in., and the circumference of the base 19 in., find the area of the curved surface.

SPHERE:

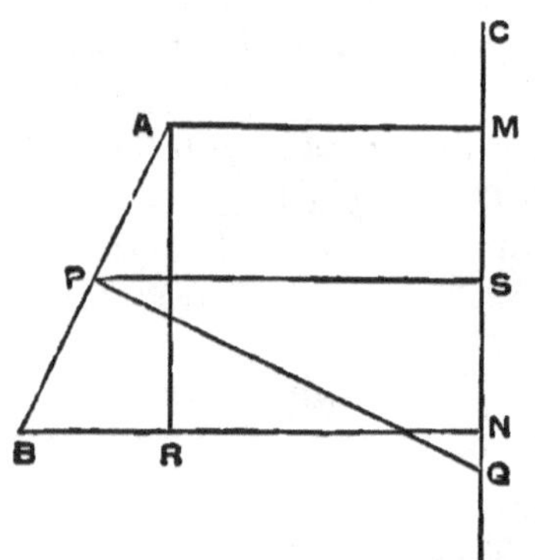

Let AB and CD represent two straight lines. P the middle point of AB.

Draw AM, PS, BN perpendicular to CD.

Draw PQ perpendicular to AB, and AR perpendicular to BN.

Revolve AB around CD as an axis.

The measure of the area of the belt which AB describes is clearly equal to the product of the measures of AB and the circumference of the circle which P describes.

∴ The measure of the area of the belt is $2\pi \times AB \times PS$, where AB and PS stand for the measures of the lengths of these lines.

The triangles ABR and PSQ are similar.

$$\therefore \frac{PS}{PQ} = \frac{AR}{AB}.$$

∴ $PS \cdot AB = PQ \cdot AR$, and $AR = MN$.

∴ The measure of the area of the belt is $2\pi \cdot PQ \cdot MN$. (MN is called the projection of AB on CD.)

1. Take half of the perimeter of a regular polygon of an even number of sides, and draw the diagonal joining its extremities.

2. Where do the perpendiculars from the middle points of the sides cut this diagonal?

3. Compare the lengths of these perpendiculars.

4. Revolve the half polygon in example 1 around the diagonal.

5. What factors in the measures of the belts in example 4 are equal?

6. What factors are unequal?

7. What is the sum of the unequal factors?

8. What is the measure of the sum of the areas of the belts in example 4, in terms of the measures of the perpendicular from the middle point of one of the sides and the diagonal of the polygon?

9. If the sides of the polygon in example 1 become very short, and increase in number, while the diagonal remains constant; what does the solid formed by its revolution resemble?

10. If the edges are smoothed off, what is the solid called?

11. What does the perpendicular from the middle point of any one of the sides in example 1 become in example 10?

12. What is the measure of the area of the surface of a sphere in terms of the measures of its radius and diameter?

13. What is the measure of the area of a sphere in terms of the measure of its radius?

14. Find the area of the surface of a sphere whose diameter is 7 ft.

15. Find the areas of the surfaces of the spheres, the radii of which are respectively 7 ft., 4 in.; 9·265 in.; 9·6̇1̇ yds.; 4 yds., 2·56 ft.

16. Find the radii of the spheres, the surfaces of which are respectively 121 sq. ft.; 5½ sq. ft.; 4 sq. yds., 3 sq. ft.; 478·96 sq. in.; 196·03̇ sq. in.

RECTANGULAR SOLID.

1. Place 5 units (cubic feet) in the form of a rectangular solid.

2. What is the length of this solid? What is the width?

3. Place an equal rectangular solid alongside of this one, so as to form with it a single rectangular solid.

4. What is the volume of the new solid, which is formed, in terms of the first one?

5. What is its volume in cubic feet?

6. Take 4 equal rectangular solids like the one in example 1, and place them on a plane so as to form a single rectangular solid.

7. What is the volume of this solid in terms of one of them?

8. What is the volume in cubic feet?

9. Place squarely on top of this solid another equal to it in all respects.

10. What is the new solid called?

11. What is its length, width, height?

12. What is its volume in terms of the solid in example 6?

13. What is its volume in terms of the solid in example 1?

14. What is its volume in cubic feet?

15. Place 8 equal cubes, edge $\frac{1}{2}$ ft., in the form of a single cube.

16. What is the name of the cube so formed?

17. Find the volume of a cube whose edges are 3 ft. each.

18. What is the name of such a cube?

19. How many cubic feet in a cubic yard?

20. Find the volumes of the rectangular solids of the following dimensions:— 3, 4, 5 ft. respectively; $2\frac{1}{2}$, $3\frac{1}{4}$, 8 yds. respectively; $2\frac{1}{4}$ yds., 4 ft., 13 in.; 3·46, 5·9, 8 ft. respectively; $2\cdot6\dot{2}$ yds., $4\cdot\dot{3}\dot{8}$ ft., $12\cdot\dot{9}$ in.

21. If the volume of a rectangular solid is 1728 cubic inches, and the height is 12 inches; find the area of the base.

22. If the base in example 21 be a square; find a side of it.

23. If the volume is 864 cub. ft., and the area of the base is 108 sq. ft.; find the height.

24. Given the volume 725 cub. yds., the height $7\frac{1}{2}$ yds., the length of the base $20\frac{1}{4}$ yds., find the width of the base.

25. Given the volume 847 cub. yds., the height $25\frac{1}{2}$ ft., one side of the base 56 inches, find the other side.

PRISM.

1. What is the volume of a solid 1 foot in height, whose base is a polygon containing 9 square feet?

2. Place another solid equal in all respects to the former exactly on top of it, so that the base of the second solid will coincide throughout with the top of the former.

3. What is the name of the new solid formed?

4. What is its volume in terms of the volume of the first?

5. What is its height?

6. What is the volume in cubic feet?

7. Place 8 solids, each identically equal to the first, on top of one another in a manner similar to that in example 2.

8. What is the new solid called?

9. What is its volume in terms of the volume of one of them?

10. What is its height?

11. What is the volume in cubic feet?

12. Find the volume of a prism 10 feet in height whose base contains 14 square feet.

13. Find the volume of a prism 3 inches in height, whose base is a square, with an edge 2 inches in length.

14. The sides of the rectangular base of a prism are 4 and 5 feet respectively, and the height is 7 feet; find its volume.

15. The sides of the base of a triangular prism are 3, 4 and 5 feet respectively, and the height is 6 feet; find its volume.

16. The sides of the base of a triangular prism are 7, 8, 9 feet respectively, and the height is 11 feet; find its volume.

17. The sides of the base of a triangular prism are 7 yds., 19 ft. and 30 inches, and the height is 7·168 ft.; find the volume.

18. If the base of a prism is a polygon of a great number of sides, but each side very short, what other solid does this prism resemble?

19. If the sharp corners, along the length of the prism, are smoothed off a very little, what is the name of the new solid formed?

20. How much of the prism has been removed by smoothing off the corners?

21. What is the difference between the volumes of the solids in examples 18 and 19?

22. What is the measure of the volume of a cylinder in terms of the measures of the area of the base and the height?

23. Find the volume of a cylinder 9 feet in height, and standing on a base whose area contains 12 square feet.

WEDGE.

1. Lay a triangular prism on one of its rectangular surfaces.

2. Find the volume of a triangular prism 9 feet in length, whose triangular end contains 12 square feet.

3. Find the volume of a triangular prism 5 ft. in length, whose triangular end has an altitude of 8 ft., and the base 3 ft. in length.

4. The rectangular base on which a triangular prism rests is 3 feet in length and 2 feet in width, and the height of the prism is 9 feet; find its volume.

5. What other name is given to this solid?

6. What is the volume of a wedge 9 feet in height, which stands on a square base containing 4 square feet?

7. A wedge 5 feet in height, stands on a square base, whose edge is 4 feet; find its volume.

8. A wedge 9 feet in height, stands on a rectangular base 4 feet in length and 3 feet in width; find its volume.

CYLINDER.

1. How many cubic feet in a right circular cylinder 1 foot in height, whose base contains 12 square feet?

2. Place a second cylinder, equal in all respects to the former, exactly on top of it?

7

3. What do the two cylinders form?

4. What is the height of the new cylinder?

5. How many cubic feet does it contain?

6. Place a third equal cylinder exactly on top of the second.

7. What do the three cylinders form?

8. What is the volume of this new cylinder in terms of the first cylinder?

9. What is its height?

10. What is its volume in cubic feet?

11. A cylinder, whose base contains 12 square feet, is 7 feet in height; find its volume.

12. The radius of the base of a cylinder is 4 inches in length and the height 7 inches; find the volume.

13. The circumference of the base of a cylinder is 14 feet and the height 8 feet; find the volume.

14. Given the volume 144 cubic feet and the height 9 feet, find the area of the base.

15. In the preceding example, find the radius of the base.

16. Given the volume 108 cubic inches and the area of the base 16 square inches, find the height.

17. Given the volume 98 cubic inches, and the radius of the base 4 inches, find the height.

18. Given the volume 396 cubic feet and the circumference of the base 35 feet, find the height.

PYRAMID.

1. If the bases of two triangular pyramids are equal in area, and the pyramids have equal altitudes, how are their volumes related to each other?

2. Divide a triangular prism into 3 triangular pyramids.

3. Compare their volumes.

4. What is the volume of the whole prism in terms of the volume of one of these pyramids?

5. What is the volume of one of the pyramids in terms of the volume of the whole prism?

6. What is the measure of the volume of the prism in terms of the measures of its base and altitude?

7. What is the measure of the volume of one of the pyramids in terms of the measures of the base and altitude of the prism?

8. What is the altitude of the first pyramid which was taken from the prism?

9. What is its base in terms of the base of the prism?

10. What is the measure of the volume of this pyramid in terms of the measures of *its* base and altitude?

11. What is the measure of the volume of any triangular pyramid in terms of the measures of its base and altitude?

12. If we have a number of equal triangular pyramids of such a shape that when resting on their bases they have a common apex, and one edge common to all of them, what complete solid do they form?

13. What is the volume of this whole pyramid in terms of the volume of one of the triangular pyramids, if it takes 6 triangular pyramids to form it?

14. What is the area of the base of the whole pyramid in terms of the base of one of the triangular pyramids?

15. What is its altitude?

16. What is the measure of the volume of the whole solid in terms of the measures of its base and altitude?

17. If its base contains 20 square feet, and its altitude is 9 feet, what is the volume?

18. Find the volume of a pyramid 6 feet in height which stands on a square base, whose edge is 4 feet.

19. The volume of a pyramid is 72 cubic inches, and its height is 12 inches; what is the area of the base?

20. The volume of a pyramid which stands on a square base is 96 cubic inches, and its height is 14 inches; find a side of the base.

21. A pyramid which stands on a rectangular base 5 feet in length and 4 feet in width, is 7 feet high; find its volume.

22. The height of a triangular prism is 9 feet, and the sides of the triangular base are 3, 4 and 5 feet respectively; find the volume.

23. The sides of the triangular base are 7, 9, 10 feet respectively, and the height of the pyramid is $4\frac{1}{2}$ yards; find the volume.

CONE.

1. Take a pyramid whose base is a polygon of a very great number of sides, and each side very short.

2. What does the perimeter of the base resemble?

3. Smooth off the slant edges of this pyramid.

4. What is it now called?

5. What is the difference between the volumes in examples 2 and 3?

6. Find the volume of a cone, given its altitude 9 ft and the area of the base 27 sq. ft.

7. Find the volume of a right circular cone 8 ft. high, the radius of whose base is 7 ft.

8. Given the volume of a right circular cone 77 cub. in. and the height 6 in., what is the area of the base?

9. In the preceding example, what is the radius of the base?

10. In a right circular cone the volume is 47 cub. yds., and height 4 yds.; find the radius of the base.

SPHERE.

Take a number of equal pyramids.

1. Place two of them side by side with their apexes at the same point.

2. What is the volume of this new solid in terms of the volume of one pyramid?

3. What is the area of the two bases in terms of the area of one base?

4. If a number of them be placed so as to form a complete solid figure, with the common apex in the centre of the solid; what is the name of this solid?

5. If it takes 20 pyramids to form this polyhedron; what is its volume in terms of the volume of one pyramid?

6. What is the area of its surface in terms of the area of the base of one pyramid?

7. What relation exists between the measures in examples 5 and 6?

8. What is the measure of the volume of this solid in terms of the measures of the altitude and the base of one pyramid?

9. What is the measure of the volume of this solid in terms of the measures of the altitude of one pyramid, and the surface of the whole solid?

10. If the bases of the pyramids become very small, what is to be noticed regarding the number required to form the solid in example 4?

11. What other solid does it now resemble?

12. What is the difference between the altitude and slant edge of a pyramid like those in example 10?

13. If the edges of the surface in example 10 are smoothed off, what is the solid then called?

14. What is the difference in volume between the solids in examples 10 and 13?

15. What are the altitudes of the pyramids in example 10 termed in example 13?

16. Find the measure of the volume in example 13 in terms of the measures of its surface and radius.

17. Find the volume of a sphere, having given the radius 7 in. and the surface 616 sq. in.

18. Find the volume of a sphere whose radius is 4 in.

19. Find the radius of a sphere whose volume is 38808 cubic inches.

20. Find the volume of a sphere whose surface is 154 square feet.

21. Find the area of the surface of a sphere whose volume is 480 cubic feet.

PROBLEMS—METRIC SYSTEM.

1. Express 1000 centimetres in metres; 1000 centimetres in decimetres; 1000 centimetres in dekametres.

2. Express 1000 metres in kilometres; 1000 metres in dekametres; 1000 metres in decimetres.

3. Express 12 kilometres in dekametres; 15 kilometres in metres; 21 kilometres in millimetres.

4. Express 123,456,789 millimetres in decimetres; in metres; in kilometres.

5. Express 8·56 Km. in centimetres; 5·632 m. in millimetres; 12468 mm. in microns.

6. Add 14·6 m., 227 cm., 162·3 Dm., 1634 Km ; express the result in metres.

7. Express in centimetres the difference between 5·678 Km. and 1364·89 Dm.

8. Multiply 12 Km. 5 m. 8 cm. by 96.

9. Given that 64 miles is very nearly equal to 103 Km., express 1 yard in centimetres.

10. A train is running at the rate of 66 Km. per hour. How many metres does it go at this rate in one second.

11. A train running at the rate of 60 Km. per hour passes over 20 spaces between telegraph poles in one minute. Find the distance between two consecutive poles in metres.

12. Using the approximate value, 1 metre = 39·37 in., express the height of a man, 5 ft. 10½ in., in centimetres.

13. Using the same approximation, express in millimetres the height of a barometer which stands at 29·5 in.

14. Find the length of the two parts into which a string 10 m. long is divided, given that one-half of one part and two-thirds of the other make a length of 600 cm.

15. How long will it take a man to walk from Toronto to Hamilton, a distance of 65 Km., at the rate of 80 m. per minute ?

16. The radius of a wheel is 1·4 m. ; how many times will it revolve in going 55 Km. ? ($\pi = \frac{22}{7}$.)

17. Taking a centimetre as $\frac{2}{5}$ of an inch, find the number of millimetres in 1 yd.

18. How many lengths each 1 m. 5 cm., can be cut from a length of 1 Km. ? Express the remainder in mm.

19. Find the number of square metres of carpet required to cover a floor, the dimensions of which are 6 m. 1·75 dm. by 4 m. 12 cm.

20. Find in metres the side of a square of which the area is 15227·56 sq. metres.

21. Express 1200 ares in hectares ; 1200 ares in centiares ; 1200 ares in dekares.

22. Express 12345 milliares in ares ; 5678 centiares in dekares ; 1 hectare in centiares.

23. Find the sum of 12·64 ares, ·0168 hectares, one million milliares. Express the result in ares.

24. Express in dekares the difference between 1 Ha., and 1 are.

25. Multiply 7 Ha. 5 Da. 6 a. 4 da. 5 ca. by 27.

26. Find in hectares the area of a rectangular field, of length 4 Hm. and of breadth 2 Hm. 7 Dm.

27. The area of a square field is 18·49 Ha.; find the perimeter in metres.

28. Express 1 square decimetre in milliares.

29. Find in centiares the whole surface of a cube, the volume of which is 27 cubic metres.

30. Express in square centimetres 1 are 5 da.

31. A parallelepiped whose edges are proportional to 2, 3 and 4 contains 3 cubic metres; find its whole surface in centiares.

32. A block of wood contains 4·5 cubic metres; find in ares the area it will cover if cut into sections 1 cm. thick.

33. Find in cubic metres the volume of a sheet of ice covering a pond of area 20 Ha. and thickness 3 cm.

34. Find the cost of painting the walls and ceiling of a room 5 m. long, 4 m. wide and 3·5 m. high, at 75 cents per milliare.

35. How many ares are there in a square field whose diagonal is $10\sqrt{2}$ metres?

36. The sides of a rectangular field are proportional to 2 and 3, its area is 24 Ha.; find the length of its diagonal in metres.

37. Using the approximation, 1 metre = 39·37 in., find the number of acres in 1 Ha.

38. A rectangular garden 15m. long and 12m. broad is surrounded by a path 1·5m. wide; find the area of the path in centiares.

39. A circular area of 7 metres radius is divided into two portions in the ratio of 3 to 4; express the area of the smaller portion in ares.

40. Using the approximation 1 Ha. = 2·5 acres, express 1 metre in inches.

41. Express in steres the volume of a cube whose edge is 3 metres.

42. The volume of a cube is 1 Ks.; find the length of its edge in centimetres.

43. How many cubic centimetres in 1 stere?

44. How many steres in one million cubic centimetres?

45. A pile of wood is 45 metres long, 1·5 metres wide and 8 metres high; find its value at \$2 per stere.

46. Find the cost of excavating a cellar 12 m. long, 10 m. broad and 3·5 m. deep, at 25 cents per stere.

47. Express 1 ds. in cubic decimetres.

48. Given 1 metre = 39·37 inches, find the number of cubic feet in 1 stere.

49. Also express the stere as the fraction of a cord.

50. How many steres of wood in a rectangular floor 40 m. long and 30 m. wide, the flooring being 2 cm. in thickness?

51. The excavation of a cellar, the dimensions of which are in the ratio 1, 4 and 5, cost \$54, at 10 cents per stere; find its dimensions.

52. If a wall 15 m. long, 1 m. thick, and 3·4 metres high contains 51000 bricks, how many bricks of the same size are required for a wall of which the volume is 45 steres?

53. Find, in milliares, the surface of a cube whose edge is equal to the side of a square, the area of which is one centiare.

54. Given that 1 cubic centimetre of water weighs 1 gramme, find in kilos. the weight of one stere of earth which is 2·5 times as heavy as water.

55. Water expands $\frac{1}{10}$ in freezing; find the weight of 1 stere of ice.

56. Each linear dimension of a block of metal is increased ·002 of itself by heating; find the volume of a block which before heating was 1 stere.

57. One centistere of metal is rolled into a rectangular sheet 10 m. by 8 m.; find its thickness in microns.

58. One stere of water is poured into an empty tank 2 m. long and 1·25 m. wide; find the depth of the water in the tank.

59. An empty tank is 4 m. long and 2·5 m. wide; water is poured in at the rate of 1 stere per minute; in what time will the water be 1 dm. deep?

60. From a cylindrical tunnel 30800 steres of earth were

removed; the diameter of the cross section is 14 metres; find the length of the tunnel. $(\pi = \frac{22}{7})$.

61. If 27 steres of wood last 3 months, what must be the length of a pile 1·5 m. wide, 2 m. high, to last 1 year?

62. How many cubic decimetres in 1 cubic metre? How many litres in 1 stere?

63. Given that 1 cu. cm. of water at 4°C. weighs 1 gramme, find the weight of 1 litre of water.

64. What must be the depth of a rectangular box 12·5 dm. long and 4 dm. wide to contain 1 Kl. of water?

65. Compare the cubic centimetre and the millilitre.

66. Express 4 cubic metres, 4 cubic decimetres, 4 cubic centimetres in steres.

67. How many litres of water at 4°C. will weigh 1000-000 grammes?

68. Given that a solid loses in weight when immersed in water, an amount equivalent to the weight of the water displaced, find the volume of a solid which weighs 1,000 grammes less when immersed in water at 4°C., than when weighed in air.

69. Using the approximation 1 metre = 39·37 inches, express the litre in terms of the cubic foot.

70. Using the above approximate value of the metre, express the gallon (277·274 cu. in.) in litres.

71. Express the difference between the quart and the litre in millilitres.

72. Compare the kilolitre and the stere.

73. The dimensions of a rectangular tank are 3 m., 2 m., 1·5 m.: in what time would it be filled from two taps, the one pouring in 5 l. in 3 sec., and the other 5 Dl. in 20 sec.?

74. A cylindrical log has radius 1 m. 4 dm.; what length of it must be cut off to contain 61·6 steres?

75. A cubical box, made of material 5 cm. in thickness, is filled with 1 stere of water; find the amount of material in the box.

76. Find the depth of a cistern, the length and breadth being equal and double the depth, which will contain four million litres.

77. Using the approximation, 1 metre = 39·37 in., find the number of litres in 1 cubic foot.

78. Find approximately the weight of a kilolitre of air under certain standard conditions, given that water is 770 times as heavy as air.

79. The area of the end of a rectangular box is 1 ca.; find its length if the volume is 1 Kl.

80. Find the number of cubic centimetres in one litre.

81. Express the kilogramme in milligrammes.

82. Find the number of grammes in 1 tonneau.

83. Find in kilos. the weight of 10 cubic metres of water at 4°C.

84. What is the weight of 1 litre of mercury in kilos., given that mercury is 13·5 times as heavy as water?

85. A litre of sulphuric acid weighs 1840 grammes: compare the weights of equal volumes of sulphuric acid and water.

86. Given that an ounce Av. = 28·35 grammes, find the number of kilos. in a ton (2000 lbs.)

87. Using the above approximation, find the number of grammes in 1 lb. Troy.

88. Find the weight of a block of iron (7 times as heavy as water) of which the dimensions are 5 dm. long, 4 dm. broad, 3·5 dm. thick.

89. A watering cart is 3 m. long, 1·5 m wide and 1 m. high; find the number of tonneaus of water it carries.

90. Read 123456789 grammes as kilogrammes; as tonneaus.

91. Given that the kilogramme is equal to 2·679 lbs. Troy, find the number of lbs. Av. in the kilo.

92. Find in centiares the area of a triangle, the sides of which are 1·3, 1·4, 1·5 metres respectively.

93. The parallel sides of a trapezium are 2·6 m., and 1·4 m.; find the perpendicular distance between these sides, given that the area is ·75 centiares.

94. Find in litres the volume of a conical vessel, the diameter of the base being 21 cm., and the height 1 dm.

95. Find in kilogrammes the weight of a pyramid of lead, 6 dm. high, having a square base the side of which is 40 cm.; 1 cubic cm. of lead weighs 11·4 grammes.

96. Find the weight of a sphere of tin of radius 3·5 dm.; 1 cu. cm. of tin weighs 7·3 grammes.

97. How long will it take a man to walk around a square field whose area is 1·44 Ha., at the rate of 5 Km. per hour?

98. A man walks 5 Km. per hour; express this rate in cm. per second.

99. The radius of the hemisphere is ·56 m.; find the height of the equivalent right circular cone, the diameter of the base being equal to the diameter of the hemisphere.

100. Express the chain (66 ft.) in metres.

101. Express in hectares the area of a farm which contains 250 acres.

102. The length of the Great Western Railway is 229 miles; find its length in kilometres.

103. Express 5280 feet in metres.

104. Express the tonneau as the decimal of a long ton (2240 lbs.)

105. The pressure of the atmosphere is about 1 Kg. on a square centimetre; express this in pounds per square inch.

106. If 3 kilometres are as much under 2 miles as 5 kilometres are over 3 miles; find the length of the metre in inches.

107. If a litre is ·22 gallons, find to the nearest penny in English money the value of a pint of liquid which is worth 10 francs the litre, 1200 francs being equivalent to £49.

108. Find the number of litres of water that will cover one hectare to the depth of one centimetre.

109. How many litres of water at temperature of maximum density will weigh one tonneau?

110. Express 1 square yard as a decimal of a centiare, and 1 acre as a decimal of a hectare, using the value 1 metre = 39·37079 inches.

111. Considering the earth spherical, with a circumference of four million metres, find the area of its surface in hectares.

112. Find in steres the volume of the largest cube that can be cut from a sphere of radius 1 metre.

MISCELLANEOUS EXERCISES IN FRACTIONS

AND THE SIMPLE RULES.

1. Divide 3·003 by 148·28; ·003003 by ·014828; and 300·3 by 1·1828.

2. Find the value of

$$\frac{\cdot 7\dot{3}25\dot{6} - \cdot 6\dot{1}\dot{9}}{\cdot 45348} \text{ of 5 miles 27 yds.}$$

3. Multiply 1·$\dot{7}$28 by 3·1$\dot{4}$6, expressing the result as a circulating decimal.

4. Express $\dfrac{\frac{1}{4} - \frac{2}{3} \text{ of } \cdot 1\dot{6}}{322\cdot 2}$ of £15 17$\frac{5}{7}\frac{9}{3}$s. as the decimal of a dollar, assuming the value of £30 to be \$146.

5. Find, by the method of abridged multiplication, the product of 26·42783 and 523·23856 true to within ·001.

6. Divide ·37848 by ·456; 3·7848 by ·0456; and 3784·8 by ·00456.

7. Reduce to decimals $\frac{33}{1056}$, $\frac{3}{17}$, and express as vulgar fractions in their lowest terms, 3·0561, 15·601$\dot{3}$78$\dot{9}$.

8. Reduce the decimals ·21$\dot{3}$1$\dot{6}$, ·312$\dot{4}\dot{9}$, and ·893$\dot{1}$ to their equivalent vulgar fractions.

9. Find the value of ·4$\dot{1}\dot{4}$ + ·0$\dot{3}$5$\dot{2}$ + 6·$\dot{1}$0$\dot{1}$, and divide 2·9$\dot{8}$0$\dot{1}$ by 7·$\dot{4}$5$\dot{0}$.

10. Divide 91·$\dot{8}$6$\dot{3}$ by 87·5$\dot{6}$.

11. Find the difference between 3·14159 and $3 + \dfrac{1}{7 + \frac{1}{16}}$.

12. Find the difference between $\sqrt{6}$ and the product of 1·732 and 1·4142.

13. What fraction of £171 16s. 6d. is $\frac{36}{37}$ of £34 4s. 6d.?

14. Add together $\frac{1}{14}$, $\frac{6}{155}$, $\frac{4}{217}$, $\frac{2}{259}$.

15. Simplify $1\frac{5}{6}$ of $\dfrac{\frac{1}{2} + \frac{1}{3} + \frac{1}{4}}{2\frac{1}{2} - 3\frac{1}{3} + 4\frac{1}{4}} \times \dfrac{\frac{2\frac{1}{2}}{3\frac{1}{7}} \times \frac{1\frac{3}{16}}{1\frac{3}{8}}}{\frac{3}{4\frac{1}{2}} \times \frac{1\frac{1}{2}}{3}}$.

16. Find the product of 325·62534 and 27·1367 as far as three places of decimals, by the method of contracted multiplication.

17. Find the value of $\dfrac{2\cdot\dot{6}\times 2\cdot 8\dot{3}}{6\cdot\dot{2}\times\cdot\dot{8}5714\dot{2}}+\dfrac{4\frac{5}{8}\times 4\cdot 0\dot{3}\dot{6}}{3\cdot 75+1\cdot\dot{7}}$.

18. Find the greatest and the least possible value of a decimal the first four digits of which are known to be ·8397.

19. Find the value in decimals of $\dfrac{1}{3+\dfrac{1}{7+\frac{1}{16}}}$.

20. Obtain the quotient when the recurring decimal ·2323 . . . ., is divided by the recurring decimal ·28752875 . . . .

21. Reduce $\frac{1}{13}$ to a decimal, multiply the decimal by 1·4, and divide the product by $\frac{7}{65}$.

22. Divide £120 7s. $10\frac{1}{4}$d. by $8\frac{7}{9}$, and multiply the result by $9\frac{7}{9}$.

23. Reduce 14 weeks 6 days 23 hrs. 45 min. to the fraction of a year of $365\frac{1}{4}$ days.

24. Simplify

$$\frac{26\frac{3}{7}-1\frac{13}{21}}{\frac{4}{9}+1\frac{1}{3}-\frac{3}{5}\text{ of }\dfrac{17\frac{1}{2}}{12}\text{ of }\frac{5}{9}\div\frac{35}{24}}\text{ of }\frac{5\frac{1}{2}}{521}.$$

25. By short division find the quotient as a whole number and recurring decimal when 1769 is divided by 105.

26. Simplify $\dfrac{\frac{17}{10}+\frac{7}{11}}{\frac{17}{10}-\frac{7}{11}}-\dfrac{9+\frac{14}{27}}{2+2\frac{1}{3}}+1761\frac{5}{48}-1650\frac{431}{432}$.

27, Reduce nine and nine-tenths inches to the decimal of a mile.

28. Add $\frac{1}{11}$, $\frac{6}{135}$, $\frac{4}{217}$, $\frac{2}{259}$, $\frac{8}{333}$, and $\frac{7}{387}$; and multiply $\frac{43089}{77508}$ by $2\frac{37}{33}$.

29. Taking the length of the year as 365·25 days, express 73·05 days as the decimal of a year.

30. Simplify $\dfrac{4\frac{1}{5}-\frac{2}{3}\text{ of }\frac{5}{6}}{1\frac{2}{3}+2\frac{3}{7}-\frac{2}{5}}$; and $\dfrac{1}{\frac{1}{2}+\frac{1}{3}+\dfrac{\frac{1}{6}}{\frac{3}{4}-\frac{1}{3}}}$.

31. Reduce $\dfrac{1\frac{7}{9}\text{ of }\frac{27}{61}}{\frac{11}{12}\text{ of }9\frac{9}{11}} \div \dfrac{1\frac{4}{7}\text{ of }\frac{21}{160}}{2\frac{5}{6}\text{ of }\frac{15}{34}}$.

32. Divide $(\frac{1}{3}+\frac{1}{5}+\frac{6}{7}+\frac{4}{15})$ by $(1-\frac{1}{2}+\frac{3}{4}+\frac{2}{3}+\frac{4}{5})$.

33. Simplify $8\frac{1}{2}-4\frac{2}{3}+\frac{3}{5}-1\frac{1}{6}$ of $\frac{2}{5}$; and $1+\cfrac{1}{2+\cfrac{1}{3+\frac{1}{4}}}$.

34. Find a decimal which shall be within $\frac{1}{10000}$ of $\frac{17}{42}$.

35. Reduce $\dfrac{\frac{97}{174}+\frac{291}{388}}{\frac{485}{582}+\frac{291}{388}}$ and $\dfrac{\frac{2\frac{1}{2}}{34}+\frac{1}{8}-\frac{1}{216}+\frac{1}{12}}{\frac{1}{2}+\frac{1}{3}-\frac{1}{6}}$.

36. Multiply $\cdot 0\dot{0}2\dot{1}$ by $18\cdot 92\dot{6}$; and divide $4\cdot 0\dot{3}$ by $\cdot 1\dot{4}0\dot{7}$

37. Express as vulgar fractions $\cdot 037\dot{1}8\dot{5}$ and $\cdot\dot{0}3718\dot{5}$.

38. Reduce to a single decimal

$$\frac{\cdot 04275}{3\cdot 0\dot{5}} \times \frac{4\cdot 21\dot{6}}{\cdot\dot{3}4\dot{2}} \times \frac{2\cdot 7}{1\cdot 5318}.$$

39. Simplify $\frac{1}{2}$ of $\frac{1}{3}\times\frac{3}{4}$ of $\frac{4}{5}\div(\frac{7}{6}+\frac{1}{3}$ of $20)$.

40. The earth's polar diameter is 7899·114 miles, and is $\dfrac{298.33}{299\cdot 33}$ of the equatorial diameter; find the equatorial diameter.

41. Find the value of

$$\frac{7\frac{1}{2}}{6\frac{2}{3}}+\frac{11\frac{1}{2}-2\frac{2}{5}}{11\frac{1}{2}+2\frac{2}{5}}\times 10\frac{9}{13}-7\frac{1}{8}.$$

42. Simplify $\dfrac{3\cdot 5-1\cdot 8\dot{3}}{4\cdot\dot{1}+5\cdot 8}$ of $\dfrac{7\cdot 25\text{ of }1\cdot 2}{3\cdot 25}+\dfrac{3\cdot 1\times\cdot 1\dot{0}\dot{1}}{2\cdot\dot{1}5}$.

43. The speed of a railway train being 65·81 fathoms in 5·6 seconds, each right to the last digit, find the limits in miles of the distance traversed in one hour.

44. Reduce to their simplest form:

$$\frac{\frac{\frac{2}{5}}{1-\frac{1}{25}}+\frac{1}{3}+\frac{1}{7}}{1-\frac{1}{7}\text{ of }\left(\frac{\frac{2}{5}}{1-\frac{1}{25}}+\frac{1}{3}\right)};\quad \frac{3}{8-\cfrac{7}{2-\frac{3}{4}}}+\frac{5}{6-\cfrac{5}{2-\frac{5}{6}}}-$$

45. What simple fraction with 20 for denominator will give the nearest approximation to $\frac{769}{998}$?

46. Find the difference between :—

$$\left(\frac{\cdot 26 + \cdot 2 \text{ of } 3\cdot 7}{\cdot 48 - \cdot 014 \text{ of } 20} - \frac{4\cdot\dot{3} + 5\cdot\dot{6}}{7\cdot\dot{4} - \cdot\dot{2} \text{ of } 11}\right) \text{ of £1 10s. 6d.}$$

and

$$\left(\frac{\frac{1}{2} \text{ of } \frac{5}{9} \text{ of } 7\frac{1}{5}}{\frac{1}{3} + 4\frac{1}{2} \text{ of } \frac{4}{27}} + \frac{\frac{4}{5} - 2\frac{1}{3} + 1\frac{8}{15}}{\frac{7}{95} + 150\frac{5}{19} - 74\frac{2}{5}}\right) \text{ of £1 5s. 6d.}$$

47. What whole number of fiftieths most nearly expresses the value of $\frac{876}{1159}$?

$$\frac{876}{1159} = \frac{\frac{876}{1159}}{1} = \frac{\frac{876}{1159} \times 50}{50} = \frac{37\frac{917}{1159}}{50} = \frac{38}{50} \text{ most nearly.}$$

48. Find the value of :

$$\frac{\cdot 2\dot{3} - \left(\frac{1}{3} - \frac{\cdot 2\frac{2}{3}}{10}\right)}{\cdot\dot{4}\dot{5} - \cdot 1\dot{1}3\dot{6} - \frac{1}{11}} \text{ of £1} + \frac{\cdot\dot{5}7142\dot{8} - \frac{3}{21}}{2\frac{2}{7} - 1\frac{1}{28} - 1\frac{1}{8}} \text{ of 1 guinea.}$$

49. Subtract $\frac{1}{3}$ of $\frac{3\frac{3}{8}}{\frac{1}{4} \text{ of } 33\frac{3}{4}} + \frac{1}{2} \text{ of } \frac{\frac{3}{4}}{1 + 1\frac{9}{16}} + \frac{\frac{3}{5} \times \frac{1}{12}}{\frac{7}{9} \text{ of } 7\frac{5}{7}}$ from 101 times the sum of $\frac{3}{10}$ and $\frac{1}{2}$ of $\frac{7}{15}$ of $\frac{7}{20}$.

50. Find the least fraction which, added to the sum of $\frac{7}{9}$, $\frac{11}{13}$ and $\frac{87}{90}$ shall make the result an integer.

51. What multiple of this fraction added to the same sum will make the result an integer?

52. Simplify $\frac{34\frac{1}{2} - 7\frac{1}{5}}{34\frac{1}{2} + 7\frac{1}{5}}$ of $\frac{139}{2}$ of $\frac{1}{6\frac{1}{2}} - 5\frac{101}{104}$.

53. Reduce $\frac{\cdot 0617 + \cdot 21605 - \cdot 005}{\cdot 0024 0623}$ to a vulgar fraction.

54. Find the value of

$3\left(1 + \frac{1}{35}\right)\left(1 + \frac{1}{143}\right)\left(1 + \frac{1}{323}\right)$, to two decimal places.

55. Which of the three quantities $\frac{106}{39}$, $\frac{2721}{1001}$ and $2\cdot 718282$ most nearly expresses the value of $2\cdot 718281828$?

56. From $\frac{6\frac{41}{88}}{6 - 4\frac{4}{11}} + \frac{7}{9} \times 1\frac{22}{27}$ of $\frac{3}{56}$ take $\frac{1}{6}$ of : $6 + \cfrac{1}{6 + \cfrac{1}{6 - \frac{1}{6}}}$.

57. Simplify

$$\frac{\frac{5}{14} - \frac{3}{7} \text{ of } \frac{1}{2}}{\frac{5}{16} + \frac{7}{12} \text{ of } 3\frac{1}{4} - \left(\frac{7}{8} \text{ of } \frac{37}{21} - \frac{1}{3}\right)} \div \frac{\frac{1}{3} \text{ of } \frac{1}{2} + \frac{3}{2} \text{ of } 5}{9\frac{1}{3} - 1\frac{2}{3}}.$$

58. Ascertain whether ·52 or ·519 more nearly represents the product of ·834 and ·623.

59. What quantity must be added to

$$\frac{1\frac{1}{2}\text{ of }3\frac{1}{3}}{3\frac{1}{2}\text{ of }2\frac{2}{3}}\text{ of }\frac{1\frac{2}{7}\text{ of }1\frac{1}{6}}{\frac{18}{49}\text{ of }32\frac{2}{3}}+\frac{2\frac{1}{5}\text{ of }6\frac{2}{3}}{3\frac{1}{3}\text{ of }4\frac{1}{2}}$$

to make it equal to $\frac{1}{28\frac{2}{7}}$ of $3\frac{3}{4}$ of $3\frac{1}{7}$ of $1\frac{5}{7}+\frac{3}{8}$?

60. Simplify

$$\left\{\frac{\frac{2}{3}+\frac{5}{6}+\frac{7}{8}+\frac{11}{12}}{\frac{3}{4}-\frac{5}{8}}\times\frac{1}{34\frac{1}{2}}\right\}\div\left\{\frac{7\frac{1}{2}}{6\frac{1}{2}}+\frac{11\frac{1}{2}-2\frac{2}{5}}{11\frac{1}{2}+2\frac{2}{5}}\times 10\tfrac{9}{15}-7\tfrac{1}{8}\right\}.$$

61. What number multiplied by 57·29577 will give a product differing from 180 by less than ·0001?

62. Simplify $\dfrac{\frac{5}{32}+\frac{7}{24}\text{ of }3\frac{1}{4}-\left(\frac{7}{15}\text{ of }\frac{37}{21}-\frac{1}{6}\right)}{\frac{5}{28}-\frac{3}{14}\text{ of }\frac{1}{2}}$.

63. Simplify $\left(\dfrac{\frac{1}{2}-\frac{1}{3}}{\frac{1}{2}+\frac{1}{3}}+\dfrac{\frac{1}{3}-\frac{1}{4}}{\frac{1}{3}+\frac{1}{4}}\right)\div\left(\dfrac{\frac{1}{4}-\frac{1}{6}}{\frac{1}{4}+\frac{1}{6}}-\dfrac{\frac{1}{6}-\frac{1}{8}}{\frac{1}{6}+\frac{1}{8}}\right)$.

64. Calculate to seven places of decimals the difference between $\sqrt{\frac{1}{2}}$ and $\frac{7}{10}+\frac{7}{985}$.

65. Find the sum of

$$\left(\tfrac{5}{7}+\tfrac{5}{2}\right),\ \left(\tfrac{5}{2}-\tfrac{5}{7}\right),\ \left(\tfrac{5}{7}\times\tfrac{5}{2}\right),\ \left(\tfrac{5}{7}\div\tfrac{5}{2}\right),\ \left(\tfrac{5}{2}\div\tfrac{5}{7}\right).$$

66 What fraction having 17 for numerator equals $\frac{6\frac{1}{3}}{7}$?

67. Simplify $\dfrac{\frac{1}{23}\text{ of }6\frac{13}{17}\text{ of }24\frac{11}{13}-4\frac{13}{18}\times 3\frac{33}{54}\div 3\frac{37}{96}}{8\frac{17}{19}\times 5\frac{11}{39}\div 4\frac{15}{32}-7\frac{19}{20}\times 5\frac{11}{65}\div 11\frac{21}{25}}\times 4\tfrac{8}{25}$.

68. Explain what is meant by the expression $3\frac{2}{5}$ of $\frac{2}{7}$ of $\frac{5}{9}$. How is it reduced to a simple fraction? Explain fully the reason of the process.

69. Simplify

(1.) $1\frac{5}{6}$ of $\dfrac{\frac{1}{2}+\frac{1}{3}+\frac{1}{4}}{2\frac{1}{2}-3\frac{1}{3}+1\frac{1}{4}}\times\dfrac{\frac{2\frac{1}{2}}{3\frac{1}{7}}+\frac{\frac{3}{16}}{1\frac{3}{8}}}{\frac{3}{4\frac{1}{2}}+\frac{4\frac{1}{2}}{3}}$; (2.) $\dfrac{6+\frac{1}{6-\frac{1}{6}}}{4-\frac{1}{4-\frac{1}{4}}}\times 10\tfrac{8}{9}$.

70. Simplify $\dfrac{8\frac{3}{5}-7\frac{3}{4}+5\frac{2}{3}-4\frac{1}{2}}{13-11\frac{9}{10}+10\frac{7}{9}-9\frac{17}{20}}\times\frac{2}{11}$ of 365.

71. Find the difference between $\frac{1}{73000}$, and ·0001 of $[1 + \frac{1}{3}\{1 + \frac{1}{10}(1 + \frac{1}{10})\}]$; what fraction of $\frac{1}{73000}$ is this difference?

72. Simplify the fractions:

(1.) $\dfrac{6\frac{3}{4} - 1\frac{5}{11}}{2\frac{1}{6} + 1\frac{3}{7}}$;

(2.) $(\frac{5}{7} \text{ of } 1\frac{6}{13}) \div \dfrac{2\frac{5}{7}}{3\frac{1}{4}}$;

(3.) $\dfrac{1}{4 - \dfrac{1}{2 - \dfrac{1}{1 - \frac{5}{13}}}}$.

73. Find, correct to four decimal places, a number which as a multiplier may be substituted for 3·14159 as divisor.

74. By what decimal, correct to five places, may any number be multiplied so as to give the same result as dividing that number by 2·302585?

75. Reduce to its simplest form—

$(3\frac{4}{5} + 5\frac{1}{9} - \frac{1}{45})(4\frac{1}{3} - 3\frac{1}{4})$, divided by $1\frac{5}{11} + 2\frac{1}{8} - (2\frac{9}{16} - \frac{1}{8} - \frac{1}{22})$.

76. Reduce the fractions $\frac{15}{119}$ and $\frac{18}{113}$ to a common numerator, and determine which is the greater.

77. Find, correct to six decimal places, the value of

$$\frac{\cdot 3472 + \cdot 03172}{6146 \cdot 38} \div \cdot 0004675.$$

78. What error do we make in taking $\frac{268}{65}$ as the value of the square root of 17?

79. Find, correct to seven decimal places, the value of

$$\tfrac{3}{5}\sqrt{5} + \tfrac{2}{3}\sqrt{2} + \tfrac{6}{7}.$$

80. The length of a seconds pendulum is 39·37079 inches; if 64 metres are equal to 70 yards, by what decimal of an inch will the length of a seconds pendulum differ from one metre?

81. Reduce to a decimal correct to six places

$$\frac{1}{1 \times 3} + \frac{1}{1 \times 3 \times 5} + \frac{1}{1 \times 3 \times 5 \times 7} + \text{\&c.}$$

82. Find the quotient of ·9840018 by ·00159982 to seven decimal places; and reduce $\cdot 700245\dot{7}$ to a vulgar fraction.

83. Find the value of

$$\frac{1}{5}+\frac{1}{3\times 5^3}+\frac{1}{5\times 5^5}+\frac{1}{7\times 5^7}+\frac{1}{9\times 5^9}.$$

Proceed thus :—

$$\frac{1}{7\times 5^7}=\frac{2^7}{7\times 5^7\times 2^7}=\frac{2^7}{7\times 10^7}$$

$$=\frac{128}{7\times 10^7}=\tfrac{1}{7}\text{ of }\cdot 0000128=\text{etc.}$$

and similarly with the other fractions.

84. Find the value of

$$1+\frac{1}{4}+\frac{1\cdot 3}{4\cdot 8}+\frac{1\cdot 3\cdot 5}{4\cdot 8\cdot 12}+\frac{1\cdot 3\cdot 5\cdot 7}{4\cdot 8\cdot 12\cdot 16}.$$

85. How many terms of the series in prob. 84 will give the square root of 2 to two decimal places?

86. Calculate to 12 places of decimals the value of the series :

$$1+1+\frac{1}{1\cdot 2}+\frac{1}{1\cdot 2\cdot 3}+\frac{1}{1\cdot 2\cdot 3\cdot 4}+\text{etc.}$$

87. Which is the more correct value of a kilometre $\frac{3}{5}$ of a mile or $\frac{5}{8}$ of a mile?

88. What number multiplied by 6391·427 will give a product coinciding with 6384·2579 to four decimal places?

89. Reduce to its simplest form

$$\frac{(\cdot 075)^3+(\cdot 025)^3}{(\cdot 075)^2-(\cdot 075)(\cdot 025)+(\cdot 025)^2}.$$

90. Express $\frac{22}{7}$ and $\frac{355}{113}$ as decimal fractions to five places; and prove that their difference is about $\frac{1}{2500}$ of either of them.

91. Find the value of

(1) $\sqrt[3]{\left(\frac{5\cdot 12}{33\cdot 75}\right)}$; (2) $\sqrt[3]{\left(\frac{5\cdot 12}{\cdot 03375}\right)}$;

(3) $\dfrac{\sqrt[3]{512}+\sqrt[3]{\cdot 003375}}{\sqrt[3]{8}-\sqrt[3]{\cdot 001}}$.

92. A vulgar fraction has for its numerator 209 and its nearest approximate value in thousandths is ·511; what is its denominator?

93. The number of Canadian dollars in a pound sterling is $\frac{109\frac{1}{2}}{100}$ of $4\cdot\dot{4}$; and the number of United States dollars in a pound sterling is $\frac{11}{12}$ of $\frac{40}{1869} \times 5760 \div \frac{9}{10}$ of 25·8; find the value of £1000 in Canadian and in U. S. money.

94. Find the value of $\frac{\sqrt{2}-1}{\sqrt{2}+1}$ to four places of decimals.

95. How close an approximation is $\frac{161}{72}$ to the square root of 5?

96. The millimètre being ·03937043 of an inch, right to the last digit, show that the metre is, to seven decimal places, most nearly 3·2808692 feet.

97. Find to four places of decimals the value of

$$\frac{3\sqrt{5}+\sqrt{3}}{3\sqrt{5}-\sqrt{3}}.$$

98. Which of the following statements is more nearly correct?

$$\frac{10}{9\cdot009} = 1\cdot11, \text{ or } \frac{10}{1\cdot11} = 9\cdot009.$$

99. Express in decimals accurately to 5 places the series

$$16 \times \left\{ \frac{1}{5} - \frac{1}{3 \times 5^3} + \frac{1}{5 \times 5^5} - \frac{1}{7 \times 5^7} + \&c. \right\} - \frac{4}{239}.$$

100. The volume of water at the boiling point is $\frac{1\cdot04315}{1\cdot0012}$ of its volume at 62°. If a pound of water at 62° occupy 25·2286 cubic inches, find its volume when heated to the boiling point.

101. State which of the following fractions will reduce to terminating and which to circulating decimals, and the limits to the repeating periods in the latter:—

$$\frac{67}{128}, \frac{28}{37}, \frac{155}{1147}, \frac{51}{102}, \frac{47}{98}, \frac{321}{375}.$$

102. Find the value, to 6 places of decimals, of

$$\frac{\sqrt{5}+\sqrt{3}}{\sqrt{5}-\sqrt{3}} - \frac{\sqrt{5}-\sqrt{3}}{\sqrt{5}+\sqrt{3}}.$$

103. Given the length of a degree $69\frac{1}{9}$ miles and the length of a metre 39·37 in., what is the error in taking the metre to be ·0000001 of the distance from the equator to the pole?

104. Find the value of $\frac{1}{7}+\frac{1}{7^3}+\frac{3}{2}\cdot\frac{1}{7^5}+\frac{5}{2}\cdot\frac{1}{7^7}$.

Find the difference between your result and the square root of $\frac{1}{47}$.

105. If 24·25 francs = £1, find the value of a franc in cents.

106. A starts from Kingston to walk to Belleville, a distance of 45 miles, at $3\frac{1}{2}$ miles an hour, and B starts from Belleville 3 hours earlier at $2\frac{1}{2}$ miles an hour. Where do they meet, and how far will B be from Kingston when A arrives at Belleville?

107. If 100 cubic inches of air weigh 31 grains, find the weight of air in a room 32 ft. long, 25 ft. wide and $11\frac{1}{2}$ ft. high.

108. Water in freezing expands one-tenth. If a cubic foot of water weighs 1000 oz., find the weight of a cubic foot of ice.

109. If the imperial gallon contains 277·2 cubic inches, and holds 10 lbs. of water, what is the error in saying that a cubic foot of water weighs 1000 oz.?

110. How many pounds of tea at 4s. 2d. per pound can be bought for £12 10s.?

111. If rain falls to the depth of $\frac{3}{4}$ in., how many gallons will have fallen on an acre of ground?

112. My farm contains exactly 184 ac. 76 sq. rd. $21\frac{1}{3}$ sq. yd. There are 3·85 ac. in garden and orchard; 9·147 ac. of green crop; 76·9 ac. of grain; 23·608 ac. of meadow; 34 ac. of pasture; and the remainder is uncleared bush; what fraction of my farm is uncleared?

113. A farmer sells to a merchant 3015 lbs. of hay at \$16 per ton, and takes in payment 6 lbs. of tea at 80 cents per lb.; $22\frac{1}{2}$ lbs. of coffee at 26 cents per lb.; 33 lbs. of sugar at 12 lb. for a dollar; $32\frac{1}{3}$ lbs. of raisins at $18\frac{3}{4}$ cents per lb.; 14 lbs. 13 oz. of bacon at 16 cents per lb.; and the balance in cash; how much cash does the farmer receive?

114. If 4 men in 5 days of 9 hrs. each can mow 15 acres of grass, how many men will mow 11 acres in 2 days of 11 hrs. each?

115. A person sold A $\frac{3}{4}$ of his land, B $\frac{1}{5}$ of the remainder, C $\frac{5}{6}$ of what then remained, and received $50 for what he had left, at $60 per acre; find the number of acres he had at first.

116. A pint contains 9000 grains of barley and each grain is one-third of an inch long; how far would the grains in 17 bush. 3 pk. 1 gal. 1 qt. 1 pt. reach if placed one after another?

117. If 3 lbs. of wheat make 2 lbs of flour, how many barrels of flour can be made from 343 bushels of wheat?

118. A certain hall 60 feet long is to be carpeted. It is found that by stretching the carpet lengthwise, any one of four pieces, width respectively $\frac{3}{4}$ yd., 1 yd., $1\frac{1}{4}$ yd., and $1\frac{1}{3}$ yd., will exactly fit the hall without cutting anything from the width of the carpet. If the narrowest piece, worth $1.10 per yard, be chosen, what will it cost to carpet the hall?

119. I bought a bush farm, 180 rods long by 96 rods wide, at $12.50 per acre. I paid $14.75 per acre for clearing, and $1.35 a rod for enclosing the whole farm with wire fencing. Taking into account that I sold the wood for $1160, and ashes for $17.20, how much has the improved farm cost me per acre?

120. A farmer sells a merchant 30 bushels of wheat at 90 cents per bushel, and makes a profit of one-fifth; the merchant sells the farmer 5 yds. of broadcloth at $3.60 per yd., 16 yds. of calico at 8 cents per yd., and 44 yds. of cotton cloth at 13 cents per yd., and makes a profit of one-fourth. Which gains the more by the transaction, and how much?

121. A train going 25 miles an hour starts at 1 o'clock p. m. on a trip of 280 miles: another going 37 miles an hour starts from the same place at 12 min. past 4 o'clock p. m.; at what time will the first train be overtaken?

122. If £1 = 10 florins = 100 cents = 1000 mils, express £13 18s. $3\frac{1}{4}$d. in pounds, florins, cents and mils.

123. Find the least sum of money that can be paid both by a whole number of farthings and by a whole number of mils.

124. A sidereal day is 23 hrs. 56 min., and the mean solar day is 24 hrs.; reduce the difference between the two to the decimal of a sidereal day.

125. A building-lot 30 feet in front by 120 feet deep is sold at $100 per foot frontage; how much is that per acre?

126. A water tank 10 ft. long and 7 ft. wide, contains 3000 gallons of water; find the depth of the water in the tank.

127. If 3 men can reap 8 acres in 5 days, working 8 hours a day, in how many days can 8 men, working 12 hours a day, reap 192 acres?

128. If marble is 2·716 times as heavy as water, find the weight of a block of marble 18 inches square and 10 feet high.

129. If gold is $19\frac{1}{4}$ times as heavy as water, find the weight of a cubic inch of gold.

130. A clock that is 3 minutes slow at noon is gaining at the rate of 11 seconds in 5 hrs.; when will it indicate true time?

131. If 2 men or 3 boys can do a piece of work in 5 days, how long will it take 6 men and 10 boys to do the same amount of work?

132. If 27 men mow a field of 90 acres in 7 days, working 8 hours a day, how many men will be required to mow 200 acres in 16 days, if they will work 10 hours a day?

133. If a cow gives 12 qt. 1 pt. of milk every day, and 1 lb. 8 oz. of butter can be made from 25 qt. of milk, how many lbs. of butter can be made in one week from the milk of 16 cows?

134. How many square feet of surface are there, if the atmospheric pressure (at 14·8 lbs. to the square inch) on a man is 29000 lbs.?

135. If 200 men can make an embankment 2 miles long in 20 days, how much over-time must 120 men work in order to finish an embankment 3 miles long in 21 days, 12 hours being a day's work?

136. A train is just 27 minutes in passing through the Mont Cenis Tunnel, the length of which is 11220 metres; find the speed of the train in miles per hour.

137. A roller 28 inches in diameter and 10 feet long makes 90 revolutions in going from one side of a field to the other; find the number of acres in the field if the roller has to cross it 70 times.

138. If 12 men, working 8 hours a day, do $\frac{4}{5}$ of a piece of work in 20 days, how many days will 15 men, working 10 hours a day, take to do $\frac{7}{8}$ of it?

139. If a barrel of salt weighs 10 quarters, and 4 such quarters make the English hundred-weight, and 20 such hundred-weights make the English ton, find the number of pounds in the English, or "long" ton, as it is called.

140. How many such quarters make a barrel of flour? Find the number of barrels of flour that will make the least whole number of long tons.

141. A merchant fails owing $10,500, and pays his creditors 35 cents on the dollar; how much will be lost by a creditor to whom one-fifth of the debt was owing?

142. Find the greatest common measure and least common multiple of 8029 and 23791; and of 157 days, 7 hrs., 4 min., 7 sec., and 243 days, 2 hrs., 11 min., 49 sec.

143. If 18 men do $\frac{2}{3}$ of a piece of work in 30 days of 10 hrs., in what time would 15 men do the work, working 9 hrs a day?

144. What is the relation between a pound Avoirdupois and a pound Troy? Express one-eighth of a ton in Troy ounces, and 5760 pounds Troy in ounces Avoirdupois.

145. From a cask of wine, worth 90 cents a gallon, a sixth part is drawn and replaced by wine worth only 50 cents a gallon; what is now the value per gallon of the wine in the cask?

146. A sold a town lot to B and gained one-eighth, B sold it to C for $306 and lost three-twentieths; how much did the lot cost A?

147. A plot of land containing 20 perches less than 5 acres is valued at £1681 17s. 6d., which is three-twentieths more than it was worth a year ago: by how much has its value increased per acre during the year?

148. If a coal dealer were to buy his coal by the long ton and sell it by the short ton at the same price per ton, what fraction of his outlay would his profits be?

149. A can beat B by 5 yards in a hundred yards race, and B can beat C by 10 yards in a 200 yards race; by how much can A beat C in a 200 yards race?

150. If 1 pound of thread makes 3 yds of linen $1\frac{1}{4}$ yds. wide, how many pounds would make 45 yds. of linen 1 yd. wide?

151. The United States silver dollar weighs $412\frac{1}{2}$ grains and is $\frac{9}{10}$ fine; Canadian silver is $\frac{37}{40}$ fine, and a 5 cent piece weighs 18 grains. If a dollar in U. S. silver is worth 85 cents, what is a dollar in Canadian silver worth?

152. A cubic foot of copper weighs 560 lbs; it is rolled into a square bar 40 feet long; an exact cube is cut from the bar: what is its weight to four decimals of a pound?

153. Find to the nearest penny in English money the value of a pint of liquid which is worth 10 francs a litre. (Franc = 19 cents.)

154. An express train leaves Marseilles at 6.35 in the evening and arrives at Paris at 9.15 the next morning: the distance is 863 kilomètres; determine the speed of the train in miles per hour, reckoning the kilomètre as equal to 3281 feet.

155. A watch is right on Monday at noon and is 10 min. slow when it indicates noon on Saturday; at what rate is it losing?

156. A grocer bought 6 cwt. of sugar for $52.10; he used 65 lbs. himself and sold the rest so as to make $1\frac{1}{8}$ cents per lb. profit on the whole quantity. How much per lb. did he sell it for?

157. What quantity of water must I add to a hogshead of wine, costing $180, to reduce its price to $2 a gallon?

158. A railway train has a journey of 65 miles to perform, and usually makes the distance in 3 hours; if the train starts 15 min. late how much must it increase its speed to arrive on time?

159. A house and lot are together worth $2100; one-fourth of the value of the house is equal to one-third of the value of the lot; find the value of each.

160. A man bought a quantity of tea supposed to be done up in packages of 1 lb. each, for which he was to pay $64; on weighing them, however, it was found that each package was 1 oz. too light; how much should he pay for the tea?

161. If 12 oxen and 35 sheep eat 6·35 tons of hay in 4 days how much will it cost per week to feed 4 oxen and 6 sheep, the price of hay being $15 per ton and 2 oxen eating as much as 5 sheep?

162. A gold dollar weighs 25·8 grains and ·9 of it is pure gold. If the value of the alloy may be disregarded what is the value in dollars of an ounce avoir. of pure gold?

163. A water tank can be filled by one tap in 3 hours and by another in 2 hours. It can be emptied by a third tap in 1 hr. 24 min. The tank being empty all three taps are opened at once, in what time will the tank be filled?

164. If three-fourths of the price of a loaf of bread depend upon the price of flour, by how much should the price of a 10 cent loaf be increased when flour rises two-fifteenths in value?

165. The sum of $3576 is to be divided among four persons in proportion to their ages, their ages being 17, 20, 24 and 30 years; how would you divide the money?

166. If a person whose income is £365 a year spend £8 16s. 3d. a week for the first 20 weeks, to what amount must he limit his daily expenditure for the remainder of the year so as to avoid being in debt at the end of it?

167. Sound travels at the rate of 1140 ft. per second: how far will a railroad engine running at the rate of 50 miles an hour have gone before the sound of its whistle can be heard 5 miles away?

168. Air is ·00122 times as heavy as salt water; salt water is ·03 heavier than fresh water; if 100 cubic inches of air weigh 31 grains, how much will 100 cubic inches of salt water weigh?

169. If a gallon measure is a quarter of a gill too small, how much does the merchant gain on each 1000 gallons sold?

170. The sum which will pay A's wages for $61\frac{1}{4}$ days, will pay B's wages for $81\frac{2}{3}$ days; for how many days will it pay the wages of A and B together?

171. In a 440 yds. race, A can give to B 20 yds. start, and to C 30 yds.; B and C run a 440 yds. race; by how much does B win?

172. A grain dealer buys 5225 bushels of wheat at \$1.05 per bushel, and pays \$125 for insurance, storage, etc; he sells ·4 of the quantity at 97 cents a bushel; at what price per bushel must he sell the remainder in order to gain \$522.50 on the whole?

173. If 40 lbs. troy of standard gold ($\frac{11}{12}$ fine) are coined into 1869 sovereigns, find the number of grains of pure gold in a sovereign.

174. A and B together can earn £3 16s. in 8 days; A and C together can earn £7 13s. in 17 days; and B and C together, £12 15s. in 30 days. How much a day can they severally earn alone?

175. The silver dollar of Central America is worth $91\frac{4}{5}$ cents and the silver dollar of Mexico $99\frac{4}{5}$ cents of our money; find the least sum in Canadian dollars that can be exactly paid by each of these coins, and also how many of each would be required.

176. I sell goods at a profit of one-tenth the prime cost, but in consequence of ready payment I throw off one-twentieth of the selling price; what fraction of the prime cost do I gain?

177. It is required to build a sidewalk a quarter of a mile in length, 8 ft. wide and 2 inches thick, supported by three continuous lines of scantling 4 inches square; what will the lumber cost at \$17 per thousand feet?

178. In a room 26 ft. 6 in. long, 16 ft. 8 in. wide, and 12 ft. 3 in. high, there are three windows each $5\frac{1}{2}$ ft. high and 3 ft. wide, and two doors each 7 ft. high and $3\frac{1}{2}$ ft. wide; the base-board is 9 in. wide; how much paper, $\frac{3}{4}$ of a yard wide, will be required to cover the walls and ceiling?

179. An orchard is $21\frac{2}{3}$ rods long and $15\frac{1}{4}$ rods wide; at $1\frac{3}{4}$ cents per cubic foot what will it cost to dig a ditch around it 3 ft. 9 in. wide and four ft. deep?

180. The square of a number differs from the square of the next number by 691: find the two numbers.

181. If the earth's equatorial diameter is 7926 miles, find the length of a degree on the equator.

182. A runs a mile race with B and loses; had his speed been a third greater he would have won by 22 yards; what fraction is A's speed of B's?

183. A boy's age now is one-fifth of his father's. In six years it will be one-third of his father's present age. How old is he?

184. The diameter of a cent is one inch. How many cents can be placed edge to edge on a surface of six inches square so as just to prevent any falling off?

185. A man having lost one-fifth of his capital is worth exactly as much as another man who has gained three-twentieths on his capital; the second man's capital was originally $9000. What was the first man's capital?

186. The hour and minute hands of a clock are together at 12 o'clock; at what intervals are they together again?

187. A laborer while working is able to save 75 cents a day. During ten days of a certain month he is unable to work, in consequence of which he is worse off by $12.50 than he expected; find his daily expenses.

188. A plot of land is sold at £1200 per acre. What is the price in francs per square metre, if £1 = 25 francs?

189. Three clocks which tick 70, 80, and 90 times, respectively, a minute, beat coincidently at a certain time; find the least time in which they will again beat together.

190. Gold sixteen carats fine is brought to the mint: what fraction of its original weight will it weigh when made into coins eleven-twelfths fine?

191. By selling cloth at $1.26 a yard I gain 11 cents more than I would lose by selling it at $1.05 a yard; what would I gain by selling 800 yards at $1.40 a yard?

192. A barn 80 ft. long and 60 ft. wide is built on a plot of ground 308 ft. long and 204 ft. wide. The rest of the plot is covered with cordwood to a depth of 8 ft.; how many cords of wood are there?

193. If 15 men be necessary to excavate 966 cubic yards in 8 days, working $10\frac{1}{2}$ hrs. daily, how many men would be required to excavate 575 cubic yards in 12 days, working $7\frac{1}{2}$ hrs. daily, 4 extra men being taken on during the last 4 days?

194. The sum of $648 is to be divided among 24 men, 36 women and 72 children, so that the shares of 2 men shall be equal to those of 3 children, and each woman's share to the shares of 2 children; what will be the share of each?

195. Bought 360 gallons of wine at $2.60 a gallon; paid for carriage $17.20, and for duties $86.50. If ·15 of it be lost by leakage, at what price must the remainder be sold to gain $50 on the whole transaction?

196. Assume that 6 men can do as much work in an hour as 7 women, and 8 women as much as 11 boys, and that 5 men can do a certain piece of work in 10 hours: how long will it take 1 man, 2 women and 3 boys together to do the same piece of work?

197. In what time would a field 80 rods by 60 rods pay for underdraining lengthwise at 2 cents a foot, if the field yield 2 bushels (at 66 cents a bushel) per acre more than before draining? The drains are 4 rods apart, and the first drain runs down the middle of the field.

198. A sells goods to B and gains one-tenth on the price he paid for them. B sells the goods to C and loses one-tenth; for what fraction of A's buying price did C buy the goods?

199. A farmer employs a number of men and 8 boys: he pays the boys $·65 and men $1.10 each per day. The amount that he paid to all was as much as if each had received $·92 per day; how many men were employed?

200. A and B start together and walk in the same direction, A at the rate of 4 and B at the rate of 3 miles an hour. At the end of 7 hours A turns and goes back; how many miles will B have gone when he meets A?

201. The real cost of an article is $\frac{14}{25}$ of the price at which it is marked for sale. It is, however, sold at a reduction of one-eighth from the marked price: what fraction of the real cost does the seller gain?

202. On a piece of work, 3 men and 5 boys are employed who do half of it in six days. After this one more man and one more boy are put on, and one-third more is done in 3 days; how many more men must be put on that the work may be completed in one day more?

203. A clock which was 1·4 minutes too fast at a quarter to 11 p.m. on Dec. 2nd, was 8 min. slow at 9 a.m. Dec. 7th; when was it exactly right?

204. The sum of £405 11s. 4d. is lent upon condition that the borrower shall owe to the lender, as interest, one-forty-sixth part of the sum borrowed for each half year that he retains the money; and if the money be not returned at the end of the first, or any subsequent half-year, the interest due, instead of being paid, is to be added to the amount lent. How much is due from the borrower to the lender, at the end of a year and a half, no interest having been paid during that time?

205. The areas of the continents in square miles are as follows: — Europe, 3,780,000; S. America, 6,700,000; N. America, 8,750,000; Africa, 11,500,000; Asia, 16,500,000; what numbers less than 20 most nearly represent their relative sizes?

206. A ditch is being dug at the rate of 81 ft. per day by 54 men; after 13 days' work, 8 of them are replaced by boys, and the work goes on for 11 days more, at the end of which the whole length dug is 1889 ft. How many feet per day do the boys do?

207. Water flows into a tank from two taps which, running separately, would in $1\frac{3}{4}$ and $2\frac{1}{2}$ hours respectively, fill it up to a certain level; at this level a waste-pipe opens; if both taps are running, in what time will the waste-pipe discharge a quantity of water equal to that in the tank?

208. Find the number of days in each four hundred consecutive years.

209. Multiply 365 days, 5hrs., 48 min., 47·5 sec., by 400.

210. What is the difference between the results in the two preceding examples? In how many years would this difference amount to one day?

211. A person buys four houses; for the second he gives half as much again as for the first; for the third, half as much again as for the second; and for the fourth, as much as for the first and third together; he pays in all $24000; what is the cost of each?

212. How many ounces of jeweller's gold, 18 carats fine, can be made from 10 ounces of gold whose fineness is nine-tenths?

213. One hundred cubic inches of air weigh 31 grains, and the U. S. silver dollar weighs $412\frac{1}{2}$ grains: find the difference between the weight of a U. S. half-dime and that of a quart of air.

214. A person sub-divides his farm of 100 acres into town lots as follows:—One-half of the farm he divides into 12 equal parts, each of these into three, and each of these last into 5 equal parts; the rest of the farm he divides into quarters, each quarter into fifths, and each fifth into eighths; find the difference in square yards between the area of one of the first lots and one of the second.

215. The inscription on the Congius of Vespasian at Dresden states that it contains 10 Roman pounds. Its weight when filled with water exceeds its weight when empty by $63460\frac{3}{5}$ French grains (French grain = ·82 Eng. grains); what fraction of a pound Avoirdupois is the Roman pound?

216. An oarsman starts to row up stream at the rate of $2\frac{1}{2}$ miles an hour; after he has done about half the distance he rests long enough to allow himself to drift down stream three-quarters of a mile; in consequence of this he is 48 minutes late in reaching his destination; find the rate of the stream.

217. Fill in the blanks in the following bill of taxes:—

Amount of Assessment.............................$900

Taxes:

Town rate, three and eight-tenths mills on the dollar....
School Debenture By-Law, one and one-fourth mills......
Harbor By-Law, two and six-tenths mills..............
Redemption of Debentures, one and four-tenths mills.....
School rate, five and one-tenth mills...................
High School rate, one and three-fourths mills...........
Total taxes..................................

218. A lot 11 rods long and 9 rods wide has a fence built round it; outside the lot at a distance of 2 ft. from the fence a sidewalk 4 ft. wide is built: how many square yards of ground does the sidewalk cover?

219. Three men are employed in a work, working respectively 8, 9 and 10 hours a day, and receiving the same daily pay; after three days each works an hour a day more and the work is finished in three days more; if the total sum paid is $114.05, how much of it should each receive?

220. Brown purchased $\frac{7}{25}$ of a mill property for $4064.55, and Smith purchased $\frac{9}{35}$ of the same property at a rate $\frac{1}{20}$ higher; what did Smith's part cost him, and what fraction of the property remains unsold?

221. Four men working eight hours a day take 22 days to pave a road a quarter of a mile long and 35 ft. broad: how many days will four men, two of whom work eight hours and two ten hours a day, take to pave a road 1575 yards long and $36\frac{1}{2}$ ft. broad?

222. Find the least number the product of which with 1500 will be a perfect square.

223. Find the least number the product of which with 14175 will be a perfect cube.

224. Find the least number the product of which with 1323 will be a perfect fourth power.

225. If the volume of a brick (2 in. by 4 in. by 8 in.) be taken as the unit, what is the measure of the volume of a cord?

226. If 27 is the measure of a cubic foot, what is the measure of the unit in cubic inches?

227. If the area of a postage stamp ($\frac{3}{4}$ in. by $\frac{7}{8}$ in.) be taken as the unit, find the measure of the area of an envelope $3\frac{1}{8}$ in. by $5\frac{1}{2}$ in.

228. Describe the units of length, surface and volume in the English and in the French system. Given the numerical value of any length in one system, express it in the other.

229. Compare, in whole numbers, the rates of speed of two locomotives, one of which travels $397\frac{5}{6}$ miles in $11\frac{2}{3}$ hrs., and the other $262\frac{4}{15}$ miles in $8\frac{4}{9}$ hrs.: what have you adopted as the unit of measurement in obtaining these numbers?

230. Find the length of the longest unit that will exactly measure both the distances 88 yd. 2 ft. 5 in. and 119 yd. 2 ft. 1 in.

231. A map is drawn to a scale of half an inch to a mile: how many acres are represented by a square inch on the map?

232. The acre being the unit, what number will express the area of a lawn 30 ft. 6 in. long, and 19 ft. 4 in. wide?

233. The yard being the unit, what number will express the volume of a rectangular solid, whose dimensions are 5 ft. 6 in.; 4 ft. 7 in.; and 3 ft. 10 in?

234. If 1 mile be the unit, what number will express the value of 5 yds. 2 ft. 6 in.?

235. If £1 be the unit, what number will express the value of 7s. 6½d.?

236. The number representing 69 bushels, 1 pk. 7 qts., when 3 bush. 2 pk. 5 qts. is the unit, is the same as that representing the volume of a solid 19 ft. long, 9 ft. wide and 3 ft. high; find the unit in the latter case.

237. When one gallon is the unit, the measure of the capacity of a cistern is 610: what is the unit when the capacity of a cistern half the size is represented by 1220.

238. If 104 is the measure of a piece of carpet 312 ft. long and 27 inches wide, express the unit of measurement in inches, in feet, and in yards.

239. If the unit is \$1.86⅔, what will be the measure of \$10.95? Of $\$18.\dot{6}$? Of $\$4.8\dot{6}$?

240. Define the terms *unit* and *measure*. The measure of a certain distance is 1760, when a yard is the unit of length; what unit of length will make the measure of this distance 320?

241. The measure of the length of a piece of fencing is given as 100; when a foot is the unit the measure is 66, what was the unit in the first case?

242. For what unit of area will the measure of an acre be 4840? 10?

What is the measure of an acre, when a square link is taken as the unit? A square inch?

243. Find the quantity whose measure is 3½ when the unit is \$1·085; also the quantity whose measure is 3·1416 when the unit is 7 ft. 4 in.

244. Find the least weight that can be exactly weighed with either the oz. Avoir., or the oz. Troy as unit.

245. A map is drawn to a scale of $\frac{1}{10}$ of an inch to a mile, on this map a township is represented by a square whose side is half an inch; how many acres are there in the township?

246. Find the smallest cask whose capacity is an exact measure of each of the following units:—3 pints, 5 quarts, $1\frac{1}{2}$ gallons, $\frac{5}{8}$ bush.

247. The map of Ontario recently issued by the Crown Lands Department is drawn on a scale of 8 miles to an inch, on this map the Township of Scott measures $1\frac{5}{16}$ inches in length and $1\frac{1}{8}$ inches in width; how many acres does it contain?

248. The map of a country is drawn on a scale of $\frac{1}{10}$ of an inch to a mile; what area on the map will represent a lake 4000 acres in extent?

249. A map of a city is drawn to a scale of 500 ft. to an inch. It is proposed to lay out as a park a block of land which measures on the map $2\frac{3}{8}$ inches by $\frac{15}{16}$ of an inch; how many acres will the park contain?

250. If the side of a square lot containing one-tenth of an acre is expressed by the number 11, find the unit of linear measurement.

251. The mass of the earth being the unit, the sun's mass is 354936; but if one-tenth of the mass of Jupiter be taken as the unit, the sun's mass is 104869; find the mass of Jupiter when the earth's mass is the unit.

252. The earth's mass is 79·89 times that of the moon. Such a unit is adopted as makes the earth's mass 1·25; what number will represent the mass of the moon?

253. The earth's equatorial diameter (7925·648 miles) being the unit, the sun's diameter is 111·454; find the diameter of the sun when a mile is the unit.

254. The power that can raise 33000 lbs. through one foot in a minute being taken as the unit, find the measure of the power of an engine that can raise 2000 tons through 50 ft. in 6 hours.

255. If 8 represents a man's, and 5 a woman's work in 3 days, where a boy's work in a day is the unit, how long will it take 3 women and 5 boys to complete a piece of work which 12 men have half finished in 10 days?

256. Taking the radius as the unit, the circumference of a circle is $6\frac{2}{7}$; how many revolutions will be made in a journey of 6 miles by a wheel whose radius is 2 ft. 4 in.?

257. If 50 cents an hour be taken as the unit, what number will represent the wages of a man who receives $50 for working 8 hours a day for 10 days?

258. If a velocity of one foot in one second be taken as the unit, what number will represent the speed of a man who can run 100 yds. in 10 seconds?

259. The number of square units in a piece of carpet is $2\frac{1}{3}$ times the number of linear units in its length; if 104 yards of carpet contain 78 square yards, find the units of measurement.

260. If the unit is the work which a man can do in an hour, what will be the measure of a piece of work which requires the work of 17 men for 9 days of 10 hours each?

261. The distance between the earth and the moon being expressed by 59·9643 with reference to the earth's radius as unit, and this radius being 3962·8 miles, each of these numbers being exact to the nearest decimal, what can be known of the moon's distance from the earth in miles?

262. Using my walking-stick as the unit, I find the distance between two telegraph poles to be 50; I then ascertain the length of the walking-stick in inches and calculate the distance between the poles to be 114 ft. 7 in.; find the measure of the length of the stick, taking half an inch as the unit of length.

263. A cubic inch of water being the unit of weight, if the weight of a cubic inch of gold be 19·258; of silver, 10·474; and of a mixture of gold and silver, 14·723, each right to the third decimal place, find the bulks and weights of gold and silver in one cubic inch of the mixture.

264. The volume of water at the boiling point is 1·043 of its volume at freezing point. If a cubic foot of water at freezing point weighs $62\frac{1}{2}$ lbs., what will it weigh at boiling point?

265. Lead weighs 11·4 times as much as water, and platinum weighs 21 times as much as water; what weight in platinum will be equal in bulk to 56 lbs. of lead?

266. If copper weighs 550 pounds, and tin 462 pounds to the cubic foot, what will be the weight of a cubic foot of a mixture of 6 parts copper to 5 parts tin?

267. What is the least number, which, being a cube, is also divisible by 4, 5, 9, and 12?

268. A cubic foot of gold weighs 19300 oz. avoir.; find the thickness of gold wire 500 ft. of which weigh one grain.

269. What is meant by the power of a number? What is the lowest power of ·4578 which is less than ·08754, the index of the power being a positive integer?

270. The link of Gunter's chain being 7·92 in., prove that 10 square chains make an acre. The Scotch ell being 37·069 inches, and 24 ells making the Scotch chain; what difference in square feet is there between 55 English and 42 Scotch acres?

271. A closed vessel formed of metal an inch thick, whose external dimensions are 8 ft. 3 in., 7 ft. 5 in., and 4 ft. 3 in., weighs 372 lbs.; how much more would a solid mass of metal of the same dimensions weigh?

272. A rectangular solid of metal has its length increased one-tenth, and its width one-seventh, by hammering; by how much has its thickness been diminished?

273. A can do as much work in two days as B can do in three, as much in three as C can in four days, and as much in four as D can in five days. B, C and D can together finish cutting 20 acres of grass in 10 days: how long would it take A to cut 10 acres?

274. Fifteen guineas weigh 4 oz., Troy, the metal consisting of 11 parts gold and 1 alloy, and the value of the alloy being $\frac{11}{239}$ that of an equal weight of gold. Find the price per pound, avoirdupois, of the alloy.

275. Out of a sum of $4,000, $217 are allotted to charity, and of the remainder A gets one-twentieth more than B, and B one-twentieth more than C; what are the shares of each?

276. A steel rod, 1 foot long and one inch square, weighs $3\frac{1}{2}$ pounds, and will just support 50 tons. What is the greatest length of steel wire which, when hung up by one end, will just not break by its own weight?

277. A vessel contains 120 gallons of wine; 20 gallons are drawn therefrom and the vessel filled with water; 15 gallons of this mixture are then drawn and the vessel again filled up with water. If this operation be performed six times, 20 and 15 gallons being drawn alternately, how much wine will the mixture contain?

278. A metre equals 39·37 inches, a cubic inch of distilled water weighs 252·458 grains, and a gallon of water weighs 10 lbs. Find the number of bushels, etc., in a hectolitre.

279. In how many ways can 1397 be separated into 3 different numbers of which the greatest common measure is 127?

280. A mass of lead ore weighed 7 tons; one portion of it yielded lead $\frac{78}{100}$, and silver 8 ounces per ton: the remaining portion yielded lead $\frac{75}{100}$, and silver $7\frac{1}{2}$ ounces per ton. The total yield of silver was 55 ounces; what fraction of the whole mass was lead?

281. In the Centigrade thermometer the freezing point is zero and the boiling point 100°; in Fahrenheit's thermometer the freezing point is 32° and the boiling point 212°; what will the Centigrade thermometer mark when Fahrenheit's marks 65°? What will Fahrenheit's mark when the Centigrade marks 4°? What reading on the Centigrade will correspond to 0° F.?

282. If the pure gold in a sovereign weighs 113·002 grains, while that in a 20-franc piece weighs 89·604 grains; find the least whole number of sovereigns that will be equal to a whole number of 20-franc pieces.

283. The driving-wheel of an engine is $27\frac{1}{2}$ ft. in circumference, and the forewheel 16 ft. Two particular spokes, one in each wheel, are observed pointing vertically upwards: how far will the engine travel before the same two spokes again point vertically upwards at the same time. How often will this happen in 9 miles?

Will they ever simultaneously point vertically downwards?

284. If 175 men and 210 boys do in 1330 days the same amount of work as 603 men and 1005 boys in 350 days, compare the average daily work done by each man with that done by each boy.

285. The old standard bushel was defined by statute to contain 2150 cubic inches, but on examination was found to contain only 2124. By the Act of 1824 the bushel was declared to contain 2218 cubic inches; find the real loss on the rental (£1075) of a farm (which was calculated on a certain fraction of the selling price of the corn grown), supposing the price per bushel to remain the same.

286. A coin consists of 11 parts gold and 1 part alloy, the alloy being worth one-fifteenth of an equal weight of gold. A new coin, of the same weight is struck, consisting of 9 parts gold and 1 alloy; compare (in the lowest whole numbers) the values of the coins.

287. If equal weights of water and air be taken, the air will occupy 814 times as much space as the water. If a quantity of air be immersed in water to a depth of 32 ft. it will be compressed to one-half its original volume; if to a depth of 64 ft., to one-third its original volume; if to 96 ft., to one-fourth; 128 ft., one-fifth; and so on. To what depth must the air be immersed to be as heavy as an equal volume of water?

288. The price of gold is £3 17s. 10½d. per oz.; a composition of gold and silver weighing 18 lbs. is worth £637 7s.; but if the proportions of gold and silver were interchanged, it would be worth only £259 1s.; find the proportion of gold and silver in the composition, and the price of silver per oz.

289. If 20 men, each earning 85 cents a day, can do, in 15 days, as much work as 28 boys, at 57 cents a day, can do in 20 days; determine whether it is more profitable to employ men or boys. If a piece of work done by men cost $1000, what would the same work cost if done by boys?

290. The external length, breadth and height of a rectangular wooden closed box are 18, 10, and 6 inches respectively, and the thickness of the wood is half an inch. When the box is empty it weighs 15 lbs. and when filled with sand, 100 lbs.; compare the weight of equal bulks of wood and sand.

291. A ditch has to be made 360 feet long, 10 feet wide at the top and 3 feet wide at the bottom, the angle of the slope of each side being 45°. Find the number of cubic yards to be excavated.

292. Into a gallon measure there are put 20000 grains of broken flint which is 2·5 times as heavy as water, and 10000 grains of granite which is 2·7 times as heavy. How many pints of water will now be required to fill the vessel?

293. A fast train leaves a place A for another place B at the same moment that a slow train leaves B for A. The fast train takes 2 hours for the journey and should meet the slow train $\frac{3}{8}$ of the distance from B. The slow train is running at reduced speed, however, and they meet at $\frac{1}{4}$ of the distance from B. How much behind time will the slow train be when it arrives at A?

294. A train 88 yards long overtook a person walking along the line at the rate of 4 miles an hour and passed him completely in 10 seconds; it afterwards overtook another person and passed him in 9 seconds. At what rate per hour was this second person walking?

295. "If eight best variegated silk scarfs, measuring each three cubits in breadth and eight in length, cost a hundred nishcas; say quickly, merchant, if thou understand trade, what a like scarf, three and a half cubits long and half a cubit wide, will cost in terms of drammas, pannas, cacinis and cowry-shells?" Lilavati.

(1 nishca = 16 drammas = 256 pannas,
1 panna = 4 cacinis = 80 cowry-shells.)

296. "If the hire of carts to convey thirty benches twelve fingers thick, the square of four wide and fourteen cubits long, a distance of one league be eight drammas, tell me, my friend, what should be the cart-hire for bringing fourteen benches, which are four less in every dimension, a distance of six leagues?" Lilavati.

(1 cubit = 24 fingers.)

297. Having three separate parcels of powders weighing respectively 81 lbs. 3 oz. 360 gr. Troy; 45 lbs. 10 oz. 252 gr. Troy; and 32 lbs. 7·232 oz. Avoirdupois; how can I sub-divide them into parcels weighing each the same integral number of grains?

298. The hour hand of a watch is $\frac{3}{7}$ of an inch long, the minute hand $\frac{4}{5}$ of an inch, and the second hand $\frac{1}{3}$ of an inch; compare the *linear* speed of their points.

299. A sum of money is divided among A B and C as follows:—A receives a certain share, B one-third and C one-sixth, and the balance is divided equally among them. It is found that B and C together receive half as much again as A. Find A's share before the balance was divided.

300. A rectangular room has its length, breadth and height, as 7, 6, 5, respectively. Its walls were decorated at a cost of 50 cents per square yard, and its ceiling finished at the same rate. The bill for the whole was $344.; what will be the expense of covering the floor with carpet at $1.40 per square yard?

301. An up train 88 yards long, travelling at the rate of 35 miles an hour, meets a down train 88 yards long at 12 o'clock, and passes it in 6 seconds. At 15 min. and 6 seconds past 12 the up train meets a second down train, 132 yards long, and passes it also in 6 seconds; at what time will the second down train overtake the first?

302. Supposing the cost of digging a trench to depend upon the depth to which it is sunk as well as the quantity of earth taken out, and that the cost of digging a trench 3 feet broad by 8 feet deep is 20 cents per yard; what would be the cost of a trench 120 yards long, 5 feet broad and 10 feet deep?

303. A workman finds that he has to spend one-half of his income on food, three-twentieths on clothes, three-twenty-fifths on rent, and he saves the remainder. He emigrates to a country where food costs half as much, clothes three times as much, and rent two and a half times as much, but wages are half as much again; find whether he is better or worse off, the test being the length of time his savings would support him in the country where he is settled.

304. "In a case of ordinary division the dividend and the divisor are always similar numbers, and the quotient is abstract." Explain this statement.

305. Explain why multiplication is regarded as a case of repeated additions.

306. Explain why division is regarded as a case of repeated subtractions.

307. Every prime number except 2 is an odd number.

308. The differences of the successive square numbers produce the odd numbers

309. The sum of any number of even numbers is even; the sum of an even number of odd numbers is even, and the sum of an odd number of odd numbers is odd.

310. If an odd number is not exactly divisible by an odd number, then when the quotient is odd the remainder is even, and when the quotient is even the remainder is odd.

311. If an odd number divides exactly an even number, it will also divide one-half of it; if an even number be divisible by an odd number, it will be divisible by double the odd number.

312. The quotient of an even by an odd number, when exact, is even; the quotient of an odd by an odd, when exact, is odd; the quotient of an even by an even, when exact, is either odd or even.

313. If an even number is not exactly divisible by an odd number, then when the quotient is even the remainder is even, and when the quotient is odd, the remainder is odd.

314. Perform the following operations by short methods:

$$479 \times 125; \quad 873294 \div 99; \quad 31687 \times 320618.$$

315. Divide 3812 by 7, and the result by 9, and explain how to get the correct remainder.

316. By subtracting $.\dot{7}$ from 10 times $.\dot{7}$ show that $.\dot{7} = \frac{7}{9}$.

317. The sum of two or more fractions may be subtracted from any number by subtracting them in succession

318. If two or more fractions are to be added to any number the result is the same in whatever order the fractions may be taken.

319. The product of any number and its reciprocal is unity.

320. In order that a fraction may be in its lowest terms it is necessary and sufficient that the numerator and denominator be prime to each other.

321. Shew that any number is divisible by four if the last two figures on the right are divisible by four.

322. Explain the following statement:—"$\frac{5}{6}$ is the measure of that magnitude which contains 5 times the sixth part of the unit magnitude."

323. Every square number is either a multiple of 3, or 1 more than a multiple of 3.

324. Shew that any number and its units' digit when divided by 2 or 5 give the same remainder.

325. Prove that no square number can end in 2.

326. Prove that no square number can end in 3.

327. Explain the difference between the operations involved in the following :—$\frac{1}{2}$ of 6 and $6 \div 2$.

328. Find the value of

$$2 + \frac{1}{2} + \frac{1}{2 \times 3} + \frac{1}{2 \times 3 \times 4} + \frac{1}{2 \times 3 \times 4 \times 5} + \frac{1}{2 \times 3 \times 4 \times 5 \times 6}$$

329. Explain the following method of adding the fractions in the previous problem :—

$$\begin{array}{r} 2{\cdot}5 \\ {\cdot}16666 \\ {\cdot}04166 \\ {\cdot}00833 \\ {\cdot}00138 \\ \hline 2{\cdot}7180\dot{5} \end{array}$$

330. How many more terms of the series, in No. 328, must be taken to give the sum 2·71828 + ?

331. Knowing that $\sqrt{2} = 1{\cdot}4142 +$, find $\sqrt{2}$ to seven decimal places without going through the ordinary process of extracting the square root.

332. Describe the respective advantages of using *vulgar* and *decimal* fractions.

333. Divide 3654 by 2·03, explaining each step of the process.

334. If two numbers are to be multiplied together, the product of the first multiplied by the second is the same as the product of the second multiplied by the first.

335. If the two whole numbers nearest to half the square of any odd number be taken as the hypothenuse and one side of a right angled triangle, show that the third side must be a whole number.

336. If a number be exactly divisible by 11, the last digit in the quotient will be the same as the last digit in the dividend.

337. "The greater part of the operations upon numbers are performed, not by operating directly upon those numbers, but upon the parts into which they can be decomposed." Ascertain how far this is true in the case of ordinary addition and multiplication.

338. In addition of numbers the operation might be performed by proceeding from left to right. What would be the disadvantage of this, as compared with the ordinary method?

339. Prove that the difference between two numbers is not altered if the same number be added to each. How if the same number be subtracted from each?

340. Prove that "multiplication and division are mere modifications of addition and subtraction respectively."

341. Shew that the number of digits in any product can never exceed the sum of the number of digits in the factors of that product.

342. Prove that if the divisor and dividend be multiplied by the same number the quotient remains the same. What about the remainder?

343. If the dividend is increased by a multiple of the divisor how is the quotient affected? How is the remainder affected?

344. How is the value of an improper fraction affected by adding the same number to both its terms?

345. "Division is rather to be regarded as the undoing of a multiplication." Explain this statement.

246. The difference between any number and its square is even.

347. The square of an even number is always divisible by 4.

348. In ordinary multiplication the last partial product is always greater than the sum of all the other partial products.

349. Which of the digits may be the last digit of a square number?

350. Examine how far the following rule holds good:—"The truth of all results in multiplication may be proved by using the multiplicand and multiplier as the multiplier and multiplicand."

351. How is the decimal system of notation extended to fractions? What change in the decimal increases its value tenfold? Why is this?

352. Explain how a sufficient degree of accuracy may be obtained in the addition and subtraction of circulating decimals to any required number of places without converting the decimals into fractions.

353. From the fact that 10 square chains make an acre, deduce the length of the link in inches.

354. Prove the rule for the division of decimals. Apply the contracted method to find $\cdot 95372843 \div 44\cdot 736546$ to eight places of decimals.

355. Shew that any proper fraction is increased by adding the same number to both its terms.

356. A fraction expresses the quotient of the numerator by the denominator.

357. Why is it necessary in the addition and subtraction of fractions to reduce the fractions to a common denominator?

358. In what sense can $\frac{1}{9}$ be said to be the value of $\cdot\dot{1}$?

359. If any number (not a multiple of 10) be divided by 9 and have no remainder then the sum of the last digit in the dividend and the last digit in the quotient will be 10.

360. In dividing any numbers by 12 shew that there will never be more than two figures between the decimal point and the repeating decimal.

361. Prove that the square of 99·9899995 differs from 9998 by little more than a unit in the eighth decimal place.

362. The following rule is sometimes given to divide by 3·14159: "Multiply by 7, divide by 11, then by 2, and add one-eighth of one-thousandth of the result." Find the error in dividing 180 by 3·14159 by this rule.

363. A person walks a feet a minute, in a quarter of an hour he goes a mile; find the value of a.

364. If a number is to be divided by two or more fractions, the result will be the same whether we divide by the fractions successively or by their product.

365. If two or more fractions are to be multiplied together, the result is the same in whatever order the fractions may be taken.

366. If two or more fractions are to be subtracted from any number, the result is the same in whatever order the fractions may be taken.

367. If the numerator and denominator of a fraction be prime to each other, the numerator and denominator of any other fraction of equal value will be equimultiples of the numerator and denominator of the given fraction.

368. The product of any two numbers is the same as the product of their l. c. m. and g. c. m.

369. The sum of two odd squares cannot be a perfect square.

370. No square number differs by more than 1 from some multiple of 5.

371. Explain how to determine by inspection whether any given fraction will produce a terminated or a repeating decimal.

372. Show that a unit of any order is always greater than the sum of the numbers expressed by all the digits which follow it.

373. Prove that every number and the sum of its digits when divided by 9 leave the same remainder.

374. Explain the following method of finding the price of $16\frac{5}{8}$ yards of cloth at $3\frac{1}{4}$ per yard:—

$$\begin{array}{r} 16\frac{5}{8} \\ 3\frac{1}{4} \\ \hline 48 \\ 4 \\ 1\frac{7}{8} \\ \frac{5}{32} \\ \hline 54\frac{1}{32} \end{array} = \$54.03.$$

375. Explain the following method of finding the approximate value of $18.75 × 1.06 × 1.06 × 1.06:—

$$\begin{array}{r} \$18.75 \\ 1\ 13 \\ \hline 19.88 \\ 1\ 19 \\ \hline 21.07 \\ 1\ 26 \\ \hline \$22.33 \end{array}$$

376. Shew by means of an example how an improper fraction is affected by subtracting the same number from its numerator and denominator.

377. Explain the following method of shortening ordinary division:—

```
324) 432578 (1335
      1085
       1137
        1658
          38
```

378. Dividing the numerator of a fraction by any number is the same in effect as multiplying the denominator by it.

379. If a number measure each of two others it will also measure their sum and their difference.

380. If a number measure two others it will measure the sum and the difference of any multiples of these numbers.

381. Shew how to convert a simple or a mixed periodic decimal into an ordinary fraction.

Examples, $\cdot\dot{3}1\dot{6}$, and $\cdot 67\dot{1}3\dot{5}$.

382. Define the terms "abstract" and "concrete" as applied to numbers. Is $6 \times 3 = 18$ a correct solution of the question: What will be the cost of six postage-stamps at three cents each?

383. Prove the rule for placing the decimal point in the division of decimals.

384. Reduce $\frac{1}{7}$ and $\frac{1}{14}$ to decimals and account for the resemblance between the repeating periods in the two results.

385. What is meant by a "mean solar day?" How does the "solar" year differ from the "civil" year? Explain fully the methods which have been made use of to correct the error arising from this difference.

386. Describe the silver and the copper coinage of Canada. For what sums respectively are silver and copper legal tenders?

387. Why should the fraction $\frac{5+8}{6+9}$ be greater than $\frac{5}{6}$ and less than $\frac{8}{9}$?

388. In ordinary division the first subtraction always takes away more than half the dividend.

389. If a square number be divided by 4, the remainder is 0 or 1.

390. If a square number is divisible by 3, the sum of its digits is divisible by 9.

391. The g. c. m. of two numbers is the l. c. m. of all their common measures.

392. Of any two common measures of two numbers, ascertain whether one is necessarily a factor of the other.

393. Every common multiple of two numbers is a multiple of their l. c. m.

394. Of any two common multiples of two numbers ascertain whether one is necessarily a factor of the other.

395. State the conditions of divisibility by 2, 3, 4, 5, 6, 8, and 9.

396. Explain clearly the difference in meaning of the quotient in the two following cases :—\$36 ÷ \$9, and \$36 ÷ 9.

397. "The place of the figure indicates its power." Explain what is meant by this statement. Shew that this principle forms the basis of our system of numeration.

398. If a number is exactly divisible by 101, the last two digits in the quotient will be the same as the last two in the dividend.

399. Every prime number when divided by 6 will leave a remainder of either 1 or 5.

400. Explain the following method of dividing a number by 9 :—Take any number, say 5374965, for dividend,

5972183
5374965

Divide the sum of the digits by 9 and place the remainder, 3, over the right-hand digit ; subtract 5 from 3 in the usual way, and place the remainder, 8, over the second digit ; subtract 6 from 8 and place the remainder, 1, over 9 ; and so on, the operation ceasing as soon as a digit has been placed over the left-hand digit of the dividend. The 597218 and the 3 are respectively the quotient and remainder obtained by dividing by 9.

401. If a number is not exactly divisible by 11, how can the remainder be found without finding the quotient?

402. How can you ascertain, without actually dividing, whether a number is exactly divisible by 11?

403. Reduce to a decimal correct to four places:

$$\frac{1}{2^2}+\frac{2}{2^3}+\frac{3}{2^4}+\frac{4}{2^5}+\frac{5}{2^6}+\frac{6}{2^7}+\frac{7}{2^8}+\frac{8}{2^9}.$$

404. Explain the following method of solving the previous problem:

$$\begin{array}{l} \quad\ \ 8 \\ 7+4 \\ 6+5\cdot5 \\ 5+5\cdot75 \\ 4+5\cdot375 \\ 3+4\cdot6875 \\ 2+3\cdot8437 \\ 1+2\cdot9218 \\ \quad\ \ \cdot9804 \end{array}$$

405. By what arithmetical operation may a decimal be produced which neither terminates nor repeats?

406. Express $\frac{1}{19}$ as a recurring decimal. Show that if any whole number be divided by 19 and the quotient expressed as a whole number and a decimal, the digits will recur in the decimal portion in the same order as in $\frac{1}{19}$, starting from the proper point.

407. If a number is to be divided continuously by two or more fractions, the result will be the same in whatever order the divisors may be taken.

408. If a number is to be multiplied by two or more fractions, the result is the same whether we multiply by these fractions in succession or by their product.

409. To divide 38743295 by 11, proceed thus:

38743295
35221178

First find the remainder 8 and place it under the right-hand digit; subtract 8 from 5 and place the remainder 7 under the second digit; subtract 7 from 9 and place the remainder under the third digit, and so on until a remainder is placed under the left-hand digit. The 3522117 and the 8 are respectively the quotient and the remainder.

Give the reason for this method.

410. The sum of two or more fractions may be added to any number by adding them in succession.

411. The product of any three numbers is the same as the product of their l.c.m. and the square of their g.c.m.

412. If a cube be divided by 9 the remainder is 0, 1, or 8.

413. How can it be ascertained, without actually dividing, whether a number is exactly divisible by 99?

414. If a number has been ascertained to be divisible by 99, the quotient may be obtained by subtraction thus:

39600
39204

where 39204 has been divided by 99, and the quotient 396 obtained by placing a cipher over each of the last two digits and subtracting.

415. Shew that no square number ends in 2, 3, 7 or 8.

416. What form of vulgar fraction will give rise to a decimal consisting of p digits which do not recur, and of q digits which are repeating?

417. What must be the denominators of those fractions which, on being reduced to decimals, give pure repetends of three figures?

418. Explain the method of proving results in multiplication known as "*casting out the nines.*" In what case will it fail as a test?

419. Explain the following method of multiplying 367 by 648:—

367
648
‾2936
23188
‾237816

420. State fully the advantages and disadvantages of decimals as compared with vulgar fractions.

421. Multiplying the numerator of a fraction by any number is the same in effect as dividing the denominator by it.

422. Explain the principle of the evaluation of circulating decimals, and reduce to vulgar fractions

·576634634 - - - -   ·28345345 - - - -   ·1654321321 - - - -

423. Whence does it appear that a vulgar fraction may always be reduced either to a terminated or a circulating decimal?

424. Explain the following method of finding the price of 1 bushel 2 pk. 1 gal. 2 qts. at \$3.60 a bushel :—

$$\begin{array}{r} 3.60 \\ 1.80 \\ .45 \\ .22\frac{1}{2} \\ \hline \$6.07\frac{1}{2} \end{array}$$

PROBLEMS ARISING FROM BUSINESS TRANSACTIONS.

PERCENTAGE.

1. Find $\frac{1}{100}$ of \$400; $\frac{1}{100}$ of \$600; $\frac{1}{100}$ of \$450; $\frac{1}{100}$ of \$350; $\frac{1}{100}$ of \$627; $\frac{1}{100}$ of \$34.

2. Find 1 per cent. of \$300; 1 per cent. of \$200; 1 per cent. of \$400; 1 per cent. of \$342; 1 per cent. of \$41.

3. Find 1% of \$700; 1% of \$850; 1% of \$625; 1% of \$35.

4. Find $\frac{2}{100}$ of \$400; 2% of \$600; 2% of \$725; 2% of \$63.

5. Find 5% of \$320; 4% of \$27; 2% of \$18; 6% of \$29.

6. Write decimally $\frac{4}{100}$, $\frac{6}{100}$, $\frac{3\frac{1}{2}}{100}$, $\frac{3\frac{3}{4}}{100}$, $\frac{10\frac{1}{2}}{100}$, 7 per cent., $5\frac{1}{2}$ per cent., 10 per cent., $4\frac{1}{2}$ per cent.

7. Express decimally 3%, 2%, $4\frac{1}{2}$%, $5\frac{3}{4}$%, $12\frac{1}{2}$%, ·5%, ·02%.

8. How many hundredths of anything is $\frac{1}{4}$ of it? $\frac{1}{5}$ of it? $\frac{1}{8}$ of it? $\frac{2}{5}$ of it? $\frac{2}{7}$ of it? ·04 of it? ·125 of it? ·0875 of it?

9. What per cent. of anything is $\frac{1}{4}$ of it? $\frac{1}{5}$ of it? $\frac{1}{8}$ of it? $\frac{3}{8}$ of it? $\frac{5}{12}$ of it? ·04 of it? ·06 of it? ·125 of it? ·0425 of it?

10. What fraction of 100 is 4? What fraction of 200 is 6? What fraction of 350 is 7?

11. What per cent. of 625 is 25? What per cent. of 12 is 9? What per cent. of 375 is 15?

12. What % of 100 is 4? What % of 200 is 6? What % of 250 is 5?

13. 7 is what % of 35? 12 is what % of 72? 95 is what % of 1900? $12\frac{1}{2}$ is what % of 225? $\frac{5}{8}$ is what % of $17\frac{1}{2}$?

14. Find $12\frac{1}{2}$% of 1728 men; of 864 bushels.

15. Find $3\frac{5}{16}$% of £175. 12s. 6d.; of $144.16.

16. Find $1\frac{1}{16}$% of $265,000,000; of 50 guineas.

17. Find $33\frac{1}{3}$% of 1260 marks; of 172·80 francs.

18. $365 is what % of $5840?

19. 36 minutes is what % of 1 day?

20. £3. 2s. 6d. is what % of £25?

21. Ten per cent. of a certain number is 13; find the number.

22. The number of boys in a school is 60% of the number of girls. The number of girls is 60; how many pupils are there in the school?

23. The average attendance at a school this term is 225, which is an increase of $12\frac{1}{2}$% on the average attendance of last term; what was the average attendance last term?

24. What number increased by 14% of itself is equivalent to 285?

25. What number diminished by $11\frac{1}{4}$% of itself is equivalent to 710?

26. On account of the increase of value of flour the price of bread is advanced 25% of itself. Formerly ten loaves were sold for one dollar; what number of loaves will now be given for fifty cents?

27. A regiment lost 20% of its men in a battle; 10% of the remainder deserted, there then remained 360 men. How many men were there originally in the regiment?

28. The earnings of a mill for two years amounted to $6560; the earnings the 2nd year were 5% more than the earnings of the 1st year. Find the amount of the 2nd year's earnings.

29. Ice expands 10% in freezing; how many cubic feet of ice will weigh 1 ton, given that 1 cub. ft. of water weighs $62\frac{1}{2}$ lbs.?

30. A bankrupt pays 30% of his debts; the amount that a creditor receives is what per cent. of that which he loses?

31. A house and lot, bought for \$4000, increased in value 120 %; what was the increased value?

32. A teacher spends $92\frac{1}{2}$ % of his salary, and has \$120 left each year; in how many years will he save one year's salary?

33. One number is double another; 10 % of the greater and $12\frac{1}{2}$ % of the less together make 39; what are the numbers?

34. One number is equal to another increased by 20 % of itself; $12\frac{1}{2}$ % of the first number is greater than 5 % of the second by 10; what are the numbers?

35. The length, breadth and thickness of a block of metal are each increased $\frac{1}{10}$ % by heating; by what per cent. has the volume increased?

36. A's money is $33\frac{1}{3}$ % more than B's; how much per cent. is B's of A's?

37. A boy changed $33\frac{1}{3}$ % of his paper money into silver; he spent 50% of this silver in buying a ball and bat, which cost \$1.50; how much money had he?

38. A metre is equal to 39·37 inches: a cubic metre is how much per cent. more than a cubic yard?

39. The number of girls in a school exceeds the number of boys by 50; the number of boys is $37\frac{1}{2}$ % of the whole; find the number of girls.

40. A can do 10 % of a piece of work in 1 day, B $12\frac{1}{2}$ %; what % of the work will remain to be done after A has worked 2 days, and B 3 days?

41. A man who owned 30 % of a mine sold $33\frac{1}{3}$ % of his share for \$12000; what is the value of the mine?

42. A farm cost \$4000; 60 % of this sum was 40 % of 3 times the value of the house that was built on the farm; what was the cost of the house?

43. Divide 1440 into three parts, so that 10% of the first part, $12\frac{1}{2}$ % of the second part, and $16\frac{2}{3}$ % of the third part may be equal.

44. One-fifth is what per cent. of one-half?

45. A owns 80 % of a farm and B owns the remainder; C sells the farm for them and receives 4 % of the selling price for his services; what per cent. of the selling price does A receive?

46. From a cask containing 126 gallons of wine, 2 gal. $1\frac{3}{5}$ qt. leaked away; what per cent. was lost?

47. How many pounds of bread can a baker make from 1 cwt. of flour, if the bread is 25% heavier than the flour used?

48. A druggist buys goods at $5 per pound Av., and sells at $5 per pound Troy; his gain is what per cent. of the cost?

49. Mr. Brown sold his farm for $15840, which was 10% less than he gave for it, and he gave 10% more than it was worth; what was the actual value of the farm?

50. Four per cent. of beer is alcohol; how much alcohol does a man swallow in one year, if he drinks 3 pints of beer a day?

51. Two per cent. of a certain number, together with 3 of half of the number, makes up 21; what is the number?

52. Five per cent. of a certain number and 6% of twice that number, together make 175; find the former number?

TRADE DISCOUNT.

1. A merchant bought a quantity of goods amounting to $600; 10% of this amount having been thrown off, how much did he pay for the goods?

2. After a discount of 15% had been taken off, a merchant paid $850 for a bill of goods; what was the amount of the discount?

3. A merchant paid $170 for a bill of goods, after a discount of $30 had been taken off; what was the rate of discount?

4. A trader bought a lot of paper marked $5 per ream, at a discount of $12\frac{1}{2}$%; he received a further reduction of 2% for cash: what did the paper cost him per ream?

5. At what price must an article which cost $12 be marked, in order that after a discount of 10% has been taken off, there may be a gain of $3?

6. An article sold for $4; a discount of $12\frac{1}{2}$% had been given; what was the marked price?

25. A commission merchant received a consignment of 2000 bbls. of flour, which he sold at $8 a barrel, on a commission of $1\frac{1}{2}$%; the expenses for freight, paid by the agent out of the proceeds, amounted to $150; he bought cotton at 15 cents a pound with the net proceeds, charging $\frac{3}{4}$% commission for buying. How many pounds of cotton did he buy?

26. A consignment of goods was sold for $12500; the agent paid $200 for freight and remitted his employer $12150. What rate of commission was charged?

27. An agent receives $6360, with instructions to invest in sugar at 5 cents a pound, retaining his commission at 2% and paying in advance the freight at 20c. per cwt. How much sugar does he buy?

28. An agent receives 1500 hams, average weight 25 pounds, which he sells at 10c. a pound; he pays freight 20 cents per cwt., and charges a commission of 2% on sales. He is instructed to buy tea at 45 cents a pound, to prepay the freight on the tea (20 cents per cwt.), and retain his commission of $1\frac{1}{2}$% on the purchase. How many pounds of tea did he buy?

29. An agent charges $2\frac{1}{2}$ commission on sales and 2% for guaranteeing payment; the sales amount to $1200. Find the amount the agent receives.

30. An agent charges 2% commission on sales and $2\frac{3}{4}$% for guaranteeing payment; he received altogether $380. What was the amount of the sales?

31. An agent charges 2% commission on sales and $2\frac{3}{4}$% for guaranteeing payment; his commission for selling was $40 less than the guaranty commission. What was the amount of the sales?

32. Sold cotton on commission of 4%, invested the net proceeds in sugar at $1\frac{1}{2}$% commission; the total commission was $220. Find the value of the cotton.

33. An agent sold a consignment of apples on a commission of 3%. After deducting his commission and reserving a sufficient sum to pay the freight at 20 cents per cwt., he bought flour at $2.80 per cwt. on a commission of $2\frac{1}{4}$%. The total commission was $63. Find the amount of flour bought.

34. An agent received a consignment of wheat, which he sold, charging 2 % commission. With the net proceeds, after deducting his commission at $1\frac{1}{2}$ %, and prepaying freight at 25 cents per cwt., he bought sugar at $4\frac{3}{4}$ cents a pound. The agent's total commission was $70. Find the number of pounds of sugar bought.

35. A commission merchant sold flour on a commission of $\frac{1}{2}$ %; with the net proceeds bought tea on a commission of $\frac{1}{3}$ %; the total commission was $50. Find the cost of the tea.

36. An agent sells on a commission of a % and buys at b %; he receives a consignment, sells, and after deducting his two commissions, buys. Find what fraction the total commission is of the selling price.

37. An agent charges the same rate of commission for buying and selling. He sells a consignment for $4060, and after deducting $120 for the two commissions, invests the remainder. Find the rate per cent. charged.

38. An agent charges the same rate of commission for buying and selling. He sells a consignment for $8140, deducts $280 for his commission, and invests the balance. Find the rate per cent. charged.

39. A commission merchant charges twice the rate of commission for selling that he does for buying. He sells a consignment of leather for $3030, deducts as total commission $90, and invests the balance in hides. Find the rates of commission charged.

Solution.—Commission for selling is calculated on $3030.
Commission for buying is calculated on $2940.
Commission on $3030 at double second rate is the same as commission on $6060 at second rate.
∴ $90 is commission on $6060 + $2940, or $9000, at second rate.
∴ second rate is 1 %.

40. An agent's rate of commission for selling is one-half more than his rate for buying. He sold a consignment of flour for $4040, and after deducting $100 for both commissions invested the remainder in tea. What were the rates charged?

41. An agent's rate of commission for selling is $\frac{1}{3}$ of his rate for buying. He sold a consignment for $1421, and after deducting $49 invested the balance. What did he charge for selling?

42. An agent charges $1 more for selling goods for $100 than for buying for $100. He sold a consignment of pork for $1734, deducted $85 and invested the remainder in oats. What rate did he charge for investment?

43. An agent charges 50 cents more for selling, than he does for buying, 100 dollars' worth of goods. He sold a consignment for $3417, deducted $51 and invested the remainder. What rates did he charge?

44. A commission merchant sells goods for an employer, and, after deducting his commissions, invests the net amount in other goods. His total commission is $100, receiving $2 more for selling than for buying. If his rate for selling is the same as the rate for buying, what is that rate?

45. An agent sold a consignment of goods, and took $81 as his commission; he used the remainder in buying goods, deducting $79 for commission. If his rate for selling is the same as the rate for buying, what is that rate?

46. A consignment of 1000 bbls. of flour was sent to a commission merchant, with instructions to sell it and remit the net proceeds by draft. The consignee pays for freight and other expenses $240.80, sells the flour at $8.50 per barrel, charges $2\frac{1}{2}$% commission, and pays $\frac{3}{8}$% premium for draft; how much does the consignor receive?

INSURANCE.

1. Find the premium paid an insurance company for a policy of $1200., the rate of insurance being $\frac{3}{4}$%.

2. An insurance company charges $\frac{7}{8}$% per annum; what is the premium paid on a policy of $4000, in force 3 years?

3. What premium is paid for a policy of $1800, in force 5 years, the rate of insurance being $\frac{5}{8}$% of the policy for each year?

4. Find the premium paid to insure a house worth \$12000, for $\frac{2}{3}$ of its value, for 3 years, the rate being $\frac{2}{3}$% of the policy for each year.

5. A premium of \$45 is paid an insurance company for one year's insurance on a house, the amount of the policy being \$6000; what is the rate?

6. A premium of \$60 is paid an insurance company for two years' insurance, the amount of the policy being \$4000; what is the yearly rate?

7. \$50 is paid to secure a policy of \$2500 on a house, to run 3 years; what is the yearly rate?

8. \$42 is paid to insure for two-thirds of its value, a house worth \$6000; the policy is to run for 2 years; what is the yearly rate?

9. A cargo worth \$1250 is insured for 75 % of its value; the premium paid was \$12.50. Find the rate.

10. A house was insured for 3 years by paying a premium of \$24; the rate was $\frac{1}{2}$% a year; find the value of the house, $\frac{3}{4}$ of its value being insured.

11. A premium of \$37.50 was paid for a two-year policy on a dwelling worth \$6000; the rate was $\frac{1}{2}$% of the policy for each year; what fraction of the value was insured?

12. An insurance company charged \$18.45 for insuring a house worth \$2160, for one year; find the rate per cent.

13. An insurance company took a risk for one year of \$10000 on a warehouse worth \$15500, at $\frac{3}{4}$%; it covered 40% of its risk in another company at $\frac{7}{8}$%; how much premium did the first company receive above that which it paid the second?

14. (a.) How much is paid to secure a policy of \$100 for one year at $\frac{3}{4}$%? Ans. \$$\frac{3}{4}$.

(b.) How much does the owner receive above the premium paid, in case of loss? Ans. \$99$\frac{1}{4}$.

(c.) What must be the value of goods insured for \$100 at $\frac{3}{4}$%, so that the owner may suffer no loss in case the goods are destroyed? Ans. \$99$\frac{1}{4}$.

15. For how much must a cargo worth \$7940 be insured at $\frac{3}{4}$%, so that the owner suffers no loss if the cargo is lost?

16. For what sum should a house worth \$3965 be insured to cover the value of the house, and the cost of the policy at $\frac{7}{8}$%?

17. For what sum must a vessel worth $15800 be insured to cover the value of the vessel, the cost of the policy at $\frac{5}{8}$%, and $100 besides?

18. What must be the amount of the policy taken on a cargo worth $5940 to cover the value of the cargo, the premium paid, and an additional sum equal to the premium, the rate being $\frac{1}{2}$%?

19. A company took a risk of $40000 at $1\frac{3}{4}$%, re-insured 40 per cent. of it at 2% and 25 per cent. at $2\frac{1}{4}$%. What rate of insurance did the company get on the amount of risk it retained?

20. A company took a risk at $1\frac{1}{2}$%, re-insured 50% of it at $1\frac{3}{4}$%, and 20% of the remainder at $1\frac{1}{4}$%. What rate did the company receive on the amount of risk it carried?

21. A block of buildings worth $1000000 was insured in company No. I. for $25,000 at $1\frac{1}{4}$%; in company No. II. for $40,000 at 1%; in company No. III. for $100,000 at $\frac{7}{8}$%. Find the premiums paid in each case. If the block be damaged to the extent of $100,000, what amount of loss will be borne by each company?

22. A house is insured for $\frac{3}{4}$ of its value, the furniture for $\frac{2}{3}$ of its value. The rate in both cases is $\frac{3}{4}$%. The house is worth 5 times as much as the furniture; the total premium paid is $12.60. Find the value of the house.

23. The premium on a vessel and its cargo is $120; the rate on the vessel is $\frac{1}{2}$% and on the cargo $\frac{3}{8}$%. The value of the cargo is double that of the vessel, and each is insured for $\frac{3}{4}$ of its value. Find the value of the cargo.

24. A vessel was insured for $20000 at $\frac{3}{4}$% in one company, and for $25000 in another at $\frac{2}{3}$%. What rate of premium is paid on the whole insurance?

25. For what sum must a vessel worth $18000 and cargo worth $24000 be insured at $\frac{3}{4}$ and $\frac{1}{2}$% respectively, to cover the total value and the premiums paid?

26. A merchant had 500 barrels of flour insured for $\frac{3}{4}$ of its value at $2\frac{1}{2}$%, paying $75 premium. At what price per barrel must he sell to gain 25% of cost, as well as premium paid?

27. What is the value of a house if the insurance premium of $\frac{3}{5}$ % on $\frac{3}{4}$ of its value, including 50 cents for the policy, equals $21.50?

28. A man built a house, costing $2500, upon a lot worth $500; the house was burned and the insurance company paid the full amount of the policy, $\frac{3}{5}$ of the value at $\frac{3}{5}$ %; the land was then sold for $750. What was the man's total gain or loss?

29. The value of a yacht is $2940; for what sum must it be insured at 2%, to cover, in case of loss, the value of the yacht and the premium paid?

30. A house worth $2962.50 is insured for $3000 to cover value and premium paid; what was the premium paid?

31. A vessel worth $4925 is insured for $4000, which sum includes $\frac{4}{5}$ of the value of the vessel, and the premium paid; what was the rate of insurance?

32. A cargo worth $22125 is insured for $15000, to cover $\frac{2}{3}$ of the value, the premium and $100 besides; what was the rate of insurance?

33. Company No. I. insured a building and its stock for $\frac{2}{3}$ of the value, charging $1\frac{3}{4}$%. They reinsured in Company No. II. $\frac{1}{4}$ of the risk at $1\frac{1}{2}$%; building and stock being destroyed by fire, the second company lost $49000 less than the first. What amount did the owners lose?

34. My house is valued at one-half more than my brother's; my house is insured at $\frac{3}{4}$% on $\frac{5}{8}$ of its value, my brother's at $\frac{5}{8}$% on $\frac{2}{3}$ of its value, I pay $12 more premium than my brother; find the value of each house.

35. A shipment of flour was insured at $\frac{1}{4}$%, to cover $\frac{1}{2}$ of the value and the premium; the premium was $15; find the value of the flour.

36. A drover is taking a herd of 400 cattle from Quebec to Liverpool; the average cost of the cattle was $45; for what sum must he have the cattle insured at $1\frac{3}{4}$%, to cover, in case of loss, the value of the cattle, the premium paid, and the cost of his passage, $84?

37. What part of the value of a house must be insured at 3% so that in case of loss the owner may receive $\frac{3}{4}$ of the value, in addition to the premium paid?

TAXES.

1. Every citizen of Borden is called upon to pay for public use 2 % of the value of his property. What tax does Jones pay whose property is worth $5000 ?

2. My property is worth, according to the assessor, $2500. The rate of taxation is $1\frac{1}{2}$ cents on the dollar. What is the amount of my taxes ?

3. Find the tax on property assessed at $12000, when the rate of taxation is 2 cents on the dollar ?

4. When the rate of taxation is 15 mills on the dollar, what is the tax on property assessed at $2000 ?

5. The total assessed value of the property in a town is $750,000. What tax will be raised when the rate is $12\frac{1}{2}$ mills on the dollar ?

6. A city requires for the expenses of one year the sum of $1,500,000 ; the taxable property of the corporation is assessed at 135 million dollars. Find the rate of taxation.

7. The property of a village is assessed at $800,000 ; the rate of taxation is 18 mills on the dollar ; it costs 2 % of the tax for collection. Find the net amount received by the village.

8. A town requires $19600 to meet expenses for the year ; they pay 2 % for collection. What must be the rate, if the taxable property is assessed at $1,200,000 ?

9. A's income is $1200 annually, of which $400 is exempted from taxation. What tax does he pay when the rate is $1\frac{4}{5}$ % ?

10. What income tax does a man pay whose income is $1500 a year, the rate being 16 mills on the dollar ; $500 being exempted from taxation ?

11. What is my net income when I receive a salary of $2000 a year, $600 of which is exempted, the rate being 15 mills on the dollar ?

12. My salary is $1500 ; my net income is $1482.40 after paying income tax on all over $400. What is the rate ?

13. Mr. Jones' annual income is 25 % of his capital ; he pays $25 taxes, at the rate of $1\frac{1}{4}$ % on income. What is his capital ?

14. My salary is $1800; my net income is $1779 after paying income tax on all over $400. What was the rate?

15. The expense of constructing a bridge was $10000, which was raised by a tax on the assessable property of a town. The rate of taxation was 2 %, and the collector's commission was $150. Find the assessed value of the property of the town.

16. Incomes of $1000 or more pay tax on all over $400; incomes less than $1000 are exempted from taxation; the rate of taxation is 15 mills on the dollar. How much better off is a man whose income is $995 than a man whose income is $1000?

17. A tax of $1250 is levied on a village of which the assessed value of the property is $255,000. What is the tax on property valued at $1800?

18. A tax of $15000 is levied on a town, the assessed valuation being $930,000. What tax does a man pay whose income is $1300, $400 of which is exempted?

19. I paid $24 income tax, $400 of my income being exempted; the rate was 16 mills on the dollar. What was my income?

20. Smith bought a house for $6000; it is assessed for $\frac{2}{3}$ of its value, the rate of taxation being $16\frac{1}{2}$ mills on the dollar. The house is insured for $\frac{3}{4}$ of its value at $\frac{1}{2}$ %. If Smith had loaned his money he might have received $300 interest on it for the year. What monthly rent was Smith really paying for his house?

21. What sum must be assessed on a school district to build a school house worth $5700, and pay 5 % for collection?

22. A tax of $4500 is levied on a village, the assessed valuation being $180,000. What tax does a man pay whose income is $1350, $400 being exempted from taxation?

DUTIES AND CUSTOMS.

1. A dealer imports for me a book which was invoiced to him at $2.40; he pays 20 cents postage, 15 % ad valorem duty, and makes a gain of 25 % on the whole outlay. What does he charge me for the book?

2. If there were no duty in the previous problem, what would the dealer charge me for the book so as to make the same *rate* of gain? What would be the charge to make the same *amount* of gain?

3. What is the duty paid on an imported book, invoiced at $5, the duty being 15 per cent. of value?

4. I import a piano on which there is a specific duty of $30 and an ad valorem duty of 20 per cent.; I pay altogether for the piano $390. What was the invoice price?

5. Find the duty paid on a hogshead of molasses, invoiced at 40 cents a gallon, at 15 per cent. ad valorem.

6. What reduction, per gallon, might be made in the price of the molasses if there were no duty, the dealer selling at 25 % above total cost?

7. Find the export duty on a pine log of uniform section, the length being 30 ft. and the diameter 2 ft. 11 in., the rate of duty being $1.50 per cord.

8. Find the export duty on a stick of timber 20 ft. long, 3 ft. broad and 2 ft. thick, the rate of duty being $2 per 1000 feet board measure.

9. The duty paid on a consignment of 50 pounds of manufactured tobacco was $17.50; the duty on such tobacco is 30 cents a pound and 12½ % on the value. Find the value of the tobacco as shown in the invoice.

10. A dealer in cabinet organs sells at an advance of 40 % on the cost, laid down in his store. I pay him $304 for an organ, on which he paid a specific duty of $30 and an ad valorem duty of 15 %. What was the invoice price of the organ?

11. Giving the dealer the same rate of profit, by how much would the price be reduced in the preceding problem if there were no duty?

12. A grocer imported 150 cases of port wine, 24 bottles in each case. After 5 % had been allowed for breakage, he paid an ad valorem duty of 20 %. The freight and cartage expenses were $100, and the whole cost was $4384; what was the invoice price per bottle?

13. If goods invoiced at $1200 cost $1800 when laid down in warehouse, the cartage and freight amounting to $75, what was the rate of duty?

14. The duty on surgical instrument cases is 35%; that on the instruments 20%; the duty paid on a case of instruments invoiced at $30 was $7.50; find the invoice price of the instruments alone.

15. The duty on 1000 boxes of raisins, each containing 15 pounds, was $270; the raisins were invoiced at 8c. per lb.; the specific duty being 1 cent a pound, determine the ad valorem duty.

16. The duty on imported axes is $2 per dozen and 10% ad valorem. The whole duty paid on a lot of axes was $56, the specific duty being $24 more than the ad valorem. Find the number of axes imported.

17. The duty on a bale of canton flannel was $3.75, the specific duty being 1 cent per square yard, and the ad valorem 15%; find the width of the flannel, given that the bale contained 100 yards, invoiced at 20 cents.

18. The duty on wine containing 26% or less of spirits is 25 cents per gallon, and 30% ad valorem; for every degree above 26%, 3 cents more per gallon. A man wishes to import 100 gallons of wine 30% strong, bought at $2 per gallon; he orders enough water to be mixed with it to reduce the strength to 26%. Does he gain or lose in the amount of duty paid, the total value of the wine being reduced 10% by the mixing?

Does he gain or lose on the whole?

19. The duty on imported window shade rollers is 30% ad valorem; on the shades 5 cents per square yard and 15% ad valorem. Each shade is worth twice the value of the roller. The duty paid on a dozen (rollers and shades) invoiced at $3 each, was $9.90; the width of each shade was 4 ft. 6 in.; find the length.

20. The duty on rubber-lined cotton fire hose is 5 cents per pound and 15% ad valorem. The duty on 100 feet of hose, invoiced at 20 cents per foot, was $15.50; find its weight per foot.

STOCKS AND INVESTMENTS.

1. Find the value of 10 shares ($100 each) of bank stock at $80 per share.

2. Find the value of 75 shares of railroad stock at $125 each.

3. Merchants' Bank Stock sells at 140; find the cost of 15 shares.

4. Find the cost of $4500 stock at 90.

5. Find the cost of $2700 stock at $108\frac{1}{4}$.

6. I instruct a broker to purchase for me $5600 insurance stock which is selling at $85\frac{1}{2}$; the broker charges $\frac{1}{8}$% of par value of stock for commission; how much must I give the broker to pay for stock and commission?

7. Sold through a broker at $\frac{1}{8}$% commission, 80 shares at 92. What do I receive for the stock?

8. Purchased $8000 stock quoted at $89\frac{7}{8}$; what did it cost, brokerage $\frac{1}{8}$%?

9. How much stock, at $80\frac{5}{8}$, may be bought for $6450?

10. How much stock, at $80\frac{5}{8}$, may be bought for $6460, brokerage $\frac{1}{8}$?

11. How much stock must be sold at $117\frac{3}{4}$, to produce $4710?

12. How much stock must be sold to produce $4710, the stock being quoted at $117\frac{7}{8}$, brokerage $\frac{1}{8}$?

13. Find the income received from $4000 stock, paying an annual dividend of 8%.

14. Find the income received from $8450 stock, paying an annual dividend of 7%.

15. Invested $9100 in bank stock at $90\frac{7}{8}$, brokerage $\frac{1}{8}$, and sold out at 92, brokerage $\frac{1}{8}$; what did I gain?

16. Bought 86 shares, quoted at $96\frac{1}{8}$, and sold when the stock had fallen to 95; how much did I lose, brokerage $\frac{1}{8}$ each way?

17. What rate per cent. do I gain on my money by investing in stock at 95 which pays a 5% annual dividend?

18. What rate per cent. do I receive on my money by investing in stock at $94\frac{7}{8}$, brokerage $\frac{1}{8}$, paying an annual dividend of 5%?

19. Find the rate of dividend paid by stock, when a man who owns $24000 of it receives $1920.

20. How much 6% stock must be bought to give an annual income of $240?

What will it cost at 75, brokerage $\frac{1}{4}$?

21. What is the price of a 7% stock which pays 5% on the money invested?

22. What is the price of a 5% stock, paying $4\frac{1}{2}$% on the money invested, brokerage $\frac{1}{8}$?

23. If $7200 stock, paying 5%, be sold at $89\frac{1}{8}$, and the proceeds invested in a 6% stock at $107\frac{7}{8}$, what is the change in income, brokerage each way $\frac{1}{8}$?

24. If $50000 of 6 % stock be sold at $104\frac{7}{8}$ and the proceeds invested in an 8 % stock at $124\frac{7}{8}$, find the alteration in income, brokerage each way $\frac{1}{8}$.

25. A man decreases his income $480 by selling out of 3 % stock at 67 and investing in 4 % at par. What amount of 4 % stock did he buy?

26. A man sells out of 3 % stock at $67\frac{1}{8}$ and invests in 4 % stock at $99\frac{7}{8}$; his income was decreased by $480. What amount of 3 % stock did he sell, brokerage $\frac{1}{8}$?

27. Bought $4800 stock at 75. At what price, per $100 share, must I sell it to gain $150?

28. A man receives a half-yearly dividend of 4 % on the amount of his stock, and invests it in the same stock at 120. His next half-yearly dividend is $496. What was the amount of the first dividend?

29. What must be the price of consols in order that after deducting an income tax of 2 % an investor may make $3\frac{1}{2}$ % on his money, the consols paying 3 %?

30. A man has $9000 invested in a 4 % stock at 90, and $12000 invested in a 5 % stock at 125; he transfers from the latter to the former a sum sufficient to make the incomes from the different stocks the same. What is the amount of money transferred?

31. The expense of constructing a railroad was $4,000,-000, of which 40 % was borrowed on mortgage at 6 %, and the remainder is held in shares. What must be the average weekly receipts to pay the shareholders 5 %, the working expenses being 65 % of the gross receipts?

32. The yearly gain of a company whose capital stock is $1,000,000, is $55,000. What rate of dividend can they declare to ordinary shareholders, after paying 8 % on the preference stock, which is one-half of the whole amount?

33. Which is the more profitable investment, 6 % stock at 128, or 5 % stock at 99, brokerage $\frac{1}{8}$?

34. The whole stock of a company is $1,000,000, the net gain is sufficient to pay 6 % on the whole amount. The company pays $2\frac{1}{2}$ % to ordinary shareholders, the balance giving 8 % on the preference stock. What was the amount of the preference stock?

35. A company with a capital of $200,000 paid 8 % dividend to its shareholders. Afterwards new stock was issued, and, with the same amount of gain, the company paid only 5 %. What was the amount of the new stock issued?

36. A man invests $6,000 in 3 % stock at 75; he sells out at 80, and invests $\frac{1}{3}$ of the proceeds in $3\frac{1}{2}$ % stock at 96, and the remainder in 5 % stock at par. Find his income from the latter investments.

37. What per cent. is gained on money by investing in a $2\frac{1}{2}$ % stock at 60?

38. I sold some stock at a discount of 10% and made $12\frac{1}{2}$% on my money; at what rate of discount did I buy?

39. When money is worth 4%, what ought to be the price of consols which pay 3%?

40. A man invests $40,000 in $5\frac{1}{2}$% stock at $79\frac{3}{4}$, and $60,000 in $7\frac{1}{2}$% stock at $119\frac{3}{4}$, brokerage in each case $\frac{1}{4}$; what is his total income, and what does the broker receive for his services?

41. A person owns $15,000 bank stock paying 5%, which he sells and invests the proceeds in 6% stock at 120, his income being increased by $60; find the price at which he sold the first stock.

42. What sum invested in a 5% stock at 115 will yield a net income of $1779 after paying an income tax of 15 mills on the dollar on all over $400?

43. Money being worth 5% per annum, what sum should be paid for a $1000 bond, bearing annual interest at 6%, to be paid off at par at the end of 5 years?

44. How much stock, at 12% discount, must be bought, and sold at 8% discount, to make a clear gain of $100, brokerage each way $\frac{1}{8}$?

45. What must be the price of 5% stock to gain 8% on money invested?

46. A man invested a certain sum of money in a 6% stock at $119\frac{7}{8}$, brokerage $\frac{1}{8}$, and half as much more in a 5% stock $99\frac{7}{8}$, brokerage $\frac{1}{8}$; his income from the two investments was $900. How much did he invest in each kind of stock?

47. Which is the better investment, 5% stock at 113, or 6% stock at 134?

48. If $1200 4% stock be sold at 90, and the proceeds invested at $4\frac{1}{2}$% per annum; find the change in income.

49. A person invests £2362 10s. in 3 per cent. consols, which he sells when they have risen $\frac{5}{8}$, thereby gaining 15 guineas; at what price did he buy?

50. A man purchases £1400 stock in three per cent. consols at $94\frac{1}{2}$, and also invests £3150 in the purchase of Russian inscribed five per cent. loan at $94\frac{1}{4}$: how much stock has he standing in his name? If he sells the consols at $95\frac{1}{8}$, and the Russians at $96\frac{1}{2}$, what does he gain or lose by the transaction? [Brokerage on Consols $\frac{1}{8}$, on Russians $\frac{1}{4}$.]

SIMPLE INTEREST.

1. If 5% of the sum lent is charged for the use of money for one year, how many dollars should be paid for the use of $100 for one year? For the use of $300 for one year? For the use of $500 for one year?

2. If 6% of the sum lent is charged for the use of money for one year, how many dollars should be paid for the use of $450 for one year? For the use of $275 for one year? For the use of $242.75 for one year?

3. If 4% of the sum lent is charged for the use of money for one year, how many dollars should be paid for the use of $100 for 2 years? For the use of $100 for 3 years? For the use of $300 for 4 years?

4. If 6% of the sum lent is charged for the use of money for one year, how many dollars should be paid for the use of $100 for one-half of a year? For the use of $100 for one-third of a year? For the use of $100 for 2 months?

5. At 6% a year, find the interest on $300 for 2 years and 6 months; on $450 for 73 days; on $720 for 3 years and 146 days.

6. At 5% per annum, find the interest on $400 for $1\frac{1}{2}$ years; on $725 for $2\frac{1}{5}$ years; on $620.40 for 219 days.

7. Find the simple interest on

(*a*)	$1200	for 3	years		at 4% per annum.
(*b*)	$1750	" $2\frac{1}{2}$	"		" 3% " "
(*c*)	$926.50	" $2\frac{1}{3}$	"		" 6% " "
(*d*)	$1827.60	" 2	"	and 4 mos.	" 6% " "
(*e*)	$925.40	" 4	"	" 5 "	" 4% " "
(*f*)	$1000	" 2	"	" 65 days	" 5% " "
(*g*)	$1263.80	" 3	"	" 73 "	" 10% " "

(*h*) £1800 12s. 6d. for 2 years and 146 days at 5% per annum.

8. I loan $100 for one year at 6%; what sum should be returned to me at the end of the year?

9. Find the amount of $1200 loaned for three years at 4% per annum.

10. Find the amounts in the different examples in number 7 above, using 8% in each case.

11. Find the interest on $1400 loaned on May 1st, 1890, and returned Nov. 14th, 1890, at 8% per annum.

12. Find the interest on $1650 from Jan. 1st, 1889, to Oct. 28th, 1890, at 6% per annum.

13. Find the amount of $1275 from July 4th, 1888, to Jan. 15th, 1890, at 8% per annum.

14. Bought a farm for $5,500 to be paid in 6 months, with interest at 4%; find the amount of the payment.

15. The interest on $550 for 2 years is $44; find the interest on $100 for 1 year at the same rate.

16. The interest on $840 for 1 year and 146 days is $58.80; find the interest on $650 for 2 years at the same rate.

17. What is the rate per cent., when the interest on $1440 for $1\frac{2}{3}$ years is $72?

18. Find the rate per cent. per annum when $400 amounts to $448 in 3 years.

19. $1200 amounts to $1290 in 2 years and 6 months; find the rate per cent. per annum.

20. $1500 amounts to $1515 from Oct. 28th to Jan. 9th; find the rate per cent. per annum.

21. In how many years will $500 amount to $600 at 4%?

22. In what time will \$1260 amount to \$1340 at 3%?

23. In what time will \$100 amount to \$200 at 6%? What would be the amount of \$579.89 in the same time?

24. In what time will \$397.19 double itself at 4%?

25. In what time will any sum of money double itself at 3%?

26. At 4% for 3 years, the interest is what fraction of the sum lent?

27. At 5% for 4 years, the interest is what fraction of the principal?

28. Calculate what fraction the interest is of the principal in the following:

(*a.*) At 7½% for 4 years.
(*b.*) At 6¾% for 2 years.
(*c.*) At 3½% for 6 years.
(*d.*) At 5% for 20 years.
(*e.*) At 8% for 6 months.
(*f.*) At 6% for 4 months.
(*g.*) At 10% for 292 days.

29. In one year the interest is $\frac{1}{12}$ of the sum lent; in how many years will the interest equal the sum lent?

30. At 5% per annum, in how many years will the interest equal the sum lent?

31. In what time will a sum of money double itself at 6¼%? At 5½%? At 2%?

32. In what time will a sum of money treble itself at 4%? At 6%? At 7¼%?

33. The rent of a house, at \$20 a month, pays the taxes at 1¼% of value of house, and 6¼% on the money invested; what is the value of the house?

34. What monthly rent will pay the taxes at 1% of value, and 8% on the money invested in a house, for which \$8000 was paid?

35. What is the rate per cent. per annum when the interest on \$511,000 for 5 days is \$280?

36. The interest on \$400 for 1 year at a certain rate, together with the interest on \$500 for the same time at double that rate, amounts to \$28; find the rates.

37. The interest on \$300 for 2 years at a certain rate, together with the interest on \$600 for 3 years at double that rate, is \$105; find the rates.

38. The interest on \$250 for 6 months at a certain rate, together with the interest on \$450 for $1\frac{1}{2}$ years at $\frac{3}{4}$ of that rate, is \$25.25; find the rates.

39. Find the amount of \$100 in 2 years at 4 %.

40. What sum amounts to \$540 in 2 years at 4 %?

41. Find the sum which in 6 months, at 5 %, amounts to \$820.

42. What sum deposited in a bank now at 4 % will, in 9 months, amount to \$1339?

43. What sum deposited at 6 % will, in 219 days, amount to enough to pay the taxes on a building worth \$50,000, taxed at 15 mills on the dollar?

44. A man has an offer of \$1500 cash, or \$1650 in 9 months; which is the better offer, money being worth 8 % per annum?

45. \$250 amounts to \$275 in a certain time. What sum will amount to \$275 in one-half of the time?

46. A person borrowed money for $3\frac{1}{2}$ years at 8 per cent., (and repaid principal and interest with \$320.) How much was borrowed?

47. A person borrowed money for two years. For the first year he paid 5 %, and the second year 6 %. At the end of the time he paid back \$166. How much was borrowed?

48. "To find the interest on a sum of money at 6 %, multiply the sum by one-half of the number of months, and remove the decimal point two places to the left." Explain the rule.

49. A offers for a house \$2180, payable at the end of 3 years; B offers \$455 cash and \$455 at the end of each year for 3 years; C offers \$1600 cash. Which of these is the best offer, money being worth $8\frac{1}{3}$ per cent.?

50. If I borrow \$1200 for 3 years at 5%, with the understanding that the interest due at the end of each year shall form part of the principal for the next year, how much shall I have to pay at the end of the 3 years?

51. A man engaged in business was making 15% each year on his capital of \$15000; he gave up his business at a sacrifice of 10% of his capital, and loaned his money at 8%; what amount of income did he lose yearly?

52. A dealer bought \$2000 worth of flour on 6 months' credit, and sold it immediately for $12\frac{1}{2}\%$ advance. If from the proceeds he deposited in the bank sufficient money to amount to the \$2000 at the end of the six months, rate 5%, what sum had he left?

53. What does a dealer gain by buying goods for \$2500 at 6 months' credit, and selling immediately for \$2800, banks paying 6% on deposits?

54. The interest on a sum of money amounts to $\frac{5}{16}$ of the sum in $7\frac{1}{2}$ years; find the rate per cent.

55. The interest on \$5000 in a certain time amounts to $\frac{2}{3}$ the interest on \$100.000 for 1 month at 6%; find the time.

56. In what time will $\$1.33\frac{1}{3}$ amount to $\$1.66\frac{2}{3}$, at 5% per annum?

BANK DISCOUNT.

1. I have in my possession the written promise of John Jones to pay me, or any person I may name, \$1200 at the end of 90 days; the Bank of Commerce gives me \$1200, less the interest on it for 93 days at 6%, for my claim against Jones. How much do I receive?

2. \$500$\frac{00}{100}$. TORONTO, *June 1st, 1890.*

Four months after date I promise to pay John Jones, or order, the sum of Five Hundred $\frac{00}{100}$ Dollars, at Dominion Bank here. Value received. HENRY SMITH.

(*a*) On what date will this note be paid by Smith?

(*b*) If Jones sells this note to a bank on June 15th, how many days' interest will the bank deduct from the amount of it?

(*c*) How much will Jones receive if the bank discounts at 6%?

(*d*) What is the discount?

Find *(a)* the *day of maturity*, *(b)* the *time* between day of discount and day of maturity, *(c)* the *discount*, *(d)* the *proceeds* of the following notes:

3. \$1000$\frac{00}{100}$. HAMILTON, *March 1st, 1890.*

Six months after date I promise to pay Oliver Bland, or order, the sum of One Thousand $\frac{00}{100}$ Dollars, at my office here. Value received. JOHN SMITH.

Discounted, June 4th, 1890, at 5%.

4. $\$2356\frac{50}{100}$. TORONTO, *Feb. 14th, 1890.*

Sixty days after date I promise to pay to the order of Frank Smith, Two Thousand Three Hundred and Fifty-six $\frac{50}{100}$ Dollars, at Imperial Bank here. Value received.

Discounted immediately at 6%. G. BROWN.

5. $\$1250\frac{00}{100}$. PARKDALE, *Jan. 15th, 1890.*

Three months after date I promise to pay to Dan. Wright, or order, the sum of One Thousand Two Hundred and Fifty $\frac{00}{100}$ Dollars, at Standard Bank here. Value received.

SAMUEL NATTRASS.

Discounted Feb. 1st, 1890, at $6\frac{1}{2}$%.

6. $\$5640\frac{75}{100}$. BARRIE, Ont., *May 23rd, 1888.*

Four months from date I promise to pay James French, or order, at my office here, the sum of Five Thousand Six Hundred and Forty $\frac{75}{100}$ Dollars. Value received.

ABRAM WILKES.

Discounted July 2nd, 1888, at 8%.

7. $\$2769\frac{00}{100}$. GRIMSBY, Ont., *Dec. 1st, 1889.*

Ninety days from date I promise to pay William Barker, or order, the sum of Two Thousand Seven Hundred and Sixty-nine $\frac{00}{100}$ Dollars. Value received.

ERNEST SMITH.

Discounted Dec. 24th, at 6%.

8. $\$275\frac{00}{100}$. HARRISTON, *April 1st, 1890.*

Four months after date I promise to pay Thomas Wright, or order, the sum of Two Hundred and Seventy-five $\frac{00}{100}$ Dollars. Value received. THOMAS JONES.

Discounted June 4th, 1890, at 8%.

9. $\$4000\frac{00}{100}$. TORONTO, *Nov. 29th, 1889.*

Three months after date I promise to pay George Holmes, or order, the sum of Four Thousand $\frac{00}{100}$ Dollars at Bank of Commerce here. Value received. SAM SMITH.

Discounted Dec. 1st, 1889, at 8%.

10. $\$1234\frac{56}{100}$. NEW YORK, *May 5th, 1890.*

Six months after date I promise to pay Henry Yorker, or order, the sum of One Thousand Two Hundred and Thirty-four and $\frac{56}{100}$ Dollars, at First National Bank here. Value received. GEORGE GOULD.

Discounted June 4th, 1890, at 6%.

11. 400\frac{00}{100}$. PORT HOPE, *Jan. 29th, 1890.*

Thirty days after date I promise to pay William James, or order, the sum of Four Hundred $\frac{00}{100}$ Dollars, at Bank of Toronto here. Value received. HENRY SCOTT.

Discounted immediately at 8%.

12. 576\frac{75}{100}$. WHITBY, *Feb. 3rd, 1888.*

Four months after date, we promise to pay to the order of Charles Beemer the sum of Five Hundred and Seventy-six and $\frac{75}{100}$ Dollars, at our office here. Value received.

A. WILMOT & CO.

Discounted Mar. 1st, 1888, at 8 %.

13. $480^{00}. PARIS, *Feb. 6th, 1887.*

Three months after date, I promise to pay Samuel Cole, or order, the sum of Four Hundred and Eighty Dollars, at the Standard Bank here, with interest at 5 %. Value received. THOMAS JOHNSON.

Discounted Feb. 18th, 1887, at 6 %.

14. $2000^{00}. TORONTO, *Mar. 4th, 1889.*

Sixty days after date, I promise to pay to the order of Henry Graham the sum of Two Thousand Dollars, at the Imperial Bank here, with interest at 6%. Value received.

ALEXANDER MCCUAIG.

Discounted immediately, at 8 %.

15. $1200^{00}. PORT HOPE, *Aug. 25th, 1887.*

Ninety days after date, I promise to pay Thomas Scott, or order, the sum of Four Thousand Two Hundred Dollars, at the Bank of Montreal here, with interest at 7 %. Value received. GEORGE KELLY.

Discounted Sept. 1st, at 8 %.

16. When must notes dated and drawn as follows be paid :—

(a) Jan. 30th, at 1 month?
(b) Jan. 29th, at 1 month?
(c) Jan. 28th, 1888, at 1 month?
(d) Dec. 31st, at 2 months?
(e) Nov. 29th, at 3 months?

17. The interest on any sum for 73 days, at 5 %, is what fraction of the sum lent?

18. The interest on a note for 95 days, at 6 %, is what fraction of the amount of the note?

19. A bank discounts a 92-day note at 6 %. Find what fraction the discount is of the face value of the note.

20. A bank charges what fraction of the face of a note, when discounted 73 days before it matures, at 10 % ?

21. What fraction of the face value of a note does one receive from a bank which discounts it at 6 %, 90 days before it is due ?

22. A note is discounted 60 days before due at 8 %, and the proceeds amount to $360.20. Find the face value of the note.

23. What must be the face value of a note made June 1st at 3 months, and discounted immediately at 8 %, to produce $870 ?

24. What must be the face value of a note made May 25th at 4 months, and discounted June 3rd at 6 %, to produce $357.98 ?

25. A note for $730 was discounted 45 days before it matured and produced $724.60. What was the rate of discount ?

26. A man received from a bank $990 for a note of $1000, 73 days before it was due; what was the rate of discount ?

27. The discount on a note for $1825, which matures on Aug. 1st, and was discounted on June 4th, was $20.30; find the rate of discount.

28. The discount on a note of $1460, discounted 10 days before it was legally due, was $20; find the rate of discount.

29. For what sum must a note be drawn on June 1st, 1890, payable in 90 days, so that when it is discounted on June 14th, at 8%, the proceeds will amount to $358.60?

30. A note for $1460, discounted on May 23rd, 1888, at 6%, yielded $1448.48. When was the note nominally due ?

31. A ninety-day note, for $292, was discounted on Dec. 20th, 1887, at 8%, and yielded $289.12. On what date was the note drawn ?

32. A sixty-day note, for $1200, with interest at 6%, is discounted on the day it is made at 6%. Find the proceeds.

PARTIAL PAYMENTS.

$3500^{00} TORONTO, *March 1st, 1888.*

On demand, I promise to pay Henry Reid, or Order, the sum of Three Thousand Five Hundred Dollars, for value received, with interest at 5 per cent per annum.

THOMAS KNOTT.

On this note the following payments were made:

May 13th, 1888, $500.
Sept. 6th, 1888, $1000.

1. What amount of interest was due on May 13th?

2. By how much did the payment on May 13th exceed the interest due?

3. By how much does the payment reduce the principal?

4. What amount of interest was due on Sept. 6th?

5. By how much did the payment on Sept. 6th, exceed the interest due?

6. What amount did Knott owe Reid after making the payment on Sept. 6th?

7. How much was due Reid on Nov. 18th, 1888?

$5000^{00} TORONTO, *May 3rd, 1889.*

On demand, I promise to pay Thomas Scott, or Order, the sum of Five Thousand Dollars, for value received, with interest at 6% per annum. GEORGE COLE.

The note was endorsed as follows:

July 15th, 1889, $40.
Sept. 28th, 1889, $1200.

8. What amount of interest was due on July 15th?

9. By how much does the interest due on July 15th exceed the payment made on that date?

NOTE.—If the payment made at any time is less than the interest due at that time, this payment is added (without interest) to the next succeeding payment, and no reduction of principal is made until the sum of the payments exceeds the interest due at the time of the last payment.

10. What is the whole amount of interest due on Sept. 28th?

11. By how much does the amount of the two payments exceed the interest due?

12. How much does Cole owe Scott on Jan. 1st, 1890?

$400.00. MIMICO, *Jan. 1st, 1890.*

On demand, I promise to pay William Hill, or Order, the sum of Four Hundred Dollars, for value received, with interest at 6 % per annum. CALVIN KEMP.

On this note the following amounts were paid :—

March 15th, 1890, $20.
July 10th, " $6.
Sept. 20th, " $150.

13. How much was due on Dec. 24th, 1890?

14. The following payments were made on a demand note for $1000, drawn March 1st, 1888, bearing interest at 8 % : June 1st, 1888, $300; Sept. 1st, 1888, $10; Jan. 1st, 1889, $100; June 1st, 1889, $400. How much was due on June 1st, 1890?

15. A man bought a city lot for $2000, giving $500 cash, and making an agreement to pay 6 % interest on the balance, with the privilege of paying off any part of the principal at any time when interest has been paid up to date. The transaction took place on April 1st, 1889. On Sept. 1st, 1889, he paid $500; on Jan. 1st, 1890, he paid $600. How much remained due on June 1st, 1890?

16. On a demand note of $950, made Jan. 25th, 1888, bearing interest at 7 % per annum, the following payments were made: March 2nd, 1888, $225; May 5th, 1888, $174.19; June 29th, 1888, $187.50. What sum was due on Jan. 1st, 1889?

17. A mortgage for $3400, dated Sept. 13th, 1886, had endorsed upon it the following sums: April 20th, 1887, $800; July 2nd, 1887, $600; July 2nd, 1888, $1000. How much would pay off the mortgage on Jan. 2nd, 1889, the mortgage bearing interest at 5 %?

$2000.00 LONDON, *Jan. 4th, 1888.*

One year after date, for value received, I promise to pay Mack Jones, or order, the sum of Two Thousand Dollars, at Molson's bank here. WILLIAM KERR.

On Jan. 7th, 1889, Jones paid on this note $1200, agreeing to pay interest at 8 % per annum until he paid the balance. He made the following payments: April 7th, 1889, $300; June 7th, 1889, $200.

18. How much remained due on Dec. 7th, 1889? (Reckon time by months).

19. I borrowed $600 on June 30th, 1888, agreeing to pay $7\frac{1}{2}$ % interest on the principal, reserving the right to pay off any part of the principal when the interest is paid up to date. On Sept. 11th, 1888, I paid $200; on June 30th, 1889, I paid $150; what payment made on Jan 31st, 1891, cancelled the debt?

EQUATION OF PAYMENTS.

1. The interest on $100 for 5 days equals the interest on what sum for 1 day?

2. On what sum does the interest for 1 day equal the interest on $50 for 10 days?

3. The interest on $125 for 4 days equals the interest on what sum for 1 day?

4. The interest on $500 for 1 day equals the interest on $50 for how many days?

5. How many days' use of $10 is equal to the use of $600 for 1 day?

6. How many days' use of $60 is equal to the use of $900 for 1 day?

7. How many days' interest on $50 is equal to 12 days' interest on $62.50?

8. Jones loans me $200 for 4 months; for how many months should I loan him $160 to balance the favour?

9. How many months' use of $1000 is equal to the use of $600 for 5 months?

10. Smith loaned me $300 for 4 months, $500 for 3 months, and $150 for 2 months; how much money loaned Smith for 1 month would balance the favour?

11. I owe $400, due in 6 months, and $100, due in 11 months; when may I pay $500 and equitably cancel the debt?

12. I owe $500, due 6 months ago; $800, due $1\frac{1}{2}$ months ago; how many months' interest should I pay on ($800 + $500), in addition to paying $1300, to cancel my indebtedness?

13. Bought from Morton & Co., goods on following terms :—Goods worth $1700, cash ; $1500 payable in 20 days ; $1700 payable in 40 days ; at what time might the $4900 be paid in one payment ?

14. Find the equated time of payment of the following : $100 due in 30 days ; $800 due in 40 days ; $600 due in 60 days.

15. Bought from Eaton & Co., goods amounting to $2400 on the following terms : $400 cash, $1200 due in 10 days, $800 due in 30 days ; find the equated time of payment.

16. Find the average term of credit of $500 due in 10 days, $600 due in 12 days, and $900 in 22 days.

17. A man owes a debt of $2400, due in six months. He pays $\frac{1}{3}$ of it in 3 months, $\frac{1}{4}$ of it in 5 months ; when does the remainder become due ?

18. A merchant bought on Jan. 2nd, goods on the following terms of credit : $1200 cash, $1800 due in 30 days, and $1000 due in 40 days ; what is the average term of credit ?

19. Bought from A. White & Co., on June 3rd, goods as follows : $1800 cash, $2400 on 30 days, $800 on 60 days. I settled by paying $4000 cash and giving my note for the balance. Find the time of the note.

20. A debt of $5000 is due in 40 days ; $2000 is paid 15 days before the debt is due, and $1500, 12 days before the debt is due ; when should the balance be paid ?

21. Find the average term of credit, and the equated time of payment from June 1st, of $400 due in 30 days, $600 due in 40 days, and $500 due in 60 days.

22. A merchant bought goods from the wholesale house as follows : Mar. 4th, $800 on 30 days' credit ; June 15th, $1200 on 35 days' credit. When may the merchant equitably pay the $2000 ?

23. Bought merchandise from Macdonald & Co., as follows : June 1st, $400 on 30 days ; June 10th, $850 on 10 days ; July 3rd, $1200 on 30 days. Find the equated time.

24. I owe a friend $400, due 40 days since ; $600, due now ; $1000, due in 30 days ; find the equated time.

25. One-fifth of a debt was due 10 days ago; one-half is due now; the balance in 20 days. Find the equated time of payment.

26. Find the equated time of the following sales:—

June 20th,	a bill of	$500	at	30 days.
July 4th,	" " "	600	"	15 "
August 1st,	" " "	450	"	60 "
" 10th,	" " "	800	"	90 "

27. Find when the balance of the following account should be paid:—

Dr. JOHN JONES. Cr.

1890.			1890.		
May 1	To mdse. at 30 days....	$800	May 20	By Cash....	$1000
May 15	" " " " "	600	June 15	" "	500
June 12	" " " 60 "	1000			

28. How much must be paid Jan. 1st, 1891, to balance this account, allowing interest at 8% per annum?

29. What is the equated time of payment of the balance of the following account, allowing 30 days' credit on all debit items?

Dr. SAM. SMITH. Cr.

1890.			1890.		
Jan. 5	To mdse	$840	Feb. 1	By Cash.........	$1500
Jan. 20	" "	900	Feb. 20	" "	500
Feb. 1	" "	750			
Feb. 15	" "	800			

30. How much must be paid to balance this account on June 1st, 1890, allowing interest at 6%?

COMPOUND INTEREST.

1. I deposit in the Standard Bank $100, on which I am to receive interest at the rate of 4% per annum.

How much is there to my credit at the end of 1 year?

2. If I leave the deposit and interest in the bank, on what sum should I receive interest during the second year?

3. How much is there to my credit at the end of 2 years?

4. On what sum should I receive interest during the third year?

5. How much is there to my credit at the end of 3 years?

6. To what sum will $100 amount if left on deposit in the Standard Bank for 3 years, interest calculated yearly at 4%?

7. What is the amount of $1 in 3 years, at 4% per annum, interest calculated yearly?

8. What is the amount of $100 in 4 years, at 5% per annum, interest calculated yearly?

9. What is the amount of $150 in 4 years, at 5% per annum, interest calculated yearly?

10. What is the compound interest on $150 in 4 years, at 5% per annum, interest calculated yearly?

11. Find the compound interest on $875.25 in 3 years, at 4% per annum, interest calculated yearly.

12. Find the compound interest on $1250 in 3 years, at 5% per annum, interest calculated yearly.

13. By what fraction must any given sum of money be multiplied to give the amount at compound interest for 3 years, at 5% per annum, interest calculated yearly?

14. By what fraction must $525.35 be multiplied to give the amount at compound interest, for 4 years, at 4% per annum, interest calculated yearly?

15. John Smith deposits $100 in a Savings' Bank at the beginning of each year, making the first deposit Jan. 1st, 1885. How much will there be to his credit Jan. 1st, 1891, the bank paying 4% per annum, calculated yearly?

16. What sum of money will give $150 interest in 3 years, at 4% per annum, compounded yearly?

17. Find the amount accumulated at the end of 4 years by a man who invests \$150 now, and the same sum at the beginning of each succeeding year, at 4% compound interest, calculated yearly.

18. What is the difference between the simple and the compound interest on \$1275 for 3 years, at 5%, compounded yearly?

19. The difference between the simple and the compound interest on a certain sum for 4 years, at 6%, compounded yearly, is \$100. Find the sum.

20. What sum of money loaned at 4% per annum, compounded yearly, will amount in 4 years to \$1200?

21. Find the amount of \$1200 in 2 years, at 6% per annum, interest calculated and added to the principal at the end of each half year.

22. Find the interest on \$1450 in 1 year and 6 months, at 5% per annum, interest calculated half yearly.

23. Find the amount of \$1460 in 2 years and 6 months, at 6% per annum, calculated yearly.

24. Find the interest on \$1 in 2 years and 73 days, at 5% per annum, calculated yearly.

25. What sum of money will yield \$400 interest in 2 years and 3 months, at 4% per annum, calculated yearly?

26. The rate of increase of the population of a town is 10% per annum; the increase in the last 4 years is 13923. What is the present population?

27. At 5% per annum, interest added yearly,

(*a*) What is the amount of \$100 at the end of 1 year?
Ans. \$105.

(*b*) What is the amount of \$1 at the end of 1 year?
Ans. \$1.05, or $\$(\frac{21}{20})$.

(*c*) What is the amount of \$169.52 at the end of 1 year?
Ans. $\$169.52 \times 1.05$, or $\$169.52 \times \frac{21}{20}$.

(*d*) What is the amount of \$A at the end of 1 year?
Ans. $\$A \times 1.05$, or $\$A \times \frac{21}{20}$.

(*e*) What is the amount of \$1 at the end of 2 years?
Ans. $\$(1.05)^2$ or $\$(\frac{21}{20})^2$.

(*f*) What is the amount of \$1 at the end of 3 years?
Ans. $\$(1.05)^3$, or $\$(\frac{21}{20})^3$.

(*g*) What is the amount of \$1 at the end of x years?
Ans. $\$(1.05)^x$, or $\$(\frac{21}{20})^x$.

28. What is the amount of \$*A* at the end of *x* years, at 4% per annum, interest added yearly?
Ans. $\$A \times (1.04)^x$.

29. What is the amount of \$1 at the end of 5 years, at $3\frac{1}{2}$% per annum, added yearly? Ans. $\$(1{\cdot}035)^5$.

30. What is the amount of \$1250 at the end of 4 years, at 3% per annum, added yearly? Ans. $\$1250 \times (1{\cdot}03)^4$.

31. What is the compound interest on \$1250 for 4 years, at 3% per annum, added yearly?
Ans. $\$1250\{(1.03)^4 - 1\}$.

32. What is the compound interest on \$1500 for 3 years, at 4% per annum, added yearly?
Ans. $\$1500 \times \{(1.04)^3 - 1\}$.

33. Find the compound interest on \$1789.25, for 3 years, at 4% per annum, added yearly?

34. At 6% per annum, compounded yearly, find the amount of \$1 in $3\frac{1}{2}$ years.

35. Find the amount of \$1 in 2 years and 9 months, at 6% per annum, payable yearly? Ans. $\$(1.06)^2 \times (1{\cdot}045)$.

36. Find the compound interest on \$1200, for 2 years and 3 months, at 8% per annum, added yearly.
Ans. $\$1200\{(1.08)^2(1.02) - 1\}$.

37. What sum, at 4% per annum, added yearly, will amount in $2\frac{1}{2}$ years to \$16989·7728?

38. Find the sum which, in 2 years, at 4% per annum, payable half yearly, amounts to \$10824·3216.

39. The compound interest on a certain sum for 2 years and 73 days, at 5% per annum, compounded yearly, is \$82.82. Find the sum.

40. In how many years will a sum of money double itself at 10% per annum, compounded yearly?

41. At a certain rate, compounded yearly, the difference between the interest for the first year and that of the second is \$1, the difference between the interest of the second year and that of the third year is \$1.05. Find the rate per cent. per annum.

42. What rate per cent. per annum, compounded yearly, is equivalent to 3% per half year, compounded half yearly?

43. What rate per cent. per half year, compounded half yearly, is equivalent to 6% per annum, compounded yearly?

44. A sum of money at compound interest, added yearly, amounts to \$129600 in 2 years, and to \$178506·25 in 4 years. Find the sum of money and the rate per cent. per annum.

45. \$10.50 is the *compound* interest on a certain sum for the second year, \$11·025 is that for the third year. Find the rate and the sum.

46. The interest on a sum of money, for one year, is \$50; the difference between the compound interest of the second year and that of the third is \$2·08; the difference between that of the third and fourth years is \$2·1632. Find the rate and the sum.

47. The compound interest on a sum of money, for 4 years, is \$254·78784. The difference between the interest of the first year and that of the third is \$4·896: the difference in the second and fourth years is \$5·09184. Find the rate and the sum.

48. The compound interest on a sum of money, for 4 years, reckoned yearly, is $\frac{34481}{160000}$ of the sum. Find the rate.

49. The difference between the interest at 10% per annum, added yearly, and that added half yearly, for two years, is \$55·06¼. Find the principal.

50. Find approximately the amount of \$1000 in 10 years, at 7% per annum, compounded yearly.

PRESENT WORTH AND TRUE DISCOUNT.

1. Find the amount of \$100 in 6 months, at 8% per annum. Ans. \$104.

2. What sum placed in a bank now, at 8% per annum, will amount to \$104 in 6 months? Ans. \$100.

3. Smith owes Jones \$104, due in 6 months. What sum paid now will cancel the debt, money being worth 8% per annum. Ans. \$100.

4. Find the amount of \$100 in 9 months, at 8% per annum. Ans. \$106.

5. What sum placed at interest now, at 8% per annum, will amount to $636 in 9 months? Ans. $600.

6. Find the present worth of $636, due in 9 months, money being worth 8% per annum. Ans. $600.

7. Find the present worth of $800, due in 8 months, money being worth 6% per annum.

8. Find the present value of $1, due in 6 months, money being worth 8% per annum.

9. Find the present value of $1, due in 1 year and 8 months, money being worth 6% per annum, compounded yearly.

10. Find the present worth of $8000, due in 2 years, banks paying 4% per annum, interest calculated yearly.

11. What sum should be deducted from $1200, due in 6 months, if the debt is paid off now, money being worth 4% per annum?

12. Find the true discount of $1350, due in 9 months, money being worth 5% per annum.

13. Find the true discount of—

(*a*) $485.50, due in 146 days, at $7\frac{1}{2}$% per annum.

(*b*) $1250.60, " " 1 year and 5 months, at 6% per annum.

(*c*) $1234.56, " " 2 years, at 4% per annum, compounded yearly.

(*d*) $17684.95, " " 3 years and 219 days, at 5% per annum, compounded yearly.

(*e*) $1 " " 5 years, at 5% per annum, compounded yearly.

14. A farmer pays a yearly rental of $400. What sum paid now would be equivalent to the next three years' rental, money being worth 5% per annum?

NOTE.—Banks pay compound interest on deposits.

15. What sum, paid at the end of 2 years, is equivalent to $400 paid at the end of each year, for three years, money being worth 4% per annum?

16. A owes B $100, due in 2 years; $150, due in 3 years; $200, due in 4 years. What sum paid now would cancel the debt, 5% per annum?

17. The rent of a house is $25 per quarter paid at the end of each quarter. What is the equivalent yearly rental paid in advance, 4% per annum?

18. An estate, valued at \$10000, was divided among three heirs, whose ages are 18, 20 and 22 years respectively, in such a manner that the value of each share when the recipient became of age was the same. Find the division, money being worth 5% per annum.

19. The present worth of \$a, due in 3 years, is \$b. Find the present worth of \$c, due in 2 years.

20. The true discount of \$a, due in 3 years, is \$b. Find the true discount of \$a, due in 2 years.

21. The interest for a certain time and rate is $\frac{a}{b}$ of the principal:—

(*a*) What is the interest on \$b? Ans. \$a.

(*b*) What is the amount of \$b? Ans. \$(a + b).

(*c*) What is the true discount of \$(a + b)? Ans. \$a.

(*d*) What fraction is the true discount of the amount?

Ans. $\frac{a}{a+b}$.

22. If the interest for a certain time and rate is $\frac{a}{b}$ of the principal, find what fraction the true discount is of the amount for the same time and rate. Ans. $\frac{a}{a+b}$.

23. Find what fraction the true discount is of a debt, due in 3 years, at 5% per annum, simple interest.

24. Find what fraction the true discount is of the amount, due in 2 years, at 4% per annum, compounded half yearly.

25. By how much does the *amount* exceed the *present worth*?

26. By how much does the interest on the amount exceed the interest on the present worth?

27. The difference between the *interest* and the *true discount* on the same sum, for 2 years at 5% per annum, compounded yearly, is \$8.20. Find the sum.

28. The true discount of a certain sum, due in $2\frac{1}{2}$ years, at 4% per annum, compounded half yearly, is \$360. What is the sum?

29. The compound interest on a certain sum, for a certain time and rate, is \$250; the true discount for the same time and rate is \$240. Find the sum.

30. The difference between the compound interest and the true discount, reckoned for the same time and at the same rate, is $\frac{41}{441}$ of the interest. The time is 2 years. Find the rate.

31. The true discount for 3 years is $\frac{1000}{1331}$ of the compound interest for the same time. Find the rate per cent. per annum, compounded yearly.

32. The true discount of \$270, for 6 months, is \$30. Find the discount on the same sum for 1 year, interest compounded half yearly.

33. The true discount of \$243, for 1 year, is \$51. Find the discount on the same sum for 6 months, interest compounded half yearly.

34. The compound interest on a certain sum for the *second* year is \$49.92, and for the *third* year \$51·9168. Find the sum and the rate per cent., compounded yearly.

35. The difference between the compound interest on a certain sum, for the first and second years, is \$2.40; the difference for the second and third years is \$2·496. Find the sum and the rate per cent., compounded yearly.

36. The difference between the compound interest on a certain sum, for the first and fourth years, is \$12.61. Find the difference between the interest for the third year and that of the seventh, at 5%, compounded yearly.

37. Find the sum of money whose true discount for one year, at 10%, is greater by \$$3\frac{58}{189}$ than the sum of the true discounts of one-half of it at 8%, and the other half at 12%, for one year.

38. Bought a farm for \$10000, payable one-half cash, the remainder in 1 year, with interest at 6%. I sell immediately for \$12000, payable in 3 months, with interest at 4%. What is my *present* gain, money being worth 5% per annum?

39. Show that the true discount is the present worth of the interest, for the same time and rate.

40. A town borrows \$12000, to be repaid, principal and interest, in 4 equal annual payments. Find the annual payment, money being worth 6% per annum.

41. A sum of money in 2 years, at compound interest, added yearly, amounts to \$1389.15; the present worth of the sum for 1 year is \$1200. Find the rate and the sum.

42. The difference between the present worth of a sum of money, due in 2 years, and that of the same sum, due in 4 years, is \$5.10; the difference in the case of 3 and 5 years is \$5.00. Find the value of money per cent. per annum.

ANNUITIES.

1. What sum of money, loaned at 4% per annum, interest payable yearly, will yield an annual income of \$300?

2. What sum of money, loaned at 5% per annum, interest payable yearly, will yield an annual income of \$350?

3. A farm yields a rental of \$400 yearly. What is the value of the farm, money being worth 6% per annum?

4. What is the present value of a *perpetual annuity* of \$250, money being worth 4% per annum, payable yearly?

5. What sum must be paid for a perpetual annuity of \$450, to secure $4\frac{1}{2}$% per annum on the money paid?

6. What is the present value of a perpetual annuity of \$400, the first instalment to be paid at the end of 5 years, calculated at 5% per annum?

Solution.—The value of a perpetuity of \$400 is \$8000.
The present worth of \$8000
due in 4 years is $\$8000 \times (\frac{20}{21})^4$ Ans.

NOTE.—An annuity begins, not at the time of the first payment, but one annuity interval before.

7. What is the present value of a perpetual annuity of \$250, the first payment to be made at the end of 6 years, calculated at 4% per annum?

8. Find the present value of a perpetual annuity of \$200, the first instalment to be paid at the end of 5 years, calculated at 4%, yearly.

9. Find how much should be paid for a perpetual annuity of \$100, deferred 7 years, calculated at 5%, yearly.

10. Find how much should be paid for a perpetual annuity of \$100, deferred 3 years, calculated at 5%, yearly.

11. Find the present value of an annuity of $100, deferred 3 years, and to run 4 years; calculated at 5%, yearly.

Solution.—Take the difference of the results in examples 9 and 10.

12. Find the present value of an annuity, deferred 1 year, and to run 4 years; calculated at 5%, yearly.

13. Find the present value of an annual payment of $80, paid at the end of each year, for 4 years, the first payment being made at the end of 6 months, money being worth 4%, payable half yearly.

14. What sum invested now, at 5%, compound interest, payable yearly, will, at the end of 4 years, provide for a perpetual annuity of $200?

15. A village can raise by taxation $4000 a year; of this sum $3200 is required for ordinary expenditure. What amount of money can they raise for a waterworks system, by paying the $800 as an annuity for 10 years, money being worth 5%, payable yearly?

16. What annuity, paid at the end of each year, for the next 6 years, can be purchased for $12000, money being worth 4% per annum, payable yearly?

17. What sum of money, deposited at the end of each year, for the next ten years, will then be sufficient to purchase a perpetual annuity of $40, money being worth 5%, yearly?

18. What sum of money, deposited at the end of each year, for the next five years, will then be sufficient to purchase a perpetual annuity of $50, deferred 2 years, money being worth 6%, yearly?

19. What sum of money, deposited at the end of each year, for the next six years, will then be sufficient to purchase an annuity of $500, deferred 2 years, to run 5 years, money being worth 4% per annum, payable yearly?

20. A town issues debentures for $12000, bearing interest at 6%, payable yearly, and to run 5 years. For what sum should they sell, money being worth 5%, yearly?

$$\text{Present value} = \$720\left\{\tfrac{20}{21}+\left(\tfrac{20}{21}\right)^2+\left(\tfrac{20}{21}\right)^3+\left(\tfrac{20}{21}\right)^4+\left(\tfrac{20}{21}\right)^5\right\}.$$
$$+\$12000\times\left(\tfrac{20}{21}\right)^5.$$

21. A village offers for sale debentures, to run 5 years, bearing interest at 6% per annum. At what rate should they sell, money being worth 4% per annum, payable yearly?

Value of \$100 debenture

$$= \$6\left\{\tfrac{25}{26} + \left(\tfrac{25}{26}\right)^2 + \left(\tfrac{25}{26}\right)^3 + \left(\tfrac{25}{26}\right)^4 + \left(\tfrac{25}{26}\right)^5\right\}$$
$$+ \$100 \times \left(\tfrac{25}{26}\right)^5.$$

22. A man agrees to pay for a farm \$1600 a year, for 5 years. What sum paid now would be equivalent to this price, money being worth 4% per annum, payable yearly?

23. Find the present value of an annuity of \$200, payable for 12 years, the first payment to be made at the end of 2 years, money at 3% per annum, yearly.

24. Find the present value of an annuity of \$400, payable half yearly, the first payment to be made at the end of 1 year, money being worth 4% per annum, compounded half yearly.

25. A mortgage of \$5000, bearing interest at 6% per annum, payable yearly, has 10 years to run. Find its present value, money being worth 5% per annum, payable half yearly.

26. A mortgage of \$4000, bearing interest at 5½% per annum, payable half yearly, has 5 years and 3 months to run, the next payment of interest being due in 3 months. Find its present value, money being worth 5% per annum, payable half yearly.

27. A town borrows \$12000, to be repaid in 20 equal annual instalments. Find the amount of the annual payment, calculated at 5% per annum.

28. Find the present value of an annuity of \$1000, to run 15 years, the first payment to be made at the end of 1 year, calculated at 4% per annum.

29. Find the present value of an annuity of \$600, to run 19 years, the first payment to be made at the end of 5 years, calculated at 5% per annum.

30. Find the present value of a perpetual annuity of \$150, to begin in 15 years, money being worth 4½% per annum.

31. A mortgage of $2500, bearing interest at 8% per annum, payable half yearly, has four years to run. Find its present value, calculated at 6% per annum, payable half yearly.

32. Find the amount accumulated at the end of 15 years, by a person who deposits in a bank at the beginning of each year the sum of $200, the bank paying 4% interest, compounded half yearly.

33. A farm bears a mortgage of $3000, at 8 %, interest payable half yearly; the mortgage has 5 years to run. What sum paid now would be equivalent to reducing the interest on the mortgage to 5%, money being worth 4% per annum, payable half yearly?

34. A father dying left four sons. He bequeathed a perpetuity which he owned in the following manner: the eldest was to have the income for the first year; the next in age, the second year, and so on. Find the comparative values of the sons' shares, money worth 4%, added yearly.

35. A man pays $150 yearly, for 15 years, for an endowment policy of $2500. Find the accumulated value of the payments, reckoning money worth 6% per annum.

36. What annual deposit, for 15 years, at 5% per annum, calculated yearly, will amount to $5000?

37. A man pays $240 yearly, for 15 years, for an endowment policy of $4000. Reckoning money worth 6% per annum, payable yearly, how much is he paying each year for the life risk?

38. What is the present value of a mortgage of $5000, bearing interest at 6%, payable yearly, having 5 years and 6 months to run, the next payment of interest being due in 6 months, money being worth 5% per annum, payable yearly?

39. Money is worth 4% per annum, payable yearly. At what price should City of Toronto Bonds sell, bearing interest at 5%, payable yearly, and having 6 years to run?

40. A man buys a piano, giving his note for $350, renewable at the end of 3 months, less the quarterly instalment of $25; for the first 9 months no interest is charged; after that time, interest is paid each quarter, at the rate of 6%. What cash price is equivalent to this, money being worth 5% per annum?

41. By paying $105 a year, for ten years, a man secured an endowment policy of $1000. How much more did he pay each year than enough to amount to the $1000, reckoning money worth 6% per annum?

42. The Equitable Life Assurance Society offer 'straight' insurance to a man, 29 years of age, for an annual premium of $22.07 on each $1000; to the same man they offer a ten-year endowment policy for $104.43, the premiums in each case being paid at the beginning of the year. The man takes out a straight life policy for $1000, and deposits in a savings' bank $82.36, which pays 4% a year, compounded yearly. The man dies at the end of five years. How much better off are his heirs than if he had taken an endowment policy?

43. If the man (in above) live the ten years, how much better off is he than if he had taken the endowment policy?

44. How much money deposited at the beginning of each year, for 15 years, at 5% per annum, compounded half yearly, will amount to $1000 at the end of the 15th year?

45. In how many years can a debt of $50000, drawing interest at 8%, paid half yearly, be paid off by a sinking fund of $7500 a year?

46. In how many years can a debt of $200000, drawing interest at 5% per annum, be paid off by a sinking fund of $10000 a year?

47. A corporation borrows $250,000, which is to be repaid in 20 equal annual instalments. Find the amount of the instalment, money being worth 4% per annum, compounded yearly.

PARTNERSHIP.

1. Jones and Smith engage in business, each furnishing $5000; at the end of one year they have made a gain of $2500. How should this gain be divided?

2. Jones invests $4000 and Smith $6000 in a joint business; they make a gain of $2800 in six months. How should this gain be divided?

3. A and B engaged in the lumber trade, with a joint capital of $12000; at the end of a year A's gain amounted to $400 and B's to $800. How much capital did A put in the business?

4. Sykes and Smith formed a partnership, with a joint capital of $8000,—Sykes to receive $1200 a year for managing the business; the total gain for the year was $3200, of which Smith receives $1500. What amount of capital did Sykes invest?

5. A, B and D form a partnership; their respective shares of one year's gain are $2000, $3000 and $5000; A invested $4000 less than B. How much did D invest?

6. A invested $4500, for 2 months, in a certain business; B invested $4000, for 3 months. If the gain is divided in proportion to the use of each man's investment, what amount should A receive out of a total gain of $2800?

7. A invested $1600, for 3 months; B $1100, for 2 months; C $3000, for $1\frac{1}{2}$ months; the total gain was $2400. Find each man's share.

8. A invested $2400 in a business for 6 months, acting as manager for that time on a yearly salary of $1200; B invested $3000 for 4 months, and during that time received $350 as bookkeeper; C invested $4000 for 12 months, and acted as manager and bookkeeper when A and B were not in the business; the total gain for the year was $8640. What was C's share?

9. At the beginning of a year A, B and C enter into a partnership, each contributing $4000. At the end of 4 months A withdraws one-half of his investment, and at the end of 6 months B withdraws $\frac{3}{4}$ of his. The gain for the year is $6000. Find C's share.

10. A, B and C entered into partnership, contributing respectively $3500, $2200, and $2500; their gains were $1120, $880, and $1200 respectively. If B's capital was in trade 2 months longer than A's, for what time was each man's money in the business?

11. A, B and C formed a partnership, their money being in the business for 2 months, $2\frac{1}{2}$ months and 4 months respectively; their gains were $600, $500 and $800 respectively; A's investment was $3000. Find B's and C's.

12. Hardy and Jones are in partnership, Hardy having invested $12000 and Jones $15000; Hardy acts as manager on a yearly salary of $2400, the salary to be reduced in proportion if the capital is reduced; at the end of 4 months Hardy takes $3000 out of the business, and at the end of 6 months Jones takes out $4000; the total gain for the year was $6000. How much of this does Hardy receive?

13. Lock, Smith and Knight formed a trading company. Lock put in $2500 for 10 months, Smith $2300 for 11 months, and Knight managed the business for 12 months, his services being considered equal to a capital of $2000; they gained $2972. What sum should each man receive?

14. B and C formed a partnership to dig a trench; B furnished 100 workmen for 40 days, C 120 workmen for 30 days; they received $12000 for the work. What was the share of each?

15. A owns $\frac{1}{4}$ of a vessel, B $\frac{1}{3}$ and C the remainder; the vessel is insured for $\frac{3}{4}$ of its value; the vessel is lost, and A, after receiving his share of the insurance, finds that he has lost $1000. What did B and C lose respectively?

16. At the end of a year, from the commencement of their business, Smith, Jones and Cook, after "taking stock," find the amount of goods on hand to be $40000; cash on hand, $22000; debts due them, $25000; amount of their indebtedness, $17000.

Make a statement of resources and liabilities, showing net capital and gain. Find each partner's share of the gain, Smith having put in the business $8000, Jones $9000 and Cook $3000.

17. A, B and C form a partnership, with capitals of $7500, $15000 and $22500, respectively. A draws out at the end of each year $750; B, $1200, and C, $1350. At the end of 5 years their capital is $42900. How much of it belongs to B?

18. The capital of three partners, A, B, C, was $1500, $1000 and $1200, respectively, and their gains were $600, $320 and $288, respectively. A's money was in the business 4 months longer than C's. How long was B's money in the business?

19. Terry rented a house for one year for \$180; at the end of three months he took in Tucker as a co-tenant; after four months more they admit Taylor; Tucker moves out one month before the year is up; how much rent did each pay?

20. A begins business on Jan. 1st, investing \$4000; on the 20th of March he admits B as a partner with a capital of \$3000. On July 3rd, they find the profits for the whole time to have been \$2400. What is B's share of this?

EXCHANGE.

1. George Cooper, of Toronto, owes James Good, of Hamilton, the sum of four thousand dollars. Cooper buys from a Toronto bank an order on a Hamilton bank, directing the latter to pay Good \$4000. How much must Cooper pay, such orders selling at $\frac{1}{4}$% premium?

2. How much must be paid for a sight draft on Hamilton for \$2500, sight drafts selling at $\frac{3}{8}$% premium?

3. Find the cost of a draft on Chicago for \$800, at $\frac{1}{2}$% discount.

4. Find the cost of a draft on New Orleans for \$12000, at $\frac{3}{4}$% premium.

5. Find the cost of a 60 day bill of exchange on London for £1200, the course of exchange being £1 for \$4.80$\frac{3}{8}$.

6. What amount of draft on Montreal can be purchased for \$7500, at $\frac{1}{4}$% premium?

7. What amount of bill of exchange on London can be bought for \$1350, the course of exchange being £1 for \$4.80?

8. Find the cost of a bill of exchange on Paris for 1500 francs at 5·16 francs for \$1.

9. Formerly the legal par of exchange between Canada and Great Britain was \$4.44$\frac{4}{9}$ for £1; the legal par at present is \$4.86$\frac{2}{3}$ for £1. The present value is what per cent increase on the old value?

10. Find the cost of a sixty-day bill on London for £3000, exchange being quoted at 8$\frac{1}{4}$.

NOTE. Exchange quotations, when not giving explicitly the value of the pound in dollars, usually give it as a certain percentage premium on the *old par*. When sterling exchange is quoted at $9\frac{1}{4}$, the cost of a bill of exchange is $\frac{109\frac{1}{4}}{100}$ of $\$4.44\frac{4}{9}$ for £1.

11. Find the cost of a demand-bill on London for £1500, exchange being at $9\frac{1}{8}$.

12. What amount of demand-bill can be bought for $2400, when exchange is at 8?

13. When $7300 is paid in Toronto for a bill of exchange on Liverpool for £1500, how is exchange quoted?

14. A merchant in Winnipeg owes $4000 in New York; exchange on New York is $\frac{1}{4}$ % premium, but exchange on Chicago is $\frac{1}{2}$ % discount, and from Chicago on New York $\frac{3}{4}$ % premium. Compare the cost of a bill on New York direct, with that of one through Chicago, which would pay the debt.

15. A sight-draft on New York for $2700 was purchased for $2673; what was the course of exchange?

16. I hold a 70-day draft on Chicago for $2750; I sell the draft at $\frac{1}{4}$ % premium, and with discount off at 8 % per annum. What do I receive?

17. A Canadian Company borrows in Paris 291,000 francs for which it pays an annual interest of $2920. This loan is transmitted through London when exchange in London is quoted at 25·30 francs, and sterling exchange is $9\frac{3}{8}$. Find what rate of interest the company pays on the money actually received.

18. The value of an ounce of the gold of which sovereigns are made is £3 17s. $10\frac{1}{2}$d. What is the weight of 1869 sovereigns?

19. The old par of exchange between the United States and Great Britain was £1 = $\$4.44\frac{4}{9}$; in 1834 the U. S. Congress reduced the weight of the eagle to 258 grains, and in 1837 fixed its fineness at 900 thousandths pure; the mint price of English standard gold (22 carats fine) is £3 17s. $10\frac{1}{2}$d. per ounce. From these facts show the truth of the statement: "By the new par of exchange sterling money is worth $9\frac{1}{2}$ per cent. more than by the old par."

20. A Glasgow merchant ships to his Montreal agents for sale goods for which he pays £616 in Glasgow; he pays an ad valorem duty of 12% upon the goods, and a commission of 7% to his agents for their services. The goods realize in Montreal $7800. Find the merchant's net gain, a pound sterling being equal to $4.86.

21. A bank in Toronto remits $10000 to Liverpool as follows: First to Paris, at 5·40 francs per $1; thence to Hamburg, at 185 francs per 100 marcs; thence to Amsterdam, at 17½ stivers per marc; thence to Liverpool, at 220 stivers per pound sterling; how much sterling money will he have in bank at Liverpool, and what will be his gain over direct exchange at 10% premium?

22. A commission agent sold goods to the amount of $12500, charging a commission of 2½%; with the net proceeds he bought a draft at ¾% discount. Find the face of the draft.

23. When exchange at New York on Paris is 5·16 francs per $1, and at Paris on Hamburg 2·12½ francs per marc banco, what will be the arbitrated price in New York of 11520 marcs banco of Hamburg?

24. A Toronto merchant owes 1800 francs in Paris, he buys a draft on London when sterling exchange is at 8; exchange between London and Paris 25·20 francs per £1. What does he pay for the draft?

25. The par of exchange between Paris and London is 25·2215 francs for £1, and between St. Petersburg and London 38·177 pence for 1 rouble. Find the par of exchange between Paris and St. Petersburg.

26. The par of exchange between London and New York is $4·866564 for £1, and between London and Amsterdam 12·1071 florins for £1. Find the par of exchange between New York and Amsterdam.

PROBLEMS IN MENSURATION.

For convenience of reference the rules for obtaining the measurement of surfaces, and the volumes of solids, are here expressed generally:

1. Where a is the measure of the side and b of the end of a rectangle, the measure of its area is $a \times b$.

2. Where a, b and c are the measures of the sides of a triangle, the measure of the area is

$$\sqrt{s(s-a)(s-b)(s-c)} \text{ where } 2s = a + b + c.$$

[This result can be obtained in a manner similar to that employed on page 81.]

3. Where h is the measure of the height of a cylinder and c is the measure of the circumference of its base, the measure of the area of the curved surface is $c \times h$.

4. Where a is the measure of the slant side of a cone, and c is the measure of the circumference of the base, the measure of the area of the surface of the cone is $\frac{1}{2} \times a \times c$

5. Where a, b, and c are the measures of the height, length, and width, respectively, of a rectangular parallelopiped, the measure of its volume is $a \times b \times c$.

6. Where h is the measure of the height of a cylinder, and A is the measure of the area of its base, its volume is measured by $h \times A$.

7. Where h is the measure of the height of a prism and A the measure of the base, the volume is measured by $h \times A$.

8. Where h is the measure of the height of a wedge, and A is the measure of the area of its base, the volume is measured by $\frac{1}{2} \times h \times A$.

9. Where h is the measure of the height of a pyramid or cone, and A is the measure of the area of the base, the volume is measured by $\frac{1}{3} \times h \times A$.

10. Where c is the measure of the circumference of a circle, and r is the measure of the radius, $c = 2 \pi r$.

11. The measure of the area of a circle is πr^2.

12. The measure of the area of the surface of a sphere is $4 \pi r^2$.

13. The measure of the volume of a sphere is $\frac{4}{3} \pi r^3$.

14. The measure of the area of a trapezium (4 sided figure, having two parallel sides), is $\frac{a+b}{2} \times h$, where a and b are the measures of the parallel sides and h the measure of the altitude.

NOTE.—If a diagonal be drawn, it divides the trapezium into two triangles, whose altitudes are equal.

1. A ladder, 30 feet long, stands upright against a wall. Find how far the bottom of the ladder must be pulled out to lower the top 6 feet.

2. A ladder, 40 feet long, is placed so as to reach a window 24 feet high on one side of a street, and from the same spot it will reach a window 32 feet high on the other side of the street. Find the breadth of the street.

3. The radius of a circle is 26 inches; the perpendicular, drawn from the centre on a chord, 10 inches. Find the length of the chord.

4. The side of a square is 8 feet. Find the area of a circle described about it.

5. The radius of a circle is 4 feet; from a point 7 feet from the centre, a tangent is drawn to the circle. Find the length of the tangent.

6. A foot path goes up the side, and then along the end of a rectangular field, 216 yards long and 195 broad. What distance will be saved by cutting right across in the direction of the diagonal?

7. A gas-jet is 11 feet above the pavement. How far must a man, who is 5 feet 10 inches high, stand from it so as to cast a shadow 7 ft. long?

8. Find, in inches, the side of the greatest square stick of timber which can be cut from a tree whose circumference is 12 feet.

9. What is the surface of a board 18 in. wide at one end, 25 in. at the other, and 16 feet long?

10. A cubic foot of gold is extended by hammering so as to cover an area of 3 acres. Find, correct to 7 places of decimals, its thickness in a decimal of an inch.

11. A room is 24 feet long, 18 feet broad, and 7 feet high. What length of string will reach from any corner of the floor to the farthest corner of the ceiling?

12. Find the expense of paving a road of the uniform breadth of 4 yards around the inside of a rectangular piece of ground, the length of which is 85 yards and breadth 56 yards, the cost of paving being 25 cents per square yard.

13. The radius of a circle is 2 feet. Find the whole perimeter of its semicircle.

14. The whole perimeter of a semicircle is 80 inches. Find its radius.

15. Find in feet, to three decimal places, the side of a square containing $2\frac{3}{4}$ acres.

16. The two sides of an isosceles triangle measure 65 feet each, and the base is 50 feet. What is the altitude?

17. A square field contains $\frac{1}{3}$ of an acre. Determine the length of a side of the field, correct to the nearest inch.

18. A line reaching from the top of a precipice 130 feet high on the bank of a river, to the opposite side, is 380 feet long. How wide is the river?

19. The end of a round stick of timber is 3 feet in diameter. What will be the size of the largest square stick that can be hewn from it?

20. A rectangular piece of ground is 60 yards long and contains $\frac{1}{3}$ of an acre. It contains a grass plot bordered by a walk 6 feet wide. Find the area of the plot.

21. The equatorial circumference of the earth is about 25,000 miles. What is the length of a degree of longitude at the equator?

22. Find how many persons can stand in a room measuring 15 feet by 9 feet, supposing each person to require a space 27 inches by 18 inches.

23. The sides of a quadrilateral field are 20, 30, 25, 32 chains respectively, and the diagonal joining the first and third corners is 40 chains. Find its area in square yards.

24. A ladder, standing upright beside a wall 50 feet high, just reaches the top. How far may the foot of the ladder be removed from the wall, and still reach within 11 inches of the top?

25. Find the length of a ladder and the width of the street, if when one end of the ladder is placed against a wall at the side of the street it reaches a height of 24 feet, and when it is turned and placed against a wall on the opposite side it reaches a height of 18 feet, and forms, with its first position, a right nagle.

26. The sides of a triangle are 21, 43, 35 feet respectively. Find its area.

27. The diameter of a carriage wheel is 30 inches; find how many turns the wheel makes in travelling one mile.

28. A road runs around a circular shrubbery; the outer circumference is 560 feet, and the inner 420 feet. Find the breadth of the road.

29. The difference between the diameter and the circumference of a circle is 12 feet. Find its area.

30. Find the area of a triangular field whose sides are 21 yds., 19 yds., 23 yds. 2 ft., respectively.

31. How many square feet in a board 12 feet long, 18 inches wide at one end and 24 inches wide at the other?

32. The distance to the top of a certain mountain is $1\frac{1}{3}$ miles, and the circumference of the base 4·7 miles. What is the area of its surface, supposing it to be a cone?

33. A square space containing 150 sq. yds. is to be lengthened by 4 feet 3 inches in one of its dimensions, and shortened by 3 feet 4 inches in the other. What will then be its area?

34. What is the diameter of a wheel which turns round 1000 times in travelling a mile?

35. Find how many trees there are in a wood half a mile long and 50 rods wide, supposing on an average four trees grow on each square chain.

36. The diagonals of a rhombus are 30 and 35 feet respectively. Find its area.

37. Find the width of a circular path containing 120 sq. yards, which surrounds a circular pond whose circumference is 220 yards.

38. The shadow of a man standing upright and 5 ft. 10 in. high, was measured and found to be 7 ft. 5 in.; the shadow of a pole, measured at the same time, was found to be 29 ft. 8 in. Determine the length of the pole.

39. Find the volume of a frustum of a cone, the radii of whose ends are 7 ft. and 12 ft. 4 in. respectively, and whose altitude is 9 feet.

40. Find the dimensions of a rectangle containing 240 sq. ft., if its length is to its width as 3 to 2.

41. Suppose that the planet Mercury describes in 88 days a circle around the sun, of which the radius is 37000000 miles; find the number of miles described by the planet in one second.

42. A ladder 42 feet long, placed with its foot 24 feet from a wall, reaches within 3 feet of the top. How near the wall must the foot of the ladder be brought that it may reach the top?

43. An electric light is 18 feet above the ground. What will be the length of the shadow of a man 6 feet in height, if he stands 16 feet from the post on which the light is placed?

44. One side of a quadrilateral field measures 25 rods, the side opposite and parallel to it measures 36 rods, and the distance between the two sides is 12 rods. Find the area.

45. Which requires the most fence, a circular field 15 rods in diameter, or a square one whose side is 14 rods?

46. The depth of water is 7 feet in a circular cistern, the circumference of whose base is 20 feet. Find the depth of the same quantity of water in another cistern, the perimeter of whose *square* base is 20 feet.

47. The parallel chords of a circular zone are 12 and 16, and its breadth 14. What is the diameter of the circle?

48. A rectangular garden is to be cut from a rectangular field so as to contain $\frac{3}{4}$ of an acre; one side of the field is taken for one side of the plot, and measures $2\frac{1}{2}$ chains. Find the length of the other side.

49. The length of one of the diagonals of a quadrilateral is 40 feet, and the lengths of the perpendiculars on this diagonal from the opposite angles of the quadrilateral are 12 ft. 1 in., and $9\cdot2\dot{5}$ ft. respectively. Find the area of the quadrilateral.

50. Find the diagonal of a square whose area is 14 square inches.

51. The area of a chess board, having 8 squares along each side, is 80 square inches. Find the length of a side of one of its squares.

52. A ladder is to be placed so as to reach a window, the sill of which is 70 feet from the ground; the foot of the ladder cannot be brought nearer than 30 feet from the wall. What length of ladder will be sufficient?

53. A rectangular field is 40 rods in length and 30 yds. in width. Find in feet the side of a square field of equal area.

54. A room is 20 feet long, 16 feet wide, and 12 feet high. What is the distance from one of the lower corners to the opposite upper corner?

55. What is the volume of a frustum of a square pyramid, the sides of whose bases are 18 and 25 feet respectively, and the altitude 15 feet?

56. A garden roller is 3 feet $7\frac{1}{2}$ inches wide, and 5 feet $10\frac{2}{3}$ inches in circumference. How much ground does it pass over in making three complete revolutions?

57. What is the length of a side of the greatest cube which can be cut from a sphere 1 inch in diameter?

58. A rectangular court measures 21 ft. 6 in. by 13 ft. 4 in. Find the cost of paving it at 6 pence a square foot.

59. Find the difference between the perimeter of a square field containing $22\frac{1}{2}$ acres, and the perimeter of a rectangular field of equal area, the length of the latter field being to its width as 4 to 3.

60. Find the area of a circular path the outer circumference of which is 110 yards, and the inner 88 yards.

61. Find the number of gallons of water in a tank which is in the form of a rectangular parallelopiped, 8 feet wide, 10 feet long, and 9 feet deep, if it is full of water.

62. Find the difference in feet between the perimeters of a circular and a square field, if each contains 2 acres.

63. Find the number of cedar blocks required to pave a street 40 feet wide and 300 yards long, supposing that a block occupies 24 sq. inches.

64. How many gallons of water will a cistern contain, the diameter of whose base is 8 feet, and which is 4 feet deep?

65. How fast must the water rise in a well whose diameter is 7 feet, so that it may remain the same depth when a pump is emptying it at the rate of $\frac{1}{4}$ of a ton of water per hour?

66. If we consider the earth a sphere whose diameter is 8000 miles, find the scale in miles per inch by which its surface must be represented on a globe whose diameter is 1 foot.

67. Find the number of cords of wood in a cylindrical stick of timber, the length being 40 feet and the circumference 22 feet.

68. Find the area of a triangle whose sides are 41·26, 39·4 and 29·2 yards respectively.

69. Find the area of the uniform circular walk 2 yards wide, surrounding a circular pond which contains $2\frac{1}{2}$ acres.

70. Find the whole surface of a hemispherical bowl whose inner diameter is 4 inches, and outer 6 inches.

71. Compare the volumes of a right circular cone and a cylinder of the same altitude and base.

72. A tree breaks off a certain distance from the ground, and has its top resting on the ground, so that the two parts of the tree and the ground form a right angled triangle. The height of the stump is 12 feet, and the distance of the top from the butt is 34 feet. Find the total length of the tree.

73. A circular pond which is 4 miles in diameter has a driveway around it. A man wishing to reach a point directly across the pond from where he stands, can drive at the rate of ten miles an hour and row at the rate of 6 miles an hour. Which is the speedier way?

74. A rectangular field containing 3 acres, is surrounded by a road of the uniform width of 66 feet, the total area of the road being 3 acres 36 sq. rods. Find the length and width of the field.

75. A side of a square field is 36 rods. Find a side of a square field 3 times as large as it.

76. Iron being nearly 8 times as heavy as an equal volume of water, find the weight of a solid sphere of iron whose radius is 3 inches.

77. Find the edge of a cube which is 5 times the volume of a cube whose edge is 2 inches.

78. Each edge of a cube is diminished by $\frac{1}{8}$ of itself. By what fraction of itself is the volume diminished?

79. Compare the volumes of two spheres whose radii are in the ratio of 2 to 3.

80. The diameter of a circle is 130 ft.; the breadth of a zone is 64 feet; and one of the parallel chords 120 feet. Find the other.

81. A house is 60 feet long and 25 feet broad. In a ground plan of that house the length is 5 inches. What ought to be the breadth?

82. A board is 8 inches broad; what length must be cut off to make a square yard of surface (on one side)?

83. What is the superficial area of the outside of a box whose dimensions are 9, 10, $7\frac{1}{2}$ feet respectively?

84. The base of a triangular field, containing 1 acre, is $90\frac{1}{3}$ yards in length. What is the altitude?

85. A room is 24 feet by 20 feet; in the central part is a carpet which measures 21 feet by 17 feet. Find the cost of painting the rest of the floor at 12 cents a square yard.

86. Find how many square flower beds 4 feet to the side, can be arranged in a square plot whose side is 7 feet.

87. The perimeters of a square and a rectangle are each 40 inches. Find the difference in their areas, if the sides of the rectangle are in the ratio 1 to 3.

88. The weight of iron is found to be about 7·7 times as heavy as that of an equal volume of water. Find the weight of a rectangular box, without a lid, full of water, if the outer dimensions of the box are 4, 5, 6 feet, respectively, and the iron is 1 inch in thickness.

89. The rent of a square field, at $12 an acre, is $132.24. Find the cost of putting a fence around it at 35 cents a yard.

90. The sides of a triangle are 13, 14 and 15 feet, respectively. Find the perpendicular from the opposite angle, on the 14 foot side.

91. The radius of the outer boundary of a ring is 14 inches, and its area is 462 sq. in. Find the circumference of the inner boundary.

92. Find the cost of paving a road of the uniform breadth of 4 yards around the inside of a rectangular piece of ground, the length of which is 85 yards and the breadth 50 yards, the cost of paving being 20 cents a square yard.

93. How many planks, 2 inches thick, can be sawed from a log 10 feet in circumference, allowing $\frac{1}{4}$ of an inch for each saw cut, and 2 slabs, each at least 5 inches thick, to be cast aside?

94. An elm tree partially broke off, and fell with its top on the ground. If the height of the break is 15 feet above the ground, and the distance from the foot of the tree to the point where the top reaches the ground is 66 feet; what was the entire height of the tree before it fell?

95. How many square yards of canvas will be required to make a conical tent 9 feet high and having a base of 4 feet radius, no allowance being made for seams?

96. How many gallons of water will a circular vat contain that measures 12 feet across the bottom, 15 feet across the top, and 6 feet deep?

97. How many yards of paper 27 inches wide are required to paper the walls of a room 18 feet long, 12 feet wide and 11 feet high?

98. A square and a rectangular field have the same perimeter, 100 yards. The length of the rectangular field is 4 times its width. Which contains the greater area, and how much?

99. If it costs $148 to fence a square field at $4.40 a rod, what would it cost to fence the same amount of land in the form of a rectangle whose sides are in the ratio of 9 to 16?

100. Find the width of a rectangular field which contains 12 acres, and which measures 24 chains in length.

101. The radius of a circle is 8 feet. Find the circumference of another circle of $\frac{1}{3}$ the area.

102. The area of a triangle is $3\frac{1}{2}$ square metres, and its altitude $2\frac{1}{4}$ linear metres. What is the length of the base in feet?

103. The sides of a triangle measure 11, 12, 9 chains, respectively. What is its area in square yards?

104. The side of a square is 40 yards, its corners are cut off so as to form a regular octagon (8 sides). What is the area of the octagon?

105. What is the volume of a cube, if the distance from one of the corners at the base to the extreme opposite corner on the top is 4 feet?

106. The inner circumference of a circular drive is 1050 yards, and the width of the drive is 35 feet. Find the outer circumference.

107. How many yards of paper 32 inches wide, with a pattern every 1 ft. 3 in., are required to paper the walls of a room 20 ft. long, 12 ft. wide, and 12 ft. high?

108. Find the cost of the material for fencing a square field containing 10 acres with a barbed wire fence, if the wire costs 3 cents a yard, and there are five wires in the fence, and if the posts cost 8 cents apiece, and are placed 8 feet apart.

109. Find the length of an arc which subtends an angle of 75° at the centre of a circle of 12 inch radius.

110. A pond whose area is 4 acres, is frozen over with ice 6 inches in thickness. If water expands 10% in freezing, find the weight of the whole of the ice in tons.

111. The earth which is excavated from a cellar is twice as heavy as an equal volume of water. Find the weight of earth removed in digging a cellar 40 feet long, 32 feet wide, and 8 feet deep.

112. Find the cost of paving with asphalt a walk 6 feet wide, around the outside of a block 30 yards wide and 50 yards long, at $1.50 per square yard.

113. A canal 12 miles long, has an average width of 7 yds., and is 5 feet deep; how soon would the excavation of it be completed by 400 men, each removing, on the average, 20 cubic yards per day?

114. A factory has 150 windows, 90 of which severally contain 16 panes, each pane being 8 inches by 12 inches, and the remaining windows severally contain 12 panes, each pane being 7 inches by 14 inches. Find the cost of glazing the whole at 25 cents per square foot.

115. Find the side of a square which is equivalent in area to a circle, whose circumference is 55 inches.

116. A square field contains 1296 sq. yds. Find the area of a rectangular field whose length is double its breadth, and which has a perimeter equal to that of the square field.

117. If the pressure of 15 pounds to the square inch be applied to a circular plate 2 feet in diameter, find the total pressure.

118. The inner diameter of a circular building is 112 feet, and the thickness of the wall is 22 inches. Find how many square feet of ground the base occupies.

119. Find the radius of a circle which is equivalent in area to a square, the side of which is 40 inches.

120. Find the area in acres of a quadrilateral, whose diagonal is 80 chains, and the perpendicular from the opposite angles to it 29 chains and 23 chains respectively.

121. Find the area of the sector of a circle whose radius is 60 yards, the arc of the sector being 280 yards in length.

122. What is the volume of a sphere whose surface is 616 square inches?

123. How many pails of water may be contained in a cylindrical cistern 4 feet deep, having a diameter of 6 feet, if each pail contains 12 quarts?

124. Find the difference in the perimeters of two fields, each containing $2\frac{1}{2}$ acres, one being square and the other rectangular, its length being 4 times its width.

125. A gardener wishes to lay out a flower bed. He first lays out a square, 4 feet to the edge, then with a radius of 2 feet, and the 4 corners as centres, he describes 4 equal circles. The part of the square enclosed by the arcs of the circles he makes into a flower bed. Find its area and its perimeter.

126. A string is wound around a square block, edge 2 inches. Supposing the string to be unwound in the plane in which it is now situated, and always kept stretched, find the length of the curve, which may be traced by the extremity of the string in one complete revolution.

127. In the preceding example, find the length in two complete revolutions.

128. Also, in $2\frac{1}{2}$ revolutions.

129. Three equal circles of radius 3 feet each, touch one another externally. Find the area of the space enclosed by the arcs between the touching points.

130. On the sides of a right angled triangle, squares are described towards the outside of the triangle. If the sides which include the right angle are 5 and 12 inches, respectively, find the total space enclosed by the outside perimeter; also, find the perimeter.

131. Over what area can a cow, which is tethered with a rope 40 feet long, graze?

132. How many rows of desks can be arranged lengthwise in a room 30 feet long and 24 feet wide, if each desk is 20 inches wide, and the spaces along the walls and between the rows are each to be, at least, 2 feet wide?

133. Find the cost of carpeting (without piecing) a room, 18 ft. 8 in. long, 17 ft. 11 in. wide, with carpet 2 ft. wide, at $1.75 a yard, the carpet running lengthwise, and without any pattern.

134. The area of a sector of a circle is 230 sq. ft.; the angle of the sector is 50°. Find the whole perimeter of the sector.

135. A conical tin vessel has a lid; the diameter of the lid is 24 inches, and the depth of the vessel is 18 inches. How many square feet of tin does the whole outer surface present?

136. A gardener lays out a flower bed in the following manner:—He lays out a square, whose side is 6 feet; on each side as diameter and towards the outside of the square, he describes semicircles. Find the area of the whole plot enclosed; also, find the perimeter of it.

137. The sides of a triangle are 13, 14, 15 inches, respectively. If squares are described on the sides towards the outside of the triangle, find the perimeter of the figure and the whole space occupied by it.

138. In the preceding example, if equilateral triangles are described instead of squares, find the perimeter and the space enclosed by it.

139. The radius of a circle is 8 feet; two parallel chords are drawn, each equal to the radius. Find the area of the zone between the chords.

140. A circular shrubbery is surrounded by a road of uniform breadth, the inner side of the road measuring 360 yds. in circumference, and the outer side 420. What quantity of ground does the road cover?

141. If the diameter of the earth is 8000 miles, what is the diameter of Saturn, which is 1000 times its bulk?

142. Find the volume of the greatest sphere which can be cut out of a cubical block, whose edge is 3 inches.

143. Find the solid contents of a rubber ball, 4 inches in diameter, which has a hollow in it 2 inches in diameter.

144. If a pipe, 3 inches in circumference, can fill a cistern in 2 hours, in what time can a pipe, $4\frac{1}{2}$ inches in circumference, fill it?

145. Find the cost of paving a walk, $2\frac{1}{4}$ feet wide, around the outside of a square court, containing 196 sq. yds., at 20 cents a square foot.

146. The radius of a circle is 12 feet; two parallel chords are drawn on opposite sides of the centre, one subtending at the centre an angle of 60°, and the other an angle of 90°. Find the area of the zone between the chords.

147. A field is ten thousand times as large as the plan which has been made of it. Find what length in the plan will represent a length of 24 yards in the field.

148. One extremity of a string is fastened to the corner of a square, edge 3 inches, and the string is then wound around the square; if the string is kept stretched in unwinding it, find the area of the space enclosed by the string after it is unwound and the line marked out by its extremity in one complete revolution.

149. In the preceding example, if the end which is fastened to the square was attached to it at the middle point of a side, find the area.

150. The area of a sector is 115 sq. in., the area of the circle is 275 sq. in. Find the arc of the sector.

151. What is the depth of a 6 gallon pail that is 12 inches across, the sides being upright?

152. A circle and an equilateral triangle have equal perimeters, each being 36 inches; find the difference of their areas.

153. A moat 30 feet wide surrounds a castle which stands on a circular piece of ground. What is the area of the space occupied by the moat, if the inner circumference of it is 796 yards?

154. Assuming the earth to be a sphere whose diameter is 7913 miles, find the length of a degree of longitude at the equator.

155. In the preceding example, find the length of a degree of longitude in 60 degrees north latitude.

156. Also in 45° north latitude.

157. A cistern is 12 ft. 4 in. long, by 8 ft. 6 in. wide. Find how many inches the surface will sink if 280 gallons are drawn off.

158. A box is without a lid; if the external length is 3 ft., width 2 ft., depth 1 ft. 6 in., and the thickness of the material is 1 in., find the number of cubic inches of the material.

159. The wall of China is 1500 miles long, 20 feet high, 15 feet wide at the top, and 25 feet at the bottom. Find how many cubic yards of material it contains.

160. Find how many cubic feet of earth must be dug out to make a well 3 feet in diameter, and 30 feet deep.

161. One extremity of a string is fastened to a corner of an equilateral triangle, side 5 inches, and the string is then wound around the triangle, being kept taut, when unwound. Find the area of the space enclosed by the string and the line marked out by its unattached extremity in one complete revolution.

162. In the preceding example, if the end which is fastened to the triangle was attached to it at the middle point of a side, find the area.

163. The area of a sector is 90 square feet; the radius of the circle is 15 feet. Find the arc of the sector.

164. The length of a triangular prism is 7 feet, and the sides of the triangular end are 3, 4 and 5 feet, respectively. What is its whole superficial area?

165. The difference between the radii of the front and hind wheels of a carriage is 6 inches. What are the lengths of these radii if the front wheel makes 50 revolutions more than the hind one in going a mile?

166. At what distance from the top must a cone, 14 inches high, be cut parallel to the base, that the volumes of the two parts may be equal?

167. A vessel in the form of a right circular cylinder is to have a capacity of 3 gallons, and the depth of the vessel is to be equal to the length of a diameter of the end. Find its depth.

168. Find the number of cubic yards of earth dug out to make a tunnel 80 yards long, whose section is a semicircle, with a radius of 10 feet.

169. How many pieces of money, $\frac{3}{4}$ of an inch in diameter and $\frac{1}{8}$ of an inch thick, can be coined from material in the form of a cube, whose edge is 3 inches?

170. Find the cost of a leaden pipe of $1\frac{1}{2}$ inch bore, which is $\frac{1}{2}$ inch thick and 20 feet long, at 8 cents per pound, supposing that a cubic foot of lead weighs 11500 ounces.

171. The altitude of an equilateral triangle is 14 feet. Find the length of a side and the area of the triangle.

172. A ladder, whose foot rests in a given position, just reaches a window on one side of a street, and when turned about its foot just reaches a window on the other side. If the two positions of the ladder be at right angles to each other, and the heights of the windows be 36 and 27 feet, respectively, find the width of the street and the length of the ladder.

173. The driving wheel of a locomotive-engine, of diameter 7 feet, makes $1\frac{1}{2}$ revolutions in 1 second. Find, approximately, the number of miles per hour at which the train is going.

174. How many inches of wire are necessary to make a figure of a circle and a square described about it, when each side of the square is 4 inches?

175. The large hand of a clock is 11 feet long. How many yards per day does its extremity travel?

176. Find the internal depth of a cubical box, which will exactly hold 3 gallons.

177. A sovereign is $\frac{7}{8}$ of an inch in diameter, and $\frac{1}{16}$ of an inch in thickness. If 80000 of them be melted down and formed into a cube, find an edge of the cube.

178. Find how many gallons are contained in a vessel, which is in the form of a right circular cone, the radius of the base being 8 feet, and the slant side 12 feet.

179. The great pyramid of Egypt was 481 feet in height when complete, and its base was a square 764 feet in length. Find its volume in cubic yards.

180. A solid is in the form of a right circular cylinder, with hemispherical ends; the extreme length is 42 inches and the diameter is 5 inches. Find the volume.

181. Find the weight of gunpowder required to fill a hollow sphere 9 inches in diameter, supposing that 30 cubic inches of gunpowder weigh one pound.

182. The radius of the base of a cylindrical vessel is 14 inches; a block of stone is placed in the vessel and is covered with water; on removing the block the level of the water sinks 4 inches. Find the weight of the block of stone, supposing that it is 8 times as heavy as an equal volume of water.

183. A water-wheel, whose diameter is 14 feet, makes 50 revolutions per minute. Find, approximately, the number of miles per hour traversed by a point on the circumference of the wheel.

184. A locomotive, running at the rate of 35 miles per hour, has a driving-wheel which makes 4 revolutions in 1 second. Find the diameter of the wheel.

185. The shaft of Pompey's pillar, which is situated near Alexandria, in Egypt, is a single stone of granite. The height is 90 feet, the diameter at one end is 60 inches, and at the other end 7 ft. 6 in. Find the volume.

186. A cask full of water weighs 480 lbs.; the cask when empty weighs 31 lbs. Find the number of gallons the cask will hold.

187. Find the area of the whole surface of a pyramid on a square base, having its other faces equal; each side of the base is 35 feet, and the height of the pyramid is 34 feet 8 inches.

188. The surface of a sphere is equal to that of a right circular cylinder; the radius of the base of the cylinder is 4 inches, and the height 1 foot. Find the volume of the sphere.

189. The volume of a sphere is equal to that of the right circular cylinder in the preceding example. Find the surface of the sphere.

190. A square enclosure has a side 40 feet in length. In front of it, and at a distance of 40 feet from each of the two nearest corners, a cow is tethered. If the tether rope is 100 feet in length, and the cow is unable to enter the enclosure, find the area of the ground from which the cow will be able to procure the grass.

191. How many times larger than the earth is the sun, if they are considered spheres, the radius of the earth being 3956 miles, and that of the sun 441,500 miles?

192. How many gallons of wine are contained in a cask weighing 450 pounds, if the cask when empty weighs 20 lbs., and a cubic foot of wine 995 ounces?

193. If the earth be regarded as a sphere 7912 miles in diameter, and the moon a sphere 2160 miles in diameter, how many times the bulk of the moon is the earth?

194. A flat roof is 17 ft. 4 in. long, and 13 ft. 4 in. wide. Find the cost of covering it with sheet lead $\frac{1}{16}$ of an inch thick, supposing that a cubic inch of lead weighs 6·5 ounces, and that 1 pound of it costs 7 cents.

195. A box with a lid is made of planking $1\frac{1}{2}$ inches thick. If the external dimensions be 3 ft. 4 in., 2 ft. 6 in., and 1 ft. 8 in., respectively, how many square feet of planking does it contain?

196. A telegraph wire is 70 kilometres long, and $2\frac{1}{4}$ millimetres in diameter. Find the volume in cubic decimetres.

197. The top of a flag staff being broken off by a blast of wind, struck the ground at a distance of 15 feet from the foot of the pole. Find the height of the whole flag-staff supposing the length of the broken piece to be 39 feet.

198. The area of the coal field of South Wales is 1000 square miles, and the average thickness of the coal is 60 feet. If a cubic yard of coal weigh 2200 pounds, and the annual consumption of coal in Great Britain be 70,000,000 tons, find the number of years for which this coal field alone would supply Great Britain with coal at the present rate of consumption.

199. Find the number of gallons of water which pass in 5 minutes under a bridge 20 ft. 6 in. wide, the stream being 12 ft. 4 in. deep, and its velocity $2\frac{1}{2}$ miles per hour.

200. Six men bought a grinding-stone, 65 inches in diameter, each paying one-sixth part of the expense. They agree to grind down their respective shares in succession. If the axle renders a space in the centre, 5 inches in diameter, useless, find the diameter of the grinding-stone when each of them has ground his share.

201. Three poles stand upright on level ground, with their lower ends in the same straight line. The heights of the two extreme poles are 31 and 25 feet. The top of the former is 20 feet, and that of the latter 26 feet, from the top of the middle pole; while the middle pole stands 12 feet distant from the higher of the other two, measured horizontally. What is the length of the middle pole, and the distance between the tops of the other two?

202. What must be the edge of a cube, in order that its total superficial area may be 2 square feet, expressing the answer in inches?

203. A vessel when empty weighs 1·75 kilogrammes, and when full of water weighs 7·3 kilogrammes. Find the capacity of the vessel in cubic decimetres.

204. If iron is 7·7 times as heavy as an equal volume of water, find the radius of a 64-pound spherical shot.

205. Find the measure of the area of a triangle, the measures of whose sides are a, b and c.

206. A vat, 4 ft. long, 3 ft. wide, and 9 inches deep, contains pulp for making paper. A percentage of the pulp is lost in drying, and a sheet of paper 2700 yards long, 2 feet 6 inches wide, and ·004 inches thick is obtained. What per cent. of the pulp was lost in drying?

GENERAL PROBLEMS.

1. A can do as much work in 9 days as B can do in 10, but A works only 10 days for every 11 B works. How should $59.70 be divided for work, under these conditions?

2. A man insures a house, worth $4000, for $\frac{3}{5}$ of its value, at 2% premium. If the house be destroyed, find the total loss sustained by the owner after one premium has been paid.

3. What principal will amount, at simple interest, to $373.75 in $2\frac{1}{2}$ years at 6% per annum?

4. Three men form a partnership, contributing $3200, $1000, and $1500, respectively. How should a gain of $526.50 be divided?

5. I mix 3 lbs. of tea, worth 40 cents a lb., with 5 lbs. of tea, worth 48 cents a lb. At what price per lb. should I sell the mixture to make a clear gain of 8%?

6. If 3% stock is selling at 84, what should be the price of 7% stock, to realize the same interest on the money invested?

7. Find an agent's charges, who sells 3500 bushels of wheat, at 85 cents a bushel, on a commission of $1\frac{1}{2}$%.

8. A merchant buys 4000 yards of carpet in England, at 4s. 6d. per yard. Find its value in Canadian currency, exchange being £1 = $4.87.

9. Find the cost of painting a floor, 16 ft. in length and 12 ft. 3 in. in width, at 18 cents per square yard.

10. Divide $201 between A and B, so that A may have $12 more than half of what B gets.

11. Two men start together to go in the same direction around a circular track. The first goes 5 yards while the second goes $4\frac{1}{4}$. Where will they first be together?

12. I sold a book for 78 cents, gaining $12\frac{1}{2}$%. Find the cost.

13. When, first after 4 o'clock, will the minute hand be midway between the figure IV and the hour hand?

14. Two men working together can perform a piece of work in 18 days. If the job is worth $126, and one of the men works 5 days less than the other, how should the money be divided?

15. How much tea, worth 54 cents a pound, must I mix with 34 lbs., worth 60 cents a pound, so as to form a mixture worth $57\frac{3}{5}$ cents a pound?

16. Three men form a partnership, and invest their capital in the proportion of 3, 4, and 5. If they gain $720, how should it be divided among them?

17. Find my income from investing $5100 in the $4\frac{1}{2}$ per cents at 85.

18. Find the actual cost of goods which were purchased for $510 on 3 months' credit, money being worth 8% per annum.

19. A boy bought apples at 5 for 6 cents, and sold them at 2 for 3 cents. What was his gain per cent?

20. When, first after 5 o'clock, will the minute hand be midway between the figure III and the hour hand?

21. Find the cost of sowing a field, 40 rods long and 35 rods wide, with wheat at 70 cents a bushel, if it takes $1\frac{1}{2}$ bushels to sow an acre.

22. Divide $318 between A and B, so that A may have $14 less than one-third of B's share.

23. Find the difference between $\frac{3}{4} \div \frac{1}{2}$ of $\frac{5}{6}$, and $\frac{3}{4} \div \frac{1}{2} \times \frac{5}{6}$.

24. A and B can do a piece of work in 8 days, B and C in 10 days, and C and A in 12 days. Find when they will finish it, all working together.

25. I sold 2 books for 75 cents each, on the one I gained 20%, and on the other I lost 20%. How much did I gain or lose on the whole transaction?

26. Find the rate of interest at which $137 will amount to $152.07 in 2 years.

27. Reduce $\cdot\dot{7}1428\dot{5}$ to a vulgar fraction in its lowest terms.

28. A is worth a certain sum, B is worth twice as much, and C is worth as much as both, which is $2700. What are A and B worth, respectively?

29. An agent remitted to his employer $1872, as the proceeds of the sale of wheat, after deducting his commission of 4% for the transaction. What was the amount of the commission?

30. If the $3\frac{1}{2}$ per cents are selling at 75, what rate of interest would they pay on an investment?

1. A grocer mixes 30 lbs. of tea, worth 45 cents a pound, with 20 lbs. of tea, worth 35 cents a pound. At what price per lb. must he sell the mixture so as to gain 20% on his outlay?

2. How much carpet is wasted in carpeting a room, 20 ft. long and 11 ft. wide, with carpet (running lengthwise) $\frac{3}{4}$ yd. wide, and having a complete pattern every 8 ft.?

3. A can do twice as much work in a day as B, but he works only $\frac{3}{5}$ of the time. How should $22, which they receive for their work, be divided?

4. When, first after 3 o'clock, will the hour hand of a clock be midway between the figure III and the minute hand?

5. A and B agree to pay their travelling expenses in the proportion of 2 to 3. A pays on the whole $164, and B $206. What has one to pay the other to settle the account?

6. Two-thirds of the selling price of certain goods is 10% less than cost. Find the gain per cent. at which the goods were sold.

7. In what time will $212 amount to $245.39, at 7% per annum?

8. Find the sum realized on the sale of $3000 stock at 86, brokerage at $\frac{1}{8}$%.

9. Divide $4669 among 3 men, in the proportion of 5, 7, and 11.

10. Find the true discount off $422.50 due in 9 months, at $7\frac{1}{2}$% per annum.

11. What is the cost of insuring property, worth $700, at 2%, so that in event of loss the owner may receive back the value of the property and the premium paid?

12. Divide $31.50 among A, B, and C, so that B may have one-third as much again as A, and C one-fourth as much again as A and B together.

13. Find the cost of fencing a $2\frac{1}{2}$ acre square field at 50 cents a rod.

14. Find the net income of a man whose total income is $900, on $500 of which he pays 18 mills on the dollar taxation.

15. Three brothers buy a farm, the first taking $\frac{2}{5}$ of it, the second $\frac{1}{3}$, and the third the remainder; the third paid $1884. What did each of the others pay?

16. Simplify $\frac{\cdot304 \times \cdot002 \times 1\cdot8}{\cdot0009 \times \cdot038}$

17. A can do a piece of work in 5 days, B in 6, and C in 7. If they all work together at it and make $21.40, how should the money be divided?

18. What principal will amount to $263.17½ in 4 years at 5¼% per annum?

19. How much must a man pay for $4000 stock at 87, brokerage ⅛%?

20. At what time after 5 o'clock are the hour and minute hands of a clock first together?

21. I insure a house for $\frac{2}{3}$ of its value at 3%, and if the house is destroyed my total loss will be $522.50. Find the value of the house.

22. A farmer bought 11 cows for $253, and after keeping them 17 weeks, at a cost of $1.75 a week each, he sold them for $48 each. How much did he gain or lose by the transaction?

23. Find the cost of sowing a field 40 rods long and 30 wide, with oats worth 35 cents a bushel, if it requires 2 bushels of oats to sow one acre.

24. A grocer mixes 25 lbs. of tea, worth 35 cents a lb., with 35 lbs., worth 40 cents a lb. At what price per lb. must he sell the mixture in order to gain 10%?

25. Find the cost of papering the walls of a room 14 ft. long, 12 ft. wide, and 10 feet high, in which are two windows, 3½ by 6 ft., and one door, 4 by 7 ft., with paper 27 inches wide, at 6 cents a yard.

26. Divide $612 into 3 parts, such that the first, put out at simple interest for 2 years at 6%, equals the second for 3 years at 5%, equals the third for 4 years at 4%.

27. A bankrupt owes A $1000, and B $2700, and his assets are $921.25. How should his assets be divided?

28. A grain merchant bought wheat at 87½ cents a bushel, and sold it at 77 cents a bushel. How much did he lose on every dollar he paid?

1. A and B can do a piece of work in 5 days, A could do it alone in 8 days. How long would it take B to finish the work if both together had worked at it for 3 days?

2. A grocer who throws off 5% for cash, sold the following: 3 lbs. raisins, at 8 cents a pound; $4\frac{1}{2}$ lbs. tea, at 50 cents a pound; 4 cans of salmon, at 25 cents a can; 8 lbs. of sugar, at $6\frac{1}{2}$ cents a pound; $5\frac{1}{2}$ lbs. of butter, at 20 cents a pound. Find his cash receipts.

3. I sold a horse for $153, and lost 15% by the transaction. Find the cost price of the horse.

4. A and B engage in trade; A puts into the business $400 for 6 months, and B puts in $300 for 7 months. How should a net gain of $450 be divided?

5. Find the income arising from investing $3370 in the 3 per cents, at 72.

6. A sold $2500 worth of goods to B, at a profit of 6%, and B sold them to C, at a loss of 5%. Find what C paid for the goods.

7. A certain sum of money in 3 years at 4% per annum amounts to $336. Find what it will amount to in 4 years at the same rate.

8. Find the cost of gravelling a walk, 1 yard in width, around the inside of a $2\frac{1}{2}$ acre square field, at 15 cents a square yard.

9. A, B, and C do a piece of work, and are paid $12 for it. The money is divided according to the efficiency and time each worked; A's efficiency is to B's as 2 to 3, B's to C's as 4 to 5; A worked 6 days, B 7 days, and C 8 days. How should the money be divided?

10. Divide $369 between A and B, so that A may have $1.20 more than 14% of B's share.

11. A merchant sold an article at a loss of 8%, but had he sold it for $1.05 more, he would have gained 7%. What price did he sell it at?

12. A man divided a farm among 3 sons; to the first he gave 110 acres, to the second $\frac{2}{9}$ of the whole, and to the third $1\frac{2}{3}$ as much as to both the others. How many acres did the farm contain?

13. A and B run a 100 yards race; A takes 8 steps while B takes 9, but 10 of A's are equal in length to 11 of B's. Who will win the race, and by how much?

14. Find the difference between the true and the bank discount off $508, due in $4\frac{1}{2}$ years, at 6% per annum.

15. How many yards of paper, 30 inches wide, with a pattern every 18 inches, are required to paper the walls of a room 18 ft. long, 12 ft. wide, and 10 ft. high?

16. A person sold a lot for $600, gaining 14% of the proceeds. What would he have sold it for, had he gained 14% of the cost?

17. A city increases 13% annually in population for 3 years, and has at the end of that time 1,442,897 of a population. What had it at the beginning of the time?

18. How much stock must be sold, at 72, to pay a debt of $7470, 9 months before it becomes due, true discount being allowed, at 5% per annum?

19. A merchant sold 135 barrels of flour, part of it at $5 a barrel, and the remainder at $4 a barrel, and realized $615. How many barrels of each kind did he sell?

20. A merchant marked his goods at an advance of 20% on cost, and in selling them he used a yard measure $\frac{1}{4}$ inch too short; his entire gain being $134.10, find the cost price of the goods.

21. A merchant marked his goods at an advance of 50%, but afterwards sold them 50% less than this price. Required his loss per cent.

22. I buy a ground rent of $32 per annum for $400. What per cent. do I realize on my investment?

23. A man has $10, $5, and $4 bills, the number of each denomination being proportional to 3, 4, and 6. If they amount to $518 in all, how many has he of each denomination?

24. A and B start together to travel in the same direction. A travels at the rate of 7 miles per hour, and always remain $\frac{1}{4}$ as far ahead of B as B has travelled. What is B's rate of travelling?

25. A speculator received a stock dividend of 5%, which increased the number of his shares to 42 ($100). How many shares had he at first?

1. A druggist gives a lb. troy of certain goods instead of a lb. avoirdupois. What is his gain and the customer's loss per cent.?

2. Find the cost of carpeting a room 20 ft. long, 17 ft. 6 in. in width, with carpet 2 ft. wide, at $1.50 a yard, the carpet running lengthwise, and without a pattern.

3. Find the cost of fencing a farm in the form of a rectangle, whose sides are in the ratio of 4 to 5, and which contains 800 acres, at $1.37½ a rod.

4. A and B work 6 and 7 hours respectively for one day and receive the same wages; on the second day they work 7 and 8 hours respectively; they receive for their whole work $18.10. How should it be divided?

5. Divide $248 earnings between A and B, when A works only 4 hrs. to every 6 hrs. B works, but is able to do as much in 5 days as B can do in 8 days.

6. A grocer, by selling 5 lbs. of tea for a certain price, gained 12%; afterwards he increased the price, giving only 4½ lbs. for the same money. What per cent. did he make at the increased price?

7. A merchant lost 25% of certain perishable goods and sold the remainder at a gain of 30%. What was his gain or loss per cent.?

8. Simplify $\frac{\cdot 0004 \times \cdot 00651 \times \cdot 03}{\cdot 0008 \times 3 \cdot 1}$.

9. A builder pays 3 times as much for material as for labor; had he paid 10% more for material, and 6% less for labor, his contract would have cost him $3637.92. What was his contract price?

10. Find the length of a side of a square field containing 10 acres.

11. I invest equal sums of money in the 4 per cents at 87, and in the 5 per cents at 102, and realize $27 a year more income from the latter than from the former. How much do I invest in each?

12. The hour, minute, and second hands of a watch revolve around the same centre. When, first after 7 o'clock, will the minute hand be midway between the other two?

13. A and B form a partnership, A puts into the business \$3000 for 4 months, and B puts in \$4200 for 3 months. How should \$1722 profits be divided between them?

14. A man insures a house, worth \$4900, so that in case of loss he can recover $\frac{3}{5}$ of the value of the house and the premium of 2% paid for insuring. What was the amount of insurance?

15. A 70-day note is discounted by a banker at 9%. What rate of interest is he charging for the money advanced?

16. Find the length of the diagonal of a square field containing 40 acres

17. A speculator invests some money in 3% stock at 72. What per cent. is he making on the money invested?

18. A merchant imported a quantity of goods, paying 25% of the invoice price for freight, duty etc. He sold them for \$2925, losing 10% thereby. What was the invoice price?

19. A merchant having bought a lot of goods, sells $\frac{1}{4}$ of them at a loss of 6%. By what increase per cent. must he raise that selling price in order that by selling the remainder at the increased rate he may gain 6% on the whole transaction?

20. A bankrupt's assets are only half of his liabilities, but one-third of the assets prove to be worth only 50 cents on the dollar. How many cents on the dollar can he pay?

21. A wine merchant mixes 8 gallons of wine worth \$1.12$\frac{1}{2}$ a gallon, 12 gallons worth \$1.25 a gallon, and 14 gallons worth \$1.50 a gallon, with 16 gallons of water, and sells the mixture at \$1 a gallon. Find his gain per cent.

22. A boy engages with a farmer a year for \$40 and a suit of clothes. He leaves at the end of 9 months, and is entitled to \$25 and the suit. What was the cost of the suit?

23. A man has \$400 of his income exempt from taxation, and on the balance he pays 2% income tax; if his net income is \$865.50, find his total income.

24. What time between 6 and 7 o'clock will the hour and minute hands of a watch be together?

1. A broker receives $25 for investing $4325 in stock worth 86. What rate did the broker charge?

2. Find the perimeter of a square field containing 10 acres.

3. A merchant sold 276 yards of cloth at a profit of 15%, and 398 yards at a profit of 7%, and found that had he sold it at a uniform profit of 11%, he would have realized $1.88 more than he did. What was the cost price of the cloth?

4. A man sold a horse at 20% profit. If the horse had cost him $40 more, and had sold for the same amount as before, he would have lost 5%. What was the cost?

5. What is the difference between 40% discount, and 20, 10, and 10% discount?

6. How much will I save annually by investing $7950 in the 3%'s, at 75, instead of in the 4%'s, at 106?

7. A merchant who gives a discount of 5% for cash, sold the following goods: 14 lbs. of sugar, at $6\frac{1}{2}$ cents a pound; 7 lbs. of tea, at 35 cents a pound; 3 brooms, at 25 cents each; 20 yards of cotton, at 8 cents a yard; 4 papers of pins, at 5 cents each, and 3 spools, at 3 cents each. He received in payment 5 dozen eggs, at 12 cents a dozen, 2 pounds of butter, at 20 cents a pound, $1.90 in cash, and booked the balance. How much did he book?

8. The true discount off $130 for a certain time is $10. What would it be for double that time?

9. A merchant marks his goods at a profit of 30%, but throws off 20% of this price in selling. At what advance on cost does he sell them?

10. A and B run a 300 yard race; A runs 12 yards while B runs 10, but after A has run one-half the distance, he runs only 8 yards in the time in which he formerly ran 12 yards, while B continues his original rate throughout the race. Which wins, and by how much?

11. Find the difference between the simple and the compound interest on $100 for 3 years at 7% per annum.

12. Find the difference between the true and the bank discount off $620 due in 3 years, at 8% per annum.

13. A teacher pays $5.40 taxes. What is his total salary, if $400 of it is exempt from taxation, and a 2% rate is levied on the remainder?

14. A broker invests \$5176.50 in stock at 76, on $\frac{1}{8}$% commission. What are his charges?

15. A merchant mixes teas worth 35, 40, 60, 75 cents a lb., in equal quantities, and sells the mixture for 80 cents a lb. What percentage does he gain?

16. A, B, and C do a piece of work and are paid in proportion to the number of hours each works; for 4 days A works 6 hours a day, B 7 hours a day, and C 8 hours a day. At the end of this time A leaves, while B and C work 2 days more, 10 hours a day each, and finish the work. If \$62 be paid for the whole work, how should it be divided?

17. A merchant sells 12 lbs. of tea for what 17 lbs. cost him. What advance per cent. on cost is he making?

18. An insurance company took a risk of \$4800 at $2\frac{1}{2}$%, and immediately re-insured $\frac{1}{3}$ of it in another company at 3%. If the property be destroyed, find the loss sustained by each company.

19. If I gain 30% of the proceeds in selling goods, what is my gain. per cent., on the cost?

20. Find how much a merchant cheats a customer who buys goods to the amount of \$120, when he gives only 35 inches for a yard.

21. Find the cost of paving a walk, 1 yard wide, around the inside of a rectangular field, 40 yards in length and 30 yards in width, at 75 cents a square yard.

22. A wholesale dealer sold goods, at 5% profit, to a retailer, who disposed of them at 5% loss. Find the average gain or loss per cent.

23. A can do a piece of work in 7 days, B in 10 days, and C in 14 days. How long will it take C to finish the work if A and B have both worked at it for 2 days?

24. How many pounds of tea, worth 35 cents a pound must be mixed with 14 lbs. at 40 cents a pound, so that a gain of 20% may be made by selling the mixture at \$4.41 per 10 pounds?

25. How much better is it to lend \$300 for 3 years at $6\frac{1}{2}$% simple interest, than to lend it at 6% compound interest?

26. What rate of discount is 40, 10, and 10% equal to?

1. A druggist marks his soda at 50 % profit, but by a mistake sells a pound by avoirdupois weight. What percentage does he really gain on the pound?

2. Bought land at $10 an acre. How much must I ask an acre that I may abate 25 % from my asking price, and still make 30 % on the purchase money?

3. Find the cost of planting a hedge around a square field, containing $2\frac{1}{2}$ acres, at $12\frac{1}{2}$ cents a yard.

4. A commission merchant charges $53.60 for buying goods on a commission of 2 %. What sum must his employer remit to him to cover his charges and to buy the goods?

5. A broker charges $\frac{1}{2}$ % commission for selling $4230 stock at $79\frac{1}{4}$. What are the net proceeds of the sale?

6. A man insures property worth $7140, so that in case of loss he may recover the value of the property and the premium paid for insuring, which was at 2 %. What was the premium paid for insuring?

7. A and B invest money in business in the ratio of 5 to 7. If they gain $1005.51, how should it be divided?

8. If a piece of silk cost 80 cents a yard, at what price must it be marked that the merchant may sell it at 10% less than the marked price, and still make 20 % profit?

9. When, first after 7 o'clock, will the hour hand be midway between the figure V and the minute hand?

10. A farmer sold his farm for $5300, and considered that he had gained a certain amount, but a note for $80, which he had accepted in part payment, proved worthless, and as a result he gained only $\frac{1}{3}$ as much as he had expected. Find what he considered the farm worth.

11. Find the compound interest on $39 for 16 years, at 4 % per annum.

12. Find the sides of a rectangular field, containing $3\frac{3}{4}$ acres, if the length is to the breadth as 3 to 2.

13. I sell $3500 stock at $65\frac{1}{4}$: find my receipts from the sale.

14. A speculator paid $1400 for two lots, the price of one being 40 % that of the other. He sold the cheaper lot at a gain of 50 %, and the dearer one at a loss of 30 %. Find his gain or loss on the whole transaction.

15. A and B form a partnership, A contributing \$2700, and B \$3700; they agree that 20 % of the total profits, shall be placed to the credit of the firm annually, and the remaining profits be divided in proportion to the capital invested by each. At the end of the year A's share of the profits was \$675. Find the percentage of profit realized on the entire capital.

16. A quantity of goods was sold at 25 % gain, but, if they had cost \$40 less, the gain at the same selling price would have been 35 %. What did the goods cost?

17. A man having lost 20% of his capital, is worth exactly as much as another, who has just gained 12% on his capital; the second man's capital was originally \$4000. What was the original capital of the first man?

18. A farmer employed a number of men and 8 boys; he pays the men \$1.40 a day and the boys 50 cents a day. The average price he paid was 92 cents a day. How many men were employed?

19. Divide \$39 between A and B, so that 9% of A's share may equal 17% of B's.

20. I bought through a broker 150 (\$100) shares, at $79\frac{1}{4}$. What did the stock cost me?

21. A merchant having bought a lot of goods, sells one-third at a loss of 4%. By what per cent. must he raise that selling price in order that by selling the remainder at the increased rate he may gain 4% on the whole transaction?

22. A grocer bought a stock of tea, intending to gain 30% on its sale. When he had sold $\frac{3}{4}$ of it, he was compelled to reduce the price 10 cents a pound, and so gained only $\frac{8}{9}$ of what he had intended. Find what price he paid for the tea.

23. A father willed his property, amounting to \$26600, to his two sons, aged 14 and 17 years, respectively, in such a way that the sums loaned at 6% per annum, simple interest, would be equal when they became of age. How did he divide it?

24. A man has a certain sum of money invested at 4 %, and 3 times that amount at 6 %. From both investments he obtains an income of $\$76.42\frac{1}{5}$. What has he invested?

1. A and B made a joint stock of $1300, by which they gained $715, of which A had for his share $275 more than B. What did each contribute of the stock?

2. Find the perimeter of a rectangular field, whose sides are in the proportion of 2 to 3, and which contains 3 acres, 120 perches.

3. What per cent. of the cost price of an article is the selling price, if the marked price is 20% advance on cost, and the selling price 20% less than the marked price?

4. A broker invests $2290 in stock at 85¾, and charges ⅛% brokerage. Find his brokerage.

5. A merchant bought 100 barrels of flour, part at $7 a barrel, and the remainder at $5 a barrel; by selling the former at 15% gain, and the latter at 14% loss, he just cleared himself on the transaction. How many barrels of each did he buy?

6. I bought a hind and a fore quarter of beef, weighing together 252 pounds; I paid 7¼ cents a pound for the hind quarter, and 5½ cents a pound for the fore quarter, and found that I had paid 17½ cents on the whole more than if I had bought both quarters at 6⅜ cents a pound. Find the weight of each quarter.

7. If a debt, after a reduction of 3%, becomes $1008.80, what would it become after a reduction of 4%?

8. Divide $700 into two parts, such that the simple interest on one part for 4 years at 5% per annum, may be equal to the simple interest on the other part for 2½ years at 6% per annum.

9. The difference between the annual income derived from a certain sum invested in 8½% stock at 187½, and that from an equal sum invested in 6% stock at 134 is $10.60. What is the amount invested in either kind of stock?

10. A grain merchant sold 435 bushels of wheat at a profit of 13%, and 325 bushels at a profit of 11%, and realized $1.10 more than he would have realized had he sold it all at a uniform profit of 12%. What was the price per bushel he paid for the wheat?

11. A speculator bought two lots, the price of one of them being 40% that of the other. He sold the dearer lot at a loss of 30%, and the cheaper one at a gain of 50%. Find his gain or loss per cent. on the whole transaction.

12. A merchant at one time asked 25% less than cost for an article, but afterwards sold it for 25% more than this price. Find his loss per cent.

13. Two men run around a circular track, 400 yards in circumference, at rates of 7 and 9 yards per second respectively. When and where will they first be together again, provided that they start together and run in the same direction?

14. A person gave 5 cents each to a number of beggars and had 14 cents left. He found that he would have required 22 cents more to enable him to give the beggars 8 cents each. How many beggars were there?

15. A certain sum of money loaned at simple interest amounts to $334.40 in 9 months, and in 7 months more to $345.60. Find the sum and rate.

16. Find the perimeter of a rectangular field, whose sides are in the ratio of 7 to 16, and which contains 17½ acres.

17. A lumberman sold 36840 feet of lumber at $21.12 per M, and gained 28%. How much would he have gained had he sold it for $17 per M?

18. Three men start together and travel in the same direction. Their rates are 5, 7, and 9 miles per hour respectively. If the course be circular, and 80 rods in circumference, find where they will first meet.

19. In building a house, the owner pays twice as much for material as for labor. Had he paid 5% more for material and 7% more for labor, the house would have cost $10144. What was its cost?

20. A merchant's wholesale price is 15% advance on cost, and his retail price 10% advance on wholesale. Find his gain per cent. on cost by his retail price.

21. A merchant sells two kinds of flour, the superior at $6 a barrel, the inferior at $5 a barrel. He sold 150 barrels in all and realized $810. How many barrels of each kind did he sell?

1. I invest \$4000 in stock at 112. Find the amount of the stock I receive.

2. The joint capital of A, B, C, and D is \$3150; A invests \$2 for every \$3 B invests, B \$4 for C's \$5, C \$6 for D's \$7. Required the amount invested by each.

3. Divide 99 into two parts, such that 5 times one part may be equal to 6 times the other.

4. A, B, and C work 6, 7, and 8 hours a day, respectively, for 2 days, and receive the same daily wages. They finish the work in 3 days more by each working one hour a day extra. If \$91.30 is paid for the work, what should each man receive?

5. An article sold, at a loss of 25%, for \$5.62½. What would be the gain or loss per cent. if it sold for \$7.00?

6. A speculator sells \$3650 of the Merchants' Bank stock at 142, which yields a semi-annual dividend of 3½%, and invests the proceeds in Bank of Hamilton stock at 146, which pays 4% semi-annual dividends. Find the change in his annual income.

7. A merchant knows neither the weight nor prime cost of a caddy of tea. He recollects that if he had sold the whole quantity at 70 cents a pound he would have gained \$7.00, and if he had sold it at 50 cents a pound he would have lost \$3.00. What was the weight and prime cost per pound of the tea?

8. A and B had the same amount of money, A lost 39% of his in speculation, and B gained 45% of his in another speculation; both together now have \$164.80. What sum had each at first?

9. An army, in a defeat, loses one-quarter of its number in killed and wounded, and 4000 prisoners; it is reinforced by 2500 men, but retreats, losing one-fifth of its number in doing so; there remain 6000 men. What was the original force?

10. A person buys a lot of land at \$40 an acre, and by selling it in allotments finds the value increased threefold, so that he clears \$200, and retains 25 acres for himself. How many acres did he buy?

11. Find the compound interest on \$48 for 8 years at 8% per annum, interest calculated half yearly.

12. When exchange at New York on Paris is 5·16 francs per $1, and at Paris on Hamburg 2·12½ francs per marc banco, what will be the arbitrated price in New York of 26880 marc banco of Hamburg?

13. What must be the face value of a note made May 15th, at 5 months, and discounted June 27th, at 8%, to produce $560?

14. A grocer imported 120 cases of port wine, 36 bottles in each case. After 5% had been allowed for breakage, he paid an ad valorem duty of 20%; the freight and cartage expenses were $80, and the whole cost was $3773.60. What was the invoice price per bottle?

15. The net amount received by a village for taxes is $9690. The rate of taxation is 17 mills on the dollar, and the collectors' charges 5% of the total taxes What is the amount of the assessment?

16. A merchant had 300 barrels of flour insured for $\frac{2}{3}$ of its value at 3%, paying $36 premium. At what price, per barrel, must he sell it to gain 15% of the prime cost as well as the premium paid?

17. A vessel contains 150 gallons of wine; 50 gallons are drawn and the vessel filled with water. If this operation be continued eight times, how much wine will the mixture contain?

18. My salary is $1200. If I pay 30% of it for board, 20% of the remainder for rent, 15% of the residue for clothes, $71.20 for books, and loan 40% of the remainder, what per cent. of my salary is unexpended?

19. A farmer sold oats for $18.49, the price per bushel being as many cents as there were bushels. What was the price per bushel?

20. What is the hour when $\frac{2}{3}$ of the time past noon is $\frac{2}{5}$ of the time till midnight?

21. A man bought $\frac{3}{8}$ of a vessel, and sold $\frac{4}{5}$ of his share for $11700, which was 30% above the cost. What was the cost of the vessel?

22. A sum of money in two years, at compound interest, added yearly, amounts to $648.96; the present worth of the sum for one year is 576\frac{4}{13}$. Find the rate per cent. and the sum.

1. If 8% of the cost price of an article is equal to 6% of its selling price, what is the gain per cent.?

2. A man increases his income $12, by transferring $3000 stock from the 4 per cents to $3\frac{1}{2}$ per cents at 75. Find the price of the former stock.

3. A offers to sell a farm for $5000, payable in 9 months; B borrows $5000, at 7% per annum, to buy it; but A afterwards, wanting cash, offers to throw off 5% of his price for cash, B accepts the offer, and invests the balance of the money at 10% per annum. At the end of the 9 months find what B has gained by A not holding to the first offer.

4. A merchant lost 14% of his goods, and sold the remainder at an advance of 25% on cost, gaining $360 on the whole transaction. What did his goods cost?

5. At what time, after 12 o'clock, will the hour and minute hands of a clock be first together?

6. Divide $500 into two parts, such that the simple interest on one part for 4 years at 6% per annum, may be $12 more than that on the other part for 6 years at 5% per annum.

7. A, B, and C, are engaged to do a piece of work. A can do as much in 3 days as B can do in 4 days, and B in 6 as C in 7, if they work equal hours each day. A works 5 hours a day for 6 days, B 7 hours a day for 9 days, and C 11 hours a day for 10 days, when the whole work is completed. How should $138.10, which they received for the work, be divided?

8. A merchant buys $600 worth of sugar, and at first determines to sell it at a profit of 8%, but afterwards decides to add $\frac{1}{12}$ of its original bulk of sand, and sell it at cost. What does he gain or lose by this method of disposing of it as compared with the former way?

9. A person invested in 3% stock, so as to receive $5\frac{1}{4}$% clear on his investment, after paying an income tax of 20 mills on the dollar? What was the market price of the stock, brokerage being $\frac{1}{2}$%?

10. A merchant sold cloth at 20% gain, but had it cost $49 more, he would have lost 15% by selling at the same price. What did the goods cost?

11. A gentleman put out $21950 on interest, in 3 separate sums, at 5 %, for the benefit of his 3 sons, aged 9, 11, and 14 years respectively, in such a manner that each, as he became of age, should receive the same amount. What sum was put out at interest for each (use simple interest)?

12. A and B play a game of chance. A has $56 and B $22 when they begin, and at the end of the game A has 3 times as much money as B. How much did A win?

13. Three persons can together complete a piece of work in 45 days. It is found that the first does $\frac{3}{4}$ of what the second does, and the second $\frac{4}{5}$ of what the third does. In what time could each alone complete it?

14. A man sold a horse at a loss of 25%. If he had sold him for $60 more he would have gained 15%. What did the horse cost him?

15. Find the area of a circle whose circumference measures 24 feet.

16. A man invests $7140 in stock at 84, and after it has advanced to 95 sells out. Find his gain.

17. A person buys a lot of land at $120 an acre, and by selling a portion he makes 90% on all he sells, so that after reserving 20 acres he finds he has realized $840 more than the entire lot cost him. How many acres did he buy?

18. A merchant bought a quantity of tweed and marked it at an advance of 25% on cost, and in selling it he used a yard measure $\frac{3}{4}$ inch too short, his entire gain being $124.80. Find the cost price of the cloth, and the amount the merchant gained by his dishonesty.

19. A and B start together to travel in the same direction. A travels at the rate of 5 miles per hour, and always remains one half as far ahead of B as B has travelled. What is B's rate of travelling?

20. A hare is 75 of her own leaps ahead of a hound, and she takes 5 leaps for every 3 the hound takes, but he covers as much in 1 leap as she does in 2. How many leaps will the hound take before he catches the hare?

21. The hour, minute, and second hands of a clock revolve on the same centre. When, first after 5 o'clock, will the hour hand be midway between the other two?

1. A man sold two lots, receiving $\frac{2}{3}$ as much for one as for the other; on the cheaper one he lost 8%, and on the dearer one he gained 8%, but gained \$16 on the whole transaction. Find the proceeds of the two lots.

2. A grocer spent equal sums on tea, coffee and sugar; he sold them, making 12% on the tea, 8% on the coffee, and losing 15% on the sugar. His total gain being \$63.50, find the cost of each commodity.

3. Jones received a stock dividend of 6%, he then had 83 (\$50) shares, and \$37 of another share. How many shares had he at first?

4. A merchant marked his goods at an advance of 20%, but afterwards sold them 40% less than this price. Required his loss per cent.

5. Find the difference between the compound interests on \$200 for $7\frac{1}{2}$ years at 7% per annum, interest calculated (1) half-yearly, (2) yearly.

6. What will it cost to gravel a walk, 2 yards wide, around the inside of a circular piece of ground whose diameter is 40 yards, at 14 cents a square yard?

7. A person had \$150; part of it he loaned at 5%, and the remainder at 7%, from which he received equal sums as interest. How much did he loan at 5%?

8. Divide 39 into two parts, such, that one part may be six-sevenths of the other.

9. What per cent. of a 12% gain on cost, is a 12% gain on that selling price?

10. Find the area of a triangle whose sides are 19, 24, 36 ft. long respectively.

11. What per cent. of his money will a man make, who invests it in Pacific Railroad stock at 105, and paying 6% dividends?

12. What is the difference between 40% discount, and 10% taken off 4 times?

13. A gentleman has \$25000 Bank of Commerce stock which pays a dividend of 8%. When money is worth 7%, he sells out and invests in Bank of Toronto stock at 205, which pays a dividend of 12%. What is the difference in his income after allowing $\frac{1}{2}$% brokerage for each transaction?

14. A sold 2 city lots, which cost the same price, to B at a loss of 15 %; B sold them to C, gaining 20 % on one, and losing 25% on the other. What did either lot cost A, if B received $153 more for one than for the other?

15. A barrel of coal oil lost 20% by leakage, and the remainder was sold at a gain of 20%. Find the gain or loss per cent.

16. Three men start together to travel in the same direction. The rates of the first two men are 4 and 6 miles per hour, respectively. What is the rate of the third man, provided he is always equi-distant from the other two?

17. A man buys 75 acres of land at $140 an acre, and sells a portion of it at an advance of 50 % on the cost. He realized on the part sold a sum équal to the cost of the entire lot. How many acres did he sell?

18. A jobber bought a bankrupt stock at 75 cents on the dollar and retailed it at 10% above the original wholesale price. His expenses were 6% of the total money received. Find his gain per cent. on the goods.

19. At an election in a constituency in which the number of votes was 1800, the votes polled by the candidates were in the ratio of 7 to 5, and the majority for the successful candidate was 240. Find the number who did not vote.

20. Find how long it will take a train 87 yards long and running at the rate of 40 miles an hour, to pass another one 100 yards long and going in the same direction at the rate of 30 miles an hour.

21. A grocer sells his goods at 40% profit. He receives butter for his goods at a time when it is worth 25 cents a pound, but before he disposes of the butter 10% of it is worthless, and the remainder brings only 15 cents a pound. Find his gain or loss per cent.

22. Two partners, A and B, gained $400 in trade. A's money was 3 months in trade, and his gain was $306 less than his stock. B's money, which was $150 more than A's, was in trade 4 months. Find A's stock.

23. Divide $220 among A, B, C, and D, so that B may have twice as much as A, C as much as A and B together, and D as much as B and C together.

1. A huckster bought a certain number of apples, half of them at 2 for a cent, and the remainder at 3 for a cent; he sold them again at the rate of 5 for 2 cents and lost 1 cent by his bargain. What was the number of apples he bought?

2. Sold $\frac{1}{2}$ of a lot of lumber for what $\frac{5}{8}$ of it cost. What per cent. was gained on the part sold?

3. Find the interest on $1022 for 93 days at $5\frac{5}{7}$% per annum.

4. What must I pay for Bank of Montreal stock, which yields annual dividends of 9%, that I may make 6% interest on my money?

5. A hare is 80 of her own leaps before a greyhound: she takes 3 leaps for every 2 that he takes, but he covers as much ground in 1 leap as she does in 2. How many leaps will the hare take before she is caught?

6. If an income tax of 7d. in the pound on all incomes below £100 a year, and of 1s. in the pound on all incomes over £100 a year, realize £18750 on £500000, how much is raised on incomes below £100?

7. The hour, minute, and second hands of the clock rotate on the same axis. When, first after 3 o'clock, will the minute hand be midway between the other two?

8. A manufacturer sold goods at 50, 25, 10% discount. What was the rate of discount?

9. When the 3 per cents are at 96, how much stock must be sold out to pay a debt of $1654, 9 months before it becomes due, real discount being allowed at $4\frac{1}{2}$% per annum?

10. The whole time occupied by a train, 90 yards long, and travelling at the rate of 18 miles an hour, in crossing a bridge is $22\frac{1}{2}$ seconds. Find the length of the bridge.

11. A man owes me $200 due in 8 months, and $200 due in 12 months. I accept his note for $400 due in 10 months. How much do I gain or lose by the transaction, money being worth 8% per annum?

12. A goes into business with $1400 capital; after 3 months B joins him putting in $1600, and after 4 months more C also joins the firm, putting in $1700. The gain at the end of the year was $1191. How should it be divided?

13. A company takes a risk of \$6000 at $1\frac{1}{4}\%$ (on a house), and immediately reinsures $\frac{1}{3}$ of the risk in another company at 2%. What would the first company lose in case the house is destroyed?

14. By selling a farm for \$4000 I gain 6% more than if I had sold it for \$3800. Find the price I paid for the farm.

15. A drover bought 50 horses and cows for \$3100. The horses cost \$150 each, and the cows \$40 each. Find the number of each bought.

16. I put \$2200 out at interest for 1 year, part of it at 6% and the remainder at 5%; the sum received as interest in each case being the same. Find the sum loaned at 6%.

17. The sum of \$310, was raised by A, B, and C. B contributed \$30 more than A, and C \$40 more than B. What did each contribute?

18. A person bought 30 pounds of sugar, of two different sorts, for \$2.28; the better sort at 10 cents a pound, and the inferior at 7 cents a pound. How many pounds of each did he buy?

19. A speculator buys stock when it is 25% below par, and sells it when it is 19% below par. What is his rate of gain?

20. The perimeter of a square field is 62 yards. Find its area.

21. A man sells a house, gaining thereon 10% of the proceeds. At what advance, per cent., on the cost does he sell?.

22. At what time, between 3 and 4 o'clock, is the minute hand 5 minute spaces in advance of the hour hand?

23. Find the difference in the perimeters of a square field, containing 4 acres, and a circular field, of the same size.

24. A builder takes a contract, and at the time the material will cost just twice as much as the labor. When one-half of the material is secured, and one-third of the work performed, the cost of material rises 5%, and at the same time the cost of labor falls 8%. As a consequence, the builder saves \$10 on his first estimate. What was the first estimate?

1. A broker sold $4850 stock at $87\frac{1}{2}$, and invested the proceeds in other stock at $96\frac{1}{2}$. If his rate of brokerage is the same in each transaction, and amounts altogether to $46, find his rate.

2. A company took a risk at 4 %, and reinsured $\frac{3}{4}$ of it at 3 %. The premium received exceeded the premium paid by $27. Find the amount of the risk.

3. A, B, and C enter into partnership, A contributes $1200 for 4 months, B $800 for 5 months, and C $400 for 6 months. How should a gain of $1960 be divided among them?

4. A merchant shipped $5100 worth of wheat to his agent, and received in return $4850 worth of tea. The agent charged a certain rate for selling the wheat, and one per cent. less for buying the tea (the latter commission being calculated on the amount invested in tea). Find the rates charged.

5. What is the average time at which the following bills become due: Feb. 12, 1890, $300 on 3 months' credit; March 6, 1890, $275 on 30 days' credit; May 17, 1890, $112 on 6 months' credit; Aug. 7, 1890, $500 on 2 months' credit?

6. A wine merchant has two qualities of wine, one sort worth $1.25 a gallon, and the other worth $1.60 a gallon. He wishes to form a mixture of 75 gallons from these worth $1.36\frac{2}{3}$ a gallon. How many gallons must he take from each sort?

7. A workman was hired for 40 days at 40 cents a day for every day he worked, but with this condition, that for every day he was idle he was to forfeit 16 cents; on the whole he made $7.60. How many days did he work?

8. Find the compound interest on $324 for $8\frac{1}{2}$ years at 12 % per annum, interest calculated half-yearly.

9. A man sold two horses for $200 each, on one he gained 25 %, and on the other he lost an equal amount. Find the loss per cent. in the latter case.

10. What must I pay for $3000 stock which is selling at $94\frac{1}{2}$, when the broker who is managing the business charges $\frac{1}{8}$% for his trouble?

11. If the hour, minute, and second hands of a clock are on the same centre, when, first after 12 o'clock, will the second hand be midway between the other two?

12. What are the liabilities of a bankrupt whose total assets are \$2700, and after 2 % has been deducted for settling the business, is only able to pay 66·15 cents on the dollar?

13. A certain article of consumption is subject to a duty of 6 cents a pound; in consequence of a reduction in the duty, the consumption increases one half, but the revenue falls one-third. Find the duty per pound after reduction.

14. Bought 72 (\$50) shares of railway stock, at $46\frac{5}{8}$, and gave in payment a draft on St. Louis for \$2500, which was at $\frac{7}{8}$ % discount, and the balance in cash. How much cash did I pay?

15. A sold B a bill of goods, May 7, on 6 months' credit, amounting to \$2000; July 14, B paid him \$750, and on Sept. 4, \$125. To what additional credit is B entitled?

16. What sum will be received for a lot valued at \$4200, based on a ground rent of 7 %.

17. I pay \$51 taxes on property worth \$4000, which is assessed at $\frac{3}{4}$ of its value. Find the rate of taxation.

18. A certain sum of money at 6 % interest, amounts to \$558.60 in a certain time. At $8\frac{1}{2}$ % for the same time it amounts to \$616.35. Find the time and sum (use simple interest).

19. Four persons rented a pew in church, and by taking in two more the expense of each was lowered 75 cents. What was the rent of the pew?

20. If the cost of an article had been 30 % less, the gain per cent. would have been $2\frac{1}{2}$ times as great. Find the gain per cent.

21. If it costs \$2 a cord to saw wood into 3 lengths, what will it cost to cut a cord of wood into 4 lengths?

22. A man rows 3 miles down a stream in 30 minutes, and back in 45 minutes. Find the rate of the stream, and of the man in still water.

23. If 36 be added to a certain number, the result is five times the original number. Find the number.

16

1. A person owns 300 acres of land, which brings him in an average rental of 3.37\frac{1}{2}$ an acre. Part of it he rents at $4 an acre, and the remainder at $2.50 an acre. Find the number of acres he rents at the former rate.

2. Find the compound interest on $187 for 4$\frac{1}{3}$ years at 12% per annum, interest calculated every 4 months.

3. Which is the better investment, the 2$\frac{1}{2}$ per cents at 65$\frac{5}{8}$, or the 4$\frac{1}{2}$ per cents at 103$\frac{3}{8}$?

4. A man sells a house, losing 20% of the proceeds. What per cent. of the cost did he lose?

5. The hour, minute and second hands of a clock are on the same centre. When will the hour hand be mid-way between the other two for the first time after 12 o'clock?

6. Three men run a race around a circular course 100 yards in circumference. At the start A is 10 yards behind B, and B 15 yards behind C. If their rates are 10, 8, and 5 yards per second, respectively, when will they all be together for the first time?

7. A workman was employed for 60 days, on condition that for every day he worked he should receive 15 pence, and for every day he was absent he should forfeit 5 pence. At the end of the time he had 20 shillings to receive; how many days was he idle?

8. A, B, and C engage in trade. A puts into the business $3200 for 6 months, B puts in $1000 for 5 months, and C puts in $2500 for 7 months. C receives 10% of the total profits for managing the business, and also his just share of the remainder of the profits. If C receives altogether $428.40, find A's and B's share of the profits.

9. A speculator had a certain amount of stock in the 3 per cents at 96, and an equal amount in the 4 per cents at 101. If he sells out and invests the proceeds of each in the other kind of stock, he finds that his annual income will be diminished by $12.15. Find the amounts of money originally invested.

10. The simple interest on a certain sum for 3 years is $600; the true discount off the same sum for the same time is $500. Find the sum and the rate per cent.

11. A farmer sold 50 geese and turkeys for \$30.00. He received 50 cents each for the geese, and 75 cents each for the turkeys. Find the number of each sold.

12. A man sold an article so as to gain 25 %. If he had sold it for \$150 less, he would have lost 5 %. Find the cost of the article.

13. A grocer bought 2000 pounds of sugar at 6 cents a pound. It loses $\frac{1}{25}$ of its weight by drying out. Find his gain or loss by selling it at the rate of 15 ounces for 6 cents.

14. How many houses 24 feet long and 18 feet wide, can be built on a plot of ground 35 feet square?

15. A man owes \$15000, bearing interest at 5% per annum; he pays at the end of each year, for interest and part payment of the principal, \$2500. Find the amount of his debt at the end of the third year.

16. An importer paid 10 % for freight and duty on \$9875 worth of goods; he sold them at a clear gain of 20 % to the retailer, who sells them at a gain of 25 % to the consumer. Find the price paid by the consumer.

17. A commission merchant has goods consigned to him to sell, and, after deducting 2 % for both selling and investing the proceeds, in the usual way, he finds that his commission for selling the goods exceeds his commission for buying by \$12. Find the value of the goods remitted to him.

18. A certain number of men, and one half as many women, were employed on a work; each man received \$1.25, and each woman 75 cents; their total wages being \$45.50, how many of each were employed?

19. A man has 11 hours at his disposal; how far may he ride in a coach, at 8 miles an hour, so as to return in time, walking back at the rate of 3 miles an hour?

20. Find the cost of gravelling, at 15 cents a square yard, a path 1 yard wide running around the inside of a square garden, whose side is 114 yards.

21. A man's income decreased \$16.87½ by transferring \$4470 of 4½% stock at 102½ to the 3 per cents.; if he paid ⅛% brokerage on each transaction, find the market value of the 3 per cents.

1. After A has received $10 from B he has as much money as B and $6 more; and between them they have $40. What money had each at first?

2. If 8 men and 5 boys can reap $7\frac{1}{4}$ acres of grass in 3 days, and 12 men and 14 boys can reap 50 acres in 12 days, how long will it take 1 man and 4 boys to reap 12 acres?

3. A merchant bought 777 yards of cloth at $1.50 per yard. In selling the first half he uses a 35 inch yard measure, and in addition had the goods marked at an advance of 25 % on the cost. At what advance per cent. on the cost should he mark the remainder so as to gain only 25 % on the whole outlay, if he use a 37 inch yard measure in selling it?

4. A house worth $8937 is insured for $9000, so as to cover the value of the house, the premium of insurance, and charges amounting to 5 % of the premium. What was the rate charged for insuring?

5. Three men take an interest in a coal mine. B invests his capital for 4 months, and claims $\frac{1}{10}$ of the profits; C's capital is in 8 months, and D invests $6000 for 6 months, and claims $\frac{2}{5}$ of the profits. How much did B and C each put in?

6. A grocer mixes 30 pounds of tea with 40 pounds of an inferior quality, and finds that the mixture is worth $41\frac{3}{7}$ cents a pound; the difference in the prices being 15 cents a pound, find the price of each kind per pound.

7. A merchant pays $633.60 duty on an invoice of goods. If 20 % of the goods be exempt from duty, and 22 % is charged on the remainder, find the invoice price of the goods.

8. A man lends $1050, part of it at 4 %, and the remainder at 7 %. He finds that the income from both investments is $1.86 a year more than it would have been had he lent the entire sum at the uniform rate of $5\frac{1}{2}$%. Find the sums lent at the different rates.

9. Remitted $1200, including commission, to my agent to purchase goods on a commission of 5%. What amount did he invest in goods?

10. If \$5 be allowed as discount off a bill of \$125 due a certain time hence, what would be the discount allowed if the bill had twice as long to run?

11. A can do one half as much work as B; B can do one half as much as C; together they can complete a piece of work in 12 days. In what time could each alone complete it?

12. Divide 930 into 3 parts, such that twice the first equals 3 times the second, or 5 times the third.

13. A dealer received an invoice of glassware, 16% of which was broken. At what per cent. above cost must the remainder be sold to clear 8% on the invoice price?

14. A mixture of black and green tea, weighing 14 pounds, cost \$5.12½. If the proportions are interchanged the mixture would be worth \$5.02½. The black tea is worth 37½ cents a pound. Find the price per pound of the green tea.

15. Find the cost of gravelling at 12½ cents per square yard, a path 2 yards wide running around the inside of a square field containing 2 acres.

16. A grocer buys a certain number of oranges at 3 for 5 cents, and an equal number at 4 for 5 cents; he retails them at the rate of 7 for 10 cents. How much does he gain or lose per cent. by the transaction?

17. Does a merchant gain or lose, and how much per cent., by selling half of a barrel of sugar, giving only 15 ounces to the pound, and the other half giving 17 ounces to the pound?

18. Goods worth \$2661.75 are insured for \$2700, so as to cover the value of the goods, the premium of insurance and \$4.50 charges. What was the rate of premium charged?

19. A and B are partners. A's capital is to B's as 5 to 8; at the end of 4 months A withdraws ½ of his capital, and B ⅔ of his; at the end of the year their whole gain is \$1000. How much of the gain is each man entitled to?

20. A wine merchant mixed 8 gallons of water, with 14 gallons of wine at 75 cents a gallon, 22 gallons at 90 cents, and 16 gallons at \$1.10. How much is a gallon of the mixture worth?

1. A young man, receiving a legacy of $64788.50, invests $\frac{1}{3}$ of it in 5% stock at $93\frac{1}{2}$, and the remainder in 6% stock at 108, paying brokerage at $\frac{1}{8}$%. Find his net annual income, after he has paid an income tax of 18 mills on the dollar.

2. The duty on an importation of bottled Bay Rum, after deducting 2% for breakage, was $823.20, and the invoice price of the rum was 25 cents a bottle. How many dozen sound bottles did the importer receive, duty at 20%?

3. If 1700 metres are equal to 1 mile, and if a cubic foot of water weighs 1000 ounces avoirdupois, and a cubic metre of water 1000 kilogrammes, find the ratio of a kilogramme to a pound troy.

4. A man lends $813, part of it at 6% and the remainder at 8%. If his annual income from both investments amounts to 61.84\frac{1}{2}$, find the sums lent at the different rates.

5. Received $1100 from my agent, who had deducted his commission of 5%, as the proceeds of a sale of goods. What were the goods sold at?

6. A sold goods to B, which cost him $10000; B sold them to C, who sold them for $11910.16. If they each gained the same rate per cent., find the rate of gain.

7. A certain number of men and women earn $27; each man earns $1.25 and each woman earns 75 cents. The number of women is 3 more than twice the number of men. Find the number of each employed.

8. A grocer wishes to mix 47 pounds of tea worth 40 cents a pound, with a certain quantity worth 35 cents a pound, so that he can sell 16 pounds of the mixture for $6.07. How many pounds of 35 cent tea must he use?

9. There are two places 126 miles apart, from which two persons start at the same time, with the design of meeting; one travels at the rate of 3 miles in 2 hours, and the other at the rate of 5 miles in 4 hours. When will they meet?

10. A rectangular field, whose sides are as 4 to 5, contains 1 acre, 20 sq. rods. What will it cost to gravel a path $1\frac{1}{2}$ yards wide around the inside of it, at 10 cents a square yard?

11. A man owns ·5̇4̇ of a certain patent; he sells ·32̇9̇ of his share for $756. What is the value of the patent?

12. The freight and mail earnings of a railroad company amount to $476285·48, the passenger earnings to $378-567.29, the total expenses to $564798.20, and the company was able to declare a dividend of $4\frac{1}{3}$%. How much stock had the company issued?

13. A grocer has 280 lbs. of tea, of which he sell 80 lbs. at 30 cents a pound, and gains only 8% at this price; he now raises the price so as to gain 10% on the whole outlay. What does he now sell at a pound?

14. Find a merchant's gain, through dishonesty, if he sell goods, which cost him $84, by a false pound weight, $\frac{1}{4}$ ounce too light.

15. A man has property insured for $6500, for which he paid $162·50 premium; he wishes to increase the insurance to $9000. What extra premium will he be required to pay if the rate for the latter is $\frac{1}{2}$% greater than that for the former?

16. A and B engage in trade; A invests $6000, and at the end of 5 months withdraws a certain sum; B puts into the business $4000, and at the end of 7 months, $6000 more; at the end of the year A's gain is $5800, and B's is $7800. Find the amount A withdrew.

17. How much water is there in a mixture of 100 gallons of wine and water, worth $1.00 a gallon, if the wine cost $1.20 a gallon?

18. A liquor dealer receives an invoice of 150 dozen of bottled porter, worth $1.40 a dozen. If 10% of them are broken, what duty does he pay on the remainder the rate being 24%?

19. A man lends $375 at a certain rate of interest, and $412 at a rate 2% higher. If his interest for one year from both investments is $47.59, find the rates at which each was lent.

20. A boy engages to work for 30 days at 75 cents a day and his board, on the understanding that for every day he is idle he will receive no wages, but will have to pay $62\frac{1}{2}$ cents a day for his board. At the end of the time he receives altogether 12.87\frac{1}{2}$. How many days was he idle?

1. The difference between the true and bank discount on a sum of money for 8 months at 9% per annum is \$1.95. Find the sum.

2. A bankrupt, whose liabilities are \$3000, can pay only 65 cents on the dollar, after the assignee receives \$50 for his services; what are the total assets?

3. A farmer borrows \$1500 at 7%, and agrees to pay both principal and interest in 2 equal annual payments. What is the amount of each payment?

4. Find the volume of a sphere whose surface measures 120 square inches.

5. The expense of collecting taxes amounts to 2% of the total taxes. If a 19 mill rate gives the net taxes as \$36750, find the total amount of assessable property.

6. What must I pay for a draft purchased at 2% discount to cancel a debt amounting to \$347.50?

7. What is the difference between the simple and compound interest on \$400 for 2 years at 5% per annum?

8. A stock company with a capital of \$100000, 10% of which is paid up, can declare a dividend of 12% on the paid up capital. They make a further call of 4%, which is devoted to increasing their office accommodation; what rate of dividend can they now declare?

9. I send to my agent \$31920, and instruct him to deduct his commission at 3⅛%, and invest the remainder in broadcloth at \$2.95 per yard. When I receive the goods I have to pay in addition \$1347.90 for freight, \$479.40 for insurance, \$169.83 for storage, and an ad valorem duty of 2½% on the invoice price of the goods. Required the number of yards of cloth shipped to me, and what I gain or lose per cent. on the whole transaction, by selling the goods for \$50000.

10. If a merchant, in buying goods, uses a pound weight ¼ oz. too heavy, and in selling them one ¼ ounce too light, and gains \$15 by his dishonesty, find what he paid for the goods.

11. For what sum was a store insured, if the rate of insurance was 7½ mills on the dollar, and the premium paid was \$25.50?

12. Three men form a partnership, each contributing the same amount of capital; one of them is appointed manager, and as such, is to receive $12\frac{1}{2}\%$ of the total profits. When the business is wound up, the manager receives altogether $6000. What were the total profits?

13. A merchant, in New York gave $1950 for a bill of exchange of £400 to remit to Liverpool. What was the rate in favor of England?

14. A sum of money together with the interest which accumulates during the time will pay the wages of 17 men for 4 days, or 12 men for 6 days. How many men in either case does the interest pay?

15. A grocer by selling 12 pounds of tea for a certain price gained 12%; afterwards he decided to give a less number of pounds for the same price, and as a result gained $49\frac{1}{3}\%$. What was the number of pounds he decided to give?

16. A person owed $2500 due in 8 months; he paid $500 in 3 months, and $800 in 6 months. When was the balance due?

17. If 7 bushels of wheat are worth 10 bushels of rye, and 5 bushels of rye are worth 14 bushels of oats, and 1 bushel of oats is worth 50 cents, how many bushels of wheat will $30 buy?

18. At a 14 mill rate a man who has $400 of his salary exempt, pays $7 taxes. What was his total salary?

19. A man employed a number of men and 6 boys; he paid the men $1.25 a day, and the boys 50 cents a day. The average price paid was $1.13 a day. How many men were employed?

20. A huckster bought a certain number of apples at the rate of 5 for 2 cents, and sold one-half of them at the rate of 3 for 1 cent, and the other half at the rate of 2 for 1 cent, gaining altogether 4 cents. How many apples did he buy?

21. A person holds a certain amount of 6% accumulating stock, and after the first dividend has been added to the original stock, he holds $1317.58 stock. If he secured the original stock when it was selling at 102, find the amount of money he invested.

1. A man borrows \$2500, and agrees to pay both principal and interest in 3 equal annual payments. What is the amount of each payment, interest being at 5%?

2. A commission merchant bought a lot, 40 feet frontage, with the commission he realized from selling wheat at $5\frac{1}{4}$%; the net proceeds of the wheat, after deducting the commission, being \$3790, find the price paid per foot for the lot.

3. A merchant buys a quantity of coal oil for \$1500; by the use of a false measure, and by marking his goods at a profit of 25%, he gains \$750. Find the size of his gallon measure.

4. A house is insured for a certain sum, $\frac{3}{4}$ of that sum at $1\frac{1}{2}$% premium, and the remainder at $1\frac{1}{8}$%. What is the average premium paid?

5. Divide \$500 among three persons in such a manner that the share of the second may be $\frac{1}{2}$ greater than that of the first, and the share of the third $\frac{1}{2}$ greater than that of the second.

6. If 50 barrels of flour in Chicago are worth 125 yards of cloth in New York, and 80 yards of the same cloth in New York are worth 13 bales of cotton in Charleston, and 13 bales of cotton in Charleston are worth $3\frac{1}{2}$ hogsheads of sugar in New Orleans; how many hogsheads of sugar in New Orleans are worth 1500 barrels of flour in Chicago?

7. What rate of trade discount deducted 4 times in succession is equivalent to $59\frac{1}{25}$% off?

8. A sum of money, together with the interest which accumulates during the time, will pay the wages of 14 men for 9 days, or 5 men for 27 days. How many men's wages will it pay for 3 days?

9. A bankrupt was able to pay 40 cents on the dollar, had not a debt of \$500 proved worthless: now he is able to pay only 24 cents on the dollar. Find the total amount of his liabilities.

10. Three men engage in trade with a joint capital of \$13200. The first puts in \$6 as often as the second puts in \$10, and as often as the third puts in \$14. Their annual gain is equal to the third man's stock. What is each partner's gain?

11. Find the difference between the true and bank discount off $558, due 3 years hence, at 8% per annum.

12. The assessable property of a corporation is $320000. What rate of taxation is levied, if the collector receives $172.80 from a 3% commission?

13. I want an alloy weighing 1092 ounces, consisting of 13 parts by weight of nickel, 12 parts of lead, and 14 parts of tin. The only nickel I can obtain is contained in an alloy consisting of 14 parts by weight of nickel, 9 parts of lead, and 12 parts of tin. How much of the latter alloy, and of lead and tin must I use?

14. Bought goods as follows: June 1, 1890, on 2 months, $350; July 15, on 3 months, $400; Aug. 10, on 4 months, $450: Sept. 12, on 6 months, $600. Find the equated time of payment.

15. Find the area of a square inscribed in a circle, whose circumference is 20 feet.

16. A certain railway company is able to declare a dividend of 6$\frac{1}{16}$% on their entire stock, but there being $150000 of preference stock, which is guaranteed 8%, the ordinary stock holders receive only 5$\frac{3}{4}$%. Find the total stock of the company.

17. A commission merchant sold a shipment of flour for $5330, and invested the proceeds, less his commission, in tea. The selling commission, exceeded the buying commission by $6.50. The rates being the same in each case, find the rate charged.

18. A merchant buys $7400 worth of goods. By the use of a false measure, and by marking his goods at a profit of 12$\frac{1}{2}$%, he gains $700. Find the length of the yard measure used.

19. Goods to the amount of $20000 were insured in 3 separate companies, for $3000, $4500 and $6000, respectively. If the goods be damaged by fire, to the extent of $9000, how much of the damage should each company sustain?

20. A grocer has a mixture of tea, weighing 80 pounds, part of it is worth 42 cents, and the remainder 37 cents a pound. If the mixture is worth 38$\frac{1}{4}$ cents a pound, find the number of pounds of each kind in the mixture.

1. A draft on Dublin for £360 costs $1736. What was the course of exchange?

2. A merchant sold a lot of goods at a gain of 10%; but had they cost $50 more, and selling them at the same price as formerly, he would have lost $8\frac{1}{3}$%. Find the cost price.

3. A person makes 7% clear on his money, after paying an income tax of 2 cents on the dollar, by investing in 5% stock. Find the market value of the stock, if a broker charged $\frac{1}{8}$% for transacting the business.

4. An agent sold a consignment of wheat for $4120, and invested this sum, less his commission, in corn. His total commission on both transactions amounted to $240. What rate did he charge, the rate in each case being the same?

5. A merchant buys $1050 worth of sugar. He uses a pound weight $\frac{1}{4}$ ounce too light, and marks his goods at a profit of 15%. Find his gain.

6. A man pays $342.50 for insuring goods. What is the total amount of insurance, if $\frac{1}{2}$ of the amount is at a premium of 2%, $\frac{1}{3}$ at $2\frac{1}{2}$%, and the remainder at $2\frac{1}{4}$%?

7. Divide $3648 among three persons, so that the share of the first to that of the second shall be as 7 to 9, and of the first to the third as 3 to 4.

8. A boy hired with a mechanic for 20 weeks, on condition that he should receive $20 and a coat; at the end of 12 weeks the boy quit work, when it was found that he was entitled to $9 and the coat. What was the value of the coat?

9. A baker's outlay for material is 60% of the receipts, and his other expenses to 20% of the receipts. The price of flour falls 30%, and his expenses increase 10%. What should he now charge for a loaf, which formerly sold at $7\frac{1}{2}$ cents, in order that he may make the same per cent. profit?

10. The simple interest on a sum of money for 3 years is $82.50, and for $5\frac{1}{2}$ years at 2% higher rate is $110. Find the sum and rates.

11. J. Brown owes $400 due in 6 months, $320 in 9 months, and $280 in 12 months. Required the average term of credit.

12. A merchant becomes bankrupt and owes \$5000. His assets amount to \$3347.50. What per cent. of his indebtedness can he pay, allowing the assignee 3%, on the amount distributed, for his services?

13. A township has assessable property amounting to \$450000 and on a 3 mill rate they raise only \$1296, after paying the collector's charges. What per cent. of the total taxes did he obtain for collecting?

14. What is the total duty charged on 2000 pounds of sugar worth $4\frac{1}{2}$ cents a pound, supposing it subject to an ad valorem duty of 20%, and also a specific duty of 14 cents per 100 pounds?

15. A jeweller is required to supply 110 ounces of alloy, consisting of 3 parts of gold to one of silver. The only gold he has in stock is an alloy consisting of 22 parts of gold to 2 of silver. How much alloy must he use, and how much silver must he add to make up the order?

16. A ladder 12 metres in length, when placed upright against a wall exactly reaches the top of it. How far must it be pulled out at the bottom so that it may just reach within 1 metre of the top?

17. A person sells out \$5210 stock in the 6 per cents. at $104\frac{1}{4}$, and invests in the $3\frac{1}{2}$ per cents., losing thereby \$21.05 in his income. Find the market price of the $3\frac{1}{2}$ per cents., brokerage $\frac{1}{8}$% on each transaction.

18. A merchant shipped 2000 bushels of wheat to his agent, which brought 85 cents a bushel; after deducting his commission of 3% for selling, he was allowed $\frac{1}{2}$% on the remainder for promptness; after then paying freight and storage to the amount of \$130.25, he (the agent) remitted the net proceeds by a draft purchased at $1\frac{1}{4}$% premium. Find the amount of the draft.

19. In buying goods a merchant uses a pound weight one-quarter ounce too heavy, and in selling them a pound weight one-quarter ounce too light. Find his gain per cent. by his dishonesty.

20. What is the value of property, if it be insured for $\frac{2}{3}$ of its value, at a premium of 2%, and 5% of premium being charged for expenses, when the total cost of insuring is \$75.60?

1. A and B engage in trade, A puts in \$5000, and at the end of 4 months takes out a certain sum. B puts in \$2500, and at the end of 5 months puts in \$3000 more. At the end of the year A's gain is \$1066$\frac{2}{3}$, and B's \$1333$\frac{1}{3}$. What sum did A withdraw at the end of 4 months?

2. A farmer sold 16 bushels of corn and 20 bushels rye for \$30; 24 bushels of corn and 10 bushels of rye for \$27. How much per bushel did he receive for each?

3. A garrison of 700 men had provisions to last for 40 days, but 12 days afterward 300 men were killed. How long will the provisions last the remainder of the garrison?

4. The simple interest on a sum of money loaned at a certain rate for 2$\frac{1}{2}$ years is \$32.50, and the interest on double the sum at 1$\frac{1}{2}$% higher rate for 4 years is \$143. Find the first sum and rate.

5. A firm become insolvent, and owe \$8100. Their assets amount to \$4981.50. What per cent. of their indebtedness can they pay, allowing the assignees 2$\frac{1}{2}$% on the amount distributed, for their services?

6. A man owes \$3200 due in 7 months; he pays \$1200 in 3 months, \$600 in 5 months, and \$500 in 8 months. How long should the balance run before it becomes due?

7. A corporation had assessable property valued at \$500000. What rate of taxation must they levy, so as to secure in taxes \$9775, after paying the collector's fees at 2$\frac{1}{4}$%?

8. A commission merchant had shipped to him 1000 barrels of flour, and 4000 bushels of wheat. He paid 12$\frac{1}{2}$ cents a barrel for the storage of the flour, 3 cents a bushel for the wheat, \$67.18 freight and duty. If he sold the flour at \$8 a barrel, on a commission of 1$\frac{1}{2}$%, and the wheat at 90 cents a bushel, on a commission of 2 cents a bushel, what sum did he remit to his employer?

9. A merchant in Chicago, owes \$4500 in New York. Which of the following methods for paying the debt is the more advantageous: To send to New York a draft on New York, purchased in Chicago at $\frac{1}{2}$% premium, or to send to New York a draft on Chicago, which sells in New York at $\frac{1}{2}$% discount?

10. A man increases his annual income \$72 by transferring \$21000 stock from the 3%'s, at 78, to the $4\frac{1}{2}$%'s. Find the price paid for the latter stock.

11. A bankrupt can pay 60 cents on the dollar; if his assets were \$750 more, he could pay $\frac{27}{40}$ of his liabilities. Find his liabilities and his assets.

12. A merchant buys 2590 yards of cloth at \$1 a yard. In selling the first half he used a 35-inch yard measure, and a 37-inch one for the remainder. At what price per yard should he mark the goods so as to gain 20% on his outlay?

13. An insurance company took a risk at $2\frac{1}{4}$%, and reinsured $\frac{3}{5}$ of the risk at 2%; the premium received exceeded the premium paid by \$42. Find the amount of the risk.

14. A, B, and C enter into partnership; A puts in \$347.80, B \$359.26, and C \$543.59. If the profits of the business amount to \$500.26, how should it be divided?

15. A, B, and C start from the same point at the same time, to travel around a lake 84 miles in circumference. A travels 7 miles and B 21 miles a day in the same direction, and C 14 miles a day in the opposite direction. In how many days will they all be together?

16. Gunpowder is composed of nitre, charcoal, and sulphur, in the proportion of 33, 7, and 5. How many pounds of each in 60 pounds of gunpowder?

17. The simple interest on a sum of money for a certain time at 6% per annum is \$62.10, and the interest on 3 times that sum for $2\frac{1}{4}$ years at 4% per annum is also \$62.10. Find the first sum and the time.

18. A merchant in Toronto has \$4800 due him in Halifax. How much less will he realize by having a draft for that sum on Halifax and selling at $\frac{1}{2}$% discount, than by having a draft on Toronto remitted to him, purchased in Halifax for this sum at $\frac{1}{2}$% premium?

19. The expense of constructing a bridge was \$873, which was defrayed by a tax on the property of the town. The rate of taxation was $2\frac{1}{4}$ mills on the dollar, and the commission for collecting, 3%. What was the assessed value of the town?

1. B owes a certain sum, $\frac{1}{2}$ payable in three months, $\frac{1}{4}$ in 5 months, $\frac{1}{5}$ in 6 months, and the balance in 10 months. Required the average term of credit.

2. A farmer has sheep worth \$4, \$5, \$6, \$8 per head. What number may he sell of each to realize an average price of \$5.50 per head?

3. A merchant had 400 barrels of flour insured for 70% of its cost, at $3\frac{1}{2}$%; the premium was \$58.80. At what price per barrel must he sell the flour to gain 20%?

4. Find the present worth of an annuity of \$300 having 4 years to run, money being worth 6% per annum.

5. Find the difference in area between a square field whose side is 10 chains, and a circular field whose circumference is equal to the perimeter of the square field.

6. If 3% bonds are at $80\frac{3}{8}$, what must be the price of 4% stock, so that the income may be unchanged, if the money be transferred from the first to the second stock, brokerage $\frac{1}{8}$ on each transaction?

7. A commission merchant's terms are 6% on sales, with guaranteed payment of sales to his employer, or $2\frac{1}{2}$ without any guarantee; his employer accepts the latter: but \$90 of the sales proving worthless, how much does his employer lose by not selecting the former method, the total sales being \$2300.

8. A and B run a race. A starts out at the rate of 150 yards a minute, and every successive minute he diminishes his pace by 2 yards. B increases his pace by the same and overtakes A in 10 minutes. What was B's rate at starting?

9. J. Brown and L. Stuart enter into partnership, and agree that Brown is to receive \$400 of the profits for managing the business, and the balance of the profits is to be divided in proportion to their investments; Brown puts into the business \$1300 and Stuart \$1500. How should a total profit of \$1100 be divided?

10. If 375 lbs. of sugar are bought at 6 cents a pound, and 10% of it be wasted, at what price, per lb., must the remainder be sold to make a gain of 11% on the whole transaction?

11. A and B take a contract together, on which, for every 6 hours A works, B works $7\frac{1}{2}$ hours; but A can do as much in 3 days as B can do in 4. If A receives \$24 for his work, what should B receive?

12. A man in New York purchased a draft for \$4680 on New Orleans, drawn at 30 days, paying \$4627.35. What was the rate of discount at which it was purchased?

13. A speculator sold a house at 50% profit, but the buyer becomes bankrupt and pays only 60 cents on the dollar. What per cent. does the speculator gain or lose by the transaction?

14. In what time will a sum of money quadruple itself at simple interest, at 8% per annum?

15. A duty on coffee at 15%, in bags of 180 pounds gross, and invoiced at $12\frac{1}{2}$ cents a pound, was \$961.87½, tare having been allowed at 5%. How many bags were there?

16. In buying goods a merchant uses a pound weight $\frac{1}{4}$ ounce too heavy, but in selling them one $\frac{1}{4}$ ounce too light. If he gains \$13 by his dishonesty in selling the goods, what does he gain by it in buying?

17. A factory is insured at 2%. If the premium, with \$1.50 for the policy, is \$362, and the insurance is on $\frac{2}{3}$ of the value of the property, what is the value of the property?

18. A party of men were employed to do a piece of work, which it was necessary to finish in 25 days. When, however, the men had worked 20 days, they had done only $\frac{2}{3}$ of the work, and 3 more men had then to be employed to complete the work in due time. How many men were employed at first?

19. A cow is tethered in a field by a rope 40 feet in length. If the tether is tied to a fence 30 feet from the corner of the field around which she can go, find the area of ground she can procure grass from?

20. A merchant buys wine at \$1.20 a pint, and dilutes it with water, so that he may make a profit of 20% by selling the mixture at 90 cents a pint. How much water does he add to every gallon of the mixture sold?

1. A man rows 3 miles down a stream in 40 minutes; without the aid of the stream it would take him an hour. How long would it take him to return the 3 miles against the stream?

2. A bankrupt's assets are \$4800, out of which he pays 60 cents on the dollar on $\frac{1}{3}$ of his debts, and 50 per cent. on the remainder. What is the amount of his liabilities?

3. A merchant lays out £1000 in buying cloth in England at 3 shillings a yard. He takes the cloth to France at an expense of 3 pence a yard for carriage, &c., and in addition pays a duty of 42 centimes a metre. He sells half the cloth at 8 francs a metre, and the remainder at 6 francs a metre. What profit does he make? (Express answer in dollars and cents, given £1 = \$4.86$\frac{2}{3}$, 1 franc = 19 cents, and 1 metre = 39$\frac{3}{8}$ inches.)

4. A man invested \$15510 in the 4 per cents; when the market price rose 3$\frac{1}{8}$ he sold out, and with the proceeds bought 3$\frac{1}{4}$% stock at 77, thereby increasing his income by \$16.10$\frac{5}{8}$. Find the price he paid for the 4% stock.

5. A commission merchant's terms are a certain rate of commission with guaranteed payment of sales, or 3% without any guarantee. His employer accepts the former method (which is better than the latter by \$19.50, owing to a bad debt of \$96). If the total amount of the sales was \$3400, what was the percentage demanded with guaranteed payment?

6. Reduce £350 5s. 7$\frac{3}{4}$d. to dollars and cents, a dollar being worth 4s. 2d.

7. A commenced business with a capital of \$12000. Four months afterwards B entered into partnership with him and put in 37$\frac{1}{2}$ acres of land. At the close of the year their profits were \$1600, of which B was entitled to \$1000. What was the value of the land per acre?

8. What per cent. advance on cost must a merchant mark his goods, so that, after allowing 5% of the sales for bad debts, an average credit of 6 months, and 7% of the prime cost of the goods for expenses, he may make a clear gain of 12$\frac{1}{2}$% on the cost, money being worth 6%?

9. A boy engages with a farmer 9 months for \$46 and a suit of clothes. But leaving at the end of 7 months he is entitled to only \$32 and the suit of clothes. Find the price of the clothes.

10. A and B can do a piece of work in 4 days; B and C in 6 days; C alone in 10 days. How long would it take A and C together to do it?

11. A person travelled from London to Loch Lomond (480 miles) by sea, rail, and coach. The distance by coach was $\frac{1}{3}$ that by rail, and the distance by rail .15 that by sea; the cost of the travel by coach was $\frac{2}{3}$ that by rail, and the cost by rail $\frac{3}{10}$ that by sea. Coach fare being 8 cents a mile, find the cost of the entire journey.

12. Find the longitude of Callao in Peru, if it is 12 minutes to 7 a.m. at Callao when it is noon at Greenwich.

13. A reduction of 20% in the price of apples would enable a purchaser to obtain 120 more for a sovereign. What is the price before reduction?

14. Sight exchange on Montreal for \$5000 cost \$5075. What was the course of exchange?

15. A man increases his capital for 4 years at the rate of $18\frac{3}{4}$% a year. At the end of the time his capital is \$1303.21. What was it at the beginning of the time?

16. A, B, and C engage in trade; A puts in \$3000, B \$4000, and C \$6000. If the total profits are \$2600, how should they be divided?

17. A ship sails at the rate of 12 miles an hour. When she is a certain distance from the shore she springs a leak, which admits 20 tons of water per hour. 70 tons would suffice to sink her, but the crew pumps out $2\frac{3}{4}$ tons of water every 10 minutes. Find the distance that she was from shore when she sprung the leak, if she is in a sinking condition on reaching the shore.

18. If a merchant in buying goods uses a pound weight $\frac{1}{4}$ ounce too heavy, but in selling them uses one which is $\frac{1}{4}$ ounce too light, and gains \$15 by his dishonesty, find what he paid for the goods.

19. What is the duty, at 12 cents a pound, and 10% ad valorem, on 450 bags of wool, each containing 110 pounds, valued at 21 cents a pound?

1. A woman sold 90 apples, some at the rate of 6 for 5 cents, and the remainder at the rate of 16 for 13 cents. She found that those sold at the latter price brought her 4 cents more than the others did. How many were sold at each rate?

2. How many litres will a cistern hold that measures on the inside 5 feet in length, 4 feet in width, and 4 feet in depth?

3. What must be the least number of soldiers in a regiment to admit of it being drawn up 2, 3, 5, 6, 8, 12 deep, and also of its being formed into a solid square?

4. When, first after 4 o'clock, will the two hands of a watch be equidistant from the figure three?

5. Two circular plates of solid gold, each one inch thick, the diameters of which are 9 and 12 inches respectively, are melted into a single circular plate one-half inch thick. Find its diameter.

6. The product of two numbers is 11154, their g.c.m. is 13. Find their l.c.m.

7. A speculator receives \$103.50 income from a sum of money invested in the 3%'s. He sells out this stock at 75, and with the proceeds buys other stock at 115, by which he increases his annual income $30\frac{10}{23}\%$. What per cent. dividend did the latter stock pay?

8. A merchant ships \$2700 worth of goods to his agent to sell. His agent's terms are $7\frac{1}{2}\%$ commission and guaranteed payment of sales, or 4% without the guarantee. The merchant accepts the latter terms, but part of the sales being worthless, he loses \$21.50 by not accepting the other plan. How much did he lose on bad debts?

9. A merchant sells goods at a profit of 60%, but his purchaser fails and pays only 60 cents on the dollar. How much per cent. does the merchant gain or lose by the transaction?

10. Three men rent a pasture for \$165. The first puts in 4 horses for 8 weeks; the second 5 horses for 11 weeks, and the third 9 horses for 5 weeks. How much ought each to pay?

11. A merchant buys a quantity of cloth for $1460, and marks it so as to gain 25%, but by using a yard measure which is too long, he gains only $340. Find the length of the yard measure.

12. A man's gross income is made up from $1100 salary, and $400 rent for a farm. His gross income is reduced to $1187.50, on account of the following expenses:—$1000 insurance at 2% on the house, 15 mills on the dollar taxation on salary and on $\frac{2}{5}$ of the value of the farm, interest on a mortgage for $\frac{2}{3}$ the value of the farm at 6%. Find the value of the farm.

13. A certain metal weighs 480 pounds per cubic foot, and is worth $50.40 per ton. What will be the cost of a quantity of metal sufficient to make a mile of piping of 9 inch bore and $\frac{3}{8}$ of an inch thick?

14. Explain clearly the difference between true and bank discount.

15. Find the longitude of the Falkland Islands, if it is 5 o'clock a.m. at them, when it is 1 o'clock p.m. at Ras el Had, the longitude of which is 60° east.

16. The l.c.m. of two numbers is 924, the g.c.m. 12; one number being 84, find the other.

17. Find, to the nearest cent, the value of £460 10s $8\frac{3}{4}$d., exchange being at par.

18. A real estate broker sold a house on a commission of $4\frac{1}{4}$%, and gave the owner $4404.50 for his share of the sale. What did the house sell for?

19. A merchant bought 20 pieces of cloth, each piece containing 25 yards, at $$4.37\frac{1}{2}$ per yard, on a credit of 9 months; he immediately sold it at $$4.62\frac{1}{2}$, on a credit of 4 months. What was his actual cash gain, money being worth 6%?

20. What principal will yield $$38.90\frac{1}{4}$ interest in $4\frac{1}{2}$ years at $3\frac{1}{5}$% per annum?

21. The assets of a bankrupt, which amount to $7290, are to be divided between two creditors, A and B, whose interests in the business are as 4 to 5. A acts as assignee, and for his work receives $3\frac{1}{2}$% of the assets. The remainder is to be divided between them in the above ratio. What share of the assets does each man receive?

1. A pound (Troy) of standard gold (22 carats fine) is coined into 45 guineas. If the value of the alloy be $\frac{11}{239}$ that of an equal weight of pure gold, find the value of the alloy per pound avoirdupois.

2. Two men starting from the same point at the same time and travelling in opposite directions are 39 miles apart in 4 hours; but if they had gone in the same direction they would have been $5\frac{1}{4}$ miles apart in 7 hours. Find their rates per hour.

3. What is the area of a right-angled triangle that has a base 32 inches, and the sum of the hypotenuse and perpendicular 50 inches?

4. A merchant every year gains 40% on his capital, of which he spends annually $1000 in expenses connected with the business. At the end of 5 years he is worth 5 times as much as he was on commencing business. What was his original capital?

5. A watch was 5 minutes fast on Tuesday noon, and on the following Saturday at midnight it was 3 minutes 10 seconds slow. What is the true time on the following Wednesday morning, when the watch indicates 15 minutes past 9?

6. If the interest on one dollar for one year at one per cent. per annum be the unit, what number will represent what a person owes for the use of $325 for $6\frac{1}{2}$ years at 6 per cent. per annum?

7. The owner of $3\frac{1}{2}$% stock sells out at 82 and realizes $10947, $\frac{2}{3}$ of which he invests in 3% stock at 75, and the remainder in 4% stock at 89. Find the alteration in his income.

8. A man is owed $2500, of which his lawyer collects 75% and charges 1% of the receipts for collecting. On sueing for the remainder he gets 40% of it, of which his lawyer's fees are 10%. What per cent. of the debt does the creditor obtain?

9. A cask contains a mixture of wine and water, in the proportion of 5 to 1. How much of the mixture must be drawn off and water substituted for it, so that the mixture in the cask may contain wine and water in the proportion of 3 to 2?

10 A and B engage in business. A puts in $1200, and B $600 at first. Three months afterwards C became a member of the firm and put in $750. After the business has been carried on one month longer A withdrew $400, and at the end of another month B put in $300 more; at the end of 13 months their gain was $1100. How should it be divided?

11. A drover shipped a car load of cattle to Toronto, and offered them for sale at an advance of 25% on the cost. The market being dull, he sold them for 14% less than his asking price, and gained $170 on the load. Find the selling price of the cattle.

12. Find how much a person will save in a year if, instead of renting a house at $30 a month, he borrows $3000 at 6% and buys one, which he insures for its full value at $1\frac{1}{4}$% annually, and which is assessed at its full value (the rate of taxation being 19 mills on the dollar). He also pays $1.25 a month water rates, and in selling the property at the end of the year receives 2% less for it than he paid.

13. If 2 pounds of tea are worth 3 pounds of coffee, and 14 pounds of coffee worth 8 pounds cocoa, and 2 pounds cocoa worth 24 pounds sugar, and 16 pounds sugar worth 11 pounds raisins, how many pounds of raisins would be worth 7 pounds of tea?

14. A can mow 6 acres of grass in 5 days, B 9 acres in 11 days, and C 12 acres in 17 days. In how many days will they jointly mow 15 acres, $75\frac{1}{3}$ sq. rods?

15. The island of Hainan is 110° east longitude. Find the longitude of St. Helena, if it is 4 p.m. at Hainan when it is 8.20 a.m. at St. Helena.

16. Prove that the difference between the interest and the true discount on a sum of money is equal to the interest on the discount.

17. What is the cost in Montreal of a bill on St. Petersburg for 2500 roubles at $1\frac{1}{2}$% discount, the par of exchange being $3.77 for 5 roubles?

18. The interest on a certain sum of money for 3 years is $11·664, and the real discount on the same sum for the same time and rate is $33.60. Simple interest being allowed in each transaction, find the rate per cent. charged.

1. A mechanic receives $2 a day for his services, and pays $4 a week for his board. At the expiration of 10 weeks he has saved $72. How many days was he idle?

2. A flour merchant bought 120 barrels of flour for $660, paying $5.75 a barrel for first quality and $5 for second quality. How many barrels were first quality?

3. A town whose property was assessed at $1500000, built a school house which cost $5092.50, the collector's commission being 3%. What was the rate of taxation?

4. Divide $10170 into 3 sums, so that the amount of the first, by simple interest, in $1\frac{1}{4}$ years at 8% per annum, may be equal to that of the second in $3\frac{1}{3}$ years at 5%, and to that of the third in $2\frac{1}{8}$ years at 6%.

5. A man invested $600 more than $\frac{2}{5}$ of his money in a house, and $1200 more than $\frac{1}{6}$ of the remainder in a lot; he has $1800 left. How much had he at first?

6. Find how much a man will plough in 10 hours in a field 40 rods in length, if he walks at the rate of 3 miles an hour, cuts 8 inches wide, and takes 2 minutes turning each time.

7. A grocer bought a quantity of tea at 40 cents a pound, and fixed a price on it to gain $23\frac{1}{2}\frac{7}{1}$%, but in selling it he used a pound weight $\frac{3}{4}$ ounce too light, thus gaining $32.50 more than he would have gained if the weight had been true. How much did he buy?

8. In running a 10-mile race, on a quarter mile track, A overlaps B for the first time at the end of the 34th round. By what distance will A win at the same rate of running?

9. Find the difference between the true and standard times of the City of Hamilton, its longitude being 79° 54′ west.

NOTE.—The standard time of any place in Canada is the same as the time indicated by the nearest of the following meridian lines: 0°, 15°, 30°, 45°, etc.

10. A merchant ships to his agent 5000 bushels of wheat, which he (the agent) sells for 70 cents a bushel on a commission of 3%: after deducting this from the proceeds, and paying $380 for freight and duty, he remits the net proceeds by a draft purchased at $\frac{1}{2}$% premium. Find the amount of the draft.

11. Two pipes can fill a cistern in 30 and 35 minutes, respectively. When the cistern is empty both pipes are opened together, but at the end of 12 minutes the second is turned off. How long does it take to fill the cistern?

12. A person invested a certain amount of money in stock, paying 3% half-yearly dividends, at 103, and after receiving his first dividend invested it in the same stock at 102½. His next half year's dividend amounted to $316.50. What amount of money did he invest at first?

13. A vessel containing 23·846 quarts of water is emptied by a pitcher which holds .04679 gallons. How many times can the pitcher be filled, and what fraction of a pint will it contain when the last quantity of water is poured into it?

14. Three men rent a pasture for $132.18¾. A puts in 3 horses for 9 weeks; B puts in 7 cows for 6 weeks; and C puts in 30 sheep for 4 weeks. If it costs as much to pasture 2 horses as 3 cows, and 4 cows as 7 sheep, how much ought each to pay?

15. A machinist sold 56 binders at $130 each. On one-half of them he gained 30%, and on the remainder he lost 30%. How much does he gain or lose on the whole transaction?

16. For what sum must a vessel, valued at $28000, be insured, so that in case of loss, the owners may recover both the value of the vessel and the premium of 16%?

17. Two trains, respectively 99 yards and 132 yards long, and moving on parallel tracks, pass each other in 6¾ seconds when running in opposite directions. When moving in the same direction the one passes the other in 17¼ seconds. Find their rates per hour.

18. A certain number of men and boys can do a certain work in 8 days, but if one-third of the number of men are absent, the same work would require 10 days. Compare the amount of work done by the men to that done by the boys.

19. A man owes $700, due in 6 months; he pays $300 down. What extension of time ought to be allowed for the payment of the latter, if money be worth 10% per annum?

1. Find the difference between the true and standard times of Goderich, its longitude being 81° 40′ west.

2. A man in Detroit wishes to draw on New Orleans for a bank stock dividend, and exchange direct on New Orleans is at $1\frac{1}{4}$% discount. How much will he gain or lose by drawing on his agent in New York at $1\frac{1}{2}$% premium, allowing his agent to draw on New Orleans at 1% discount, brokerage $\frac{1}{2}$%?

3. A merchant paid $268.80 duty on 24 casks of wine worth $1.75 a gallon, a leakage of $11\frac{1}{9}$% being allowed at the custom's house. What was the invoiced number of gallons in each cask, the rate of duty being 20%?

4. In what time will a sum of money, at simple interest double itself at $6\frac{3}{4}$% per annum?

5. What is the hour, when the time past noon is equal to $\frac{2}{3}$ of the time to midnight?

6. I buy a ground rent of $25 per annum for 333.33\frac{1}{3}$. What per cent. do I realize on my investment?

7. Took a risk for $20000 at 2%; reinsured $\frac{1}{2}$ of it at $2\frac{1}{4}$%, and $\frac{1}{4}$ at $2\frac{1}{2}$%. Find my net loss in case of fire.

8. In a poor house, where there is always the same number of inmates to be fed, the contract price of meat rises 25%, and the daily allowance to each person, at the same time, is reduced from 9 ounces to 7 ounces. If the yearly charges for meat are now $714, what were they before the rise in price?

9. A cistern can be filled by two taps running separately in 20 and 30 minutes respectively, and emptied by 2 more in 24 and 18 minutes respectively. If the cistern be full and all 4 taps opened, in what time will the cistern be emptied?

10. If 69 German thalers, of which 9 parts in 10 are fine silver, weigh 41 ounces, what is the value of a thaler in English money, when standard silver, of which 37 parts in 40 are pure, is worth 5s. $1\frac{1}{2}$d an ounce?

11. The stocks of 3 partners, A, B, and C, are $4000, $6500, and $3500, respectively; their gains are $960, $2340, and $1400, respectively. If B's stock is in trade 3 months longer than A's, what time was each stock in trade?

12. If it is 2 p.m. on July 7th at a place, the longitude of which is 178° east, what is the time at a place situated in longitude 178° west?

13. If a cask contains 4 parts of wine and 3 parts of water, what fraction of the mixture must be drawn off and water substituted, for the mixture to become 3 parts of wine and 4 parts of water?

14. A speculator holding \$11771.25, 6% stock, finds that by selling out and investing the proceeds in 7% stock at $107\frac{1}{2}$, he can increase his annual income by \$6.57; before he can effect the transfer each stock has increased in value 2. By how much would his annual income be now increased?

15. A merchant ships 2500 barrels of apples to his agent, with instructions to sell the apples, deduct his commission of 4% for selling, pay the freight which amounted to \$300, and remit the proceeds by a draft. If the draft was purchased at $1\frac{1}{4}$% premium, and amounted to \$8000, what did the apples bring per barrel?

16. I have to be at a certain place at a certain time and I find that if I walk at the rate of $3\frac{1}{2}$ miles an hour I shall be 10 minutes late, if at the rate of $4\frac{1}{4}$ miles an hour I shall be 14 minutes too early. How far have I to go?

17. Four men form a partnership, the second puts in twice as much capital as the first, the third as much as the first and second, and the fourth as much as the other three. How should a profit of \$4800 be divided among them?

18. If I buy a piece of land, which increases each year at the rate of 50%, on the value of the previous year, for 4 years, and then is worth \$4500, how much did it cost?

19. A banker discounts a 70 day note at 8% per annum. What rate of interest is he making on his money?

20. If silver is worth \$1.10 an ounce, and gold is \$17 an ounce, what will be the weight of a \$10 coin which contains 20% (by weight) silver, and the remainder gold?

21. Prove that the product of the l.c.m. and g.c.m of two numbers is equal to the product of the numbers.

1. From a cask containing wine one-third is drawn off, and the cask filled with water, one-third of the mixture is then drawn off, and the cask is again filled with water. After this has been done 5 times altogether, what portion of the original quantity of wine will be left in the cask?

2. The longitude of Tokio is 140° east, that of New York 74° west. What is the time at Tokio, when it is 5.10 p.m. on October 7th in New York?

3. A does $\frac{3}{4}$ of a piece of work in 15 days, and then B joins him. They work together for $1\frac{1}{2}$ days, when B leaves and A finishes the work in $1\frac{1}{4}$ days more. How long would it take B to do the whole work?

4. My agent remitted to me a draft for \$1000 as the net proceeds of a sale. He charged me 2% commission for effecting the sale, and $\frac{1}{4}$% for buying the draft, which was at $\frac{1}{2}$% premium. Find the total proceeds of the sale?

5. A merchant marks his sugar at 10% advance on cost, and afterwards mixes sand with it, equal in weight to $\frac{1}{3}$ of the original quantity. What per cent. profit does he really make, taking sand as worthless?

6. Two mechanics work together; for 15 days' work of the first and 8 days' work of the second they receive \$61, and for 6 days' work of the first and 10 days' work of the second they receive \$38. How much does each man earn?

7. Two brothers residing in different towns have gross salaries which differ by \$20; the one who has the larger salary has to pay an income tax of $2\frac{1}{2}$ cents on the dollar, and as a consequence his net salary is \$1.40 less than his brother's who pays an income tax of only 6 mills on the dollar. Find each of their salaries.

8. The invoice of goods purchased on 4 months' credit was \$510, on which there was allowed a discount of 20 and 10%. What was the cash price of the goods, money being worth 6% per annum?

9. A merchant fails, owing A \$12500, and B \$17000. His assets are \$8850; the assignee charges 3% on the assets for closing up the business, and incurs other expenses to the amount of \$59 while engaged in the business. Find what each creditor receives.

10. The numerator of a certain fraction is a fifth as much again as the denominator, and the sum of the numerator and denominator is 352. Find the fraction.

11. A merchant buys goods at $2.40 per pound troy, and sells them at a gain of 20%. If he sells them by avoirdupois weight, what must he ask per ounce for them?

12. The longitude of Calcutta is 88° east, and that of Rome 12° 30′ east. What is the time at Rome when it is 2.15 p.m. at Calcutta?

13. Find the rates, in miles per hour, of two trains 50 and 60 yards long, respectively, which pass each other going in the same direction in 15 seconds, and in opposite direction in 3 seconds.

14. A speculator holding $9144 of 4% stock, finds that by selling it at 77½ and investing the proceeds in 5% stock, he can increase his annual income by $6.24; before he can effect the change the former stock has decreased 2 in value and the latter has risen ¾. If the change be then effected, find by how much his first income is altered?

15. A flour merchant shipped 3200 barrels of flour to his agent, who sold it at $7.25 a barrel, deducted his commission, paid $232.50 for duty and freight, and remitted a draft for $22500, purchased at ½% discount. What rate of commission for selling the flour did the agent charge?

16. A ditch is being dug at the rate of 32 yards a day by 20 men; after 7 days 4 men are replaced by 15 boys, and the work goes on 6 days more, at the end of which the whole length dug is 452 yards. Compare the work of a man with that of a boy.

17. The stock of 3 partners, A, B, and C, are $640, $370, $520, respectively, and their gains $192, $140.60, and $249.60, respectively; C's stock was in trade 4½ months longer than A's. How long was the money of each in trade?

18. A speculator sold a house for 30% profit, and with the money purchased another, which he sold for $3640, losing 12½%. What did the first house cost him?

19. The amount of a certain principal for a certain time at a certain rate per cent. is $416, and the amount of double the same principal at the same rate per cent. for one-half the former time is $741. Find the principal.

1. A merchant buys flannel at 32 cents a yard. At what profit per cent. must he sell it in order that the money he receives for 220 yards may be equal to his gain on $480 outlay?

2. Find the duty on 7200 pounds of sugar worth 6 cents a pound, the specific duty being $\frac{1}{2}$ cent per pound, and the ad valorem duty 25%.

3. Find the standard time of the city of Quebec, longitude 71°, 18′ west, when it is 2 p.m. at Vienna, longitude 16°, 24′ east.

4. A banker in discounting a bill legally due in 3 months at 6% per annum, charges 50 cents more than the true discount. Find the amount of the bill.

5. A merchant sent his agent $6150, with instructions to deduct his commission at $2\frac{1}{2}$%, and invest the remainder in flour at $6 a barrel. If the cost of freight and insurance amounts to $250, at what must the flour be sold a barrel so as to make a clear profit of 15%?

6. My agent in Montreal sells a house and lot for me for $8000, on a commission of $1\frac{1}{4}$%, and remits to me the proceeds by a draft purchased at $\frac{1}{4}$% premium. What sum do I receive from the sale of the property?

7. A has $8000, and B $5000 in a business; they take in as a partner C, and so arrange it that each has an equal share in the business. If the original capital is not increased, find what C pays to each of the others for his share.

8. A and B run a 100 yard race. A takes 10 steps while B takes 11, but 9 of A's are equal in length to 10 of B's. Who wins the race, and by how much?

9. Find the interest on $146, for 67 days, at 5% per annum

10. A man bought a farm for $4000; at the end of 3 months he paid his taxes, levied on $\frac{3}{5}$ of the purchase value at 18 mills on the dollar; in another 3 months he spent $500 in improvements, and at the end of the year he sold for $5500. Find his gain, money being worth 6% per annum.

11. Took a risk at $1\frac{3}{4}$%; reinsured $\frac{2}{3}$ of it at $2\frac{1}{4}$%. My net premium was $4.30. What was the amount of the risk?

12. What is the cost of 3825 pounds of sugar, bought at 4 cents a pound, on which is paid \$36.25 for freight, and $2\frac{1}{2}$ cents a pound for duty after deducting 12% for waste?

13. A cistern can be filled by two pipes in 20 and 24 minutes respectively, and can be emptied by another pipe in 30 minutes; in what time will it be filled if all are running together?

14. How many ounces of coinage gold are equal in value to 112 ounces of coinage silver, 1869 sovereigns weighing 40 pounds troy and 66 shillings weighing 1 pound troy?

15. In an examination of 120 candidates $8\frac{1}{3}$% of the whole obtain honors, and 80% of the remainder pass. How many fail to pass, and what percentage are they of the total number?

16. When it is 12 o'clock noon at Greenwich, what o'clock is it at St. Petersburg, the longitude of which is 30° east?

17. A contractor sends in a tender of \$11000 for a certain work, with nothing to be advanced till the work is completed; a second sends in a tender for \$10000, but stipulates to be paid \$2000 every 4 months. Find the difference between the tenders, supposing that the work cannot be completed before the end of 2 years, and money to be worth 2% per term of 4 months.

18. A man holds $\$6806.18\frac{3}{4}$ of stock in the 3 per cents; by selling out and investing the proceeds in $3\frac{1}{2}$% stock at $82\frac{1}{4}$, he can increase his annual income by $\$4.34\frac{7}{16}$. Find what the increase would be if he employs a broker who charges $\frac{1}{2}$% on each transaction to effect the change.

19. A merchant shipped a number of barrels of salt to his agent, who sold it for 75 cents a barrel, on a commission of $2\frac{1}{3}$%, he also realized a further commission of $\frac{1}{2}$% (calculated on the remainder of the proceeds) for prompt payment, and after paying duty to the amount of \$179.10 he remitted \$2736.25 to his employer. How many barrels of salt were in the consignment?

20. Two drovers together sold 400 sheep, the one at \$8.25 and the other at \$7.50 a head. They received altogether \$54.75 more than if they had both sold at the uniform rate of $\$7.87\frac{1}{2}$ a head. Find the number sold by each.

1. A and B start at the same time to travel from X to Y and Y to X, respectively, A walking at the rate of 5 miles per hour and B at the rate of $4\frac{1}{2}$ miles per hour. After meeting, A increases his speed 1 mile per hour and reaches Y in 2 hours more, and B reduces his speed to 4 miles per hour. Find how long it takes A and B each to travel the whole distance, also find the distance.

2. Bought 5 hogsheads of molasses, each containing 84 gallons, at $12\frac{1}{2}$ cents a gallon, and paid $25 for freight. Allowing 10% for leakage and waste, 5% of the sales for bad debts, and $2\frac{1}{2}$% of the remainder for collecting, for how much a gallon must I sell it to make a gain of 20% on the whole cost?

3. The difference between the interest and discount on a sum of money for 1 year 9 months at 8% per annum, is $9.80. Find the sum.

4. If 6% more be gained by selling a horse for $210 than by selling it for $199.50, what must the original price have been?

5. When it is 7.15 p.m at Greenwich, what o'clock is it at Madras, the longitude of which is 80° east?

6. The product of two numbers is 35643, and their quotient is 3. Find the numbers.

7. A retailer marks his goods at 3 prices, a cash price, a six months' credit price, and a 12 months' credit price. From his 12 months' price he deducts $12\frac{1}{2}$% for cash, and $7\frac{1}{2}$% for 6 months' credit. His 6 months' price is 40% in advance of the cost of the goods. Find at how much advance on cost he sells on 12 months' credit goods which sell for $6.66 cash.

8. I remit a $1000 draft on Toronto to my agent in Montreal. He disposes of the draft at $\frac{1}{2}$% discount on a commission of $\frac{1}{4}$%. After deducting a second commission of 2% (calculated on the amount invested), he invests the net proceeds in sugar at 6 cents a pound. Find, to the nearest ounce, the quantity of sugar I receive.

9. A certain piece of work can be completed in 30 days by 48 men. If 2 men drop off at the end of every 3 days, how long will it take to finish the work?

10. A man sold a lot at a gain of 9%; had he sold it for $12.25 more he would have gained $12\frac{1}{2}\%$. Find the cost price.

11. At what rate will a sum of money, loaned at simple interest, double itself in 25 years?

12. Took a risk at $1\frac{3}{4}\%$; reinsured $4000 of it at 2%, and $6000 more at $2\frac{1}{4}\%$. My net premium was $26.50. What amount was insured?

13. Find the present worth of an annuity of $700, having 4 years to run, money being worth 5%.

14. A, B, and C engage in trade with a joint stock of $1534; A's stock is $456, and his time of investment 7 months; B's stock is $546, and his gain $132.60; C's gain is $108.80, and his time 8 months. Required A's gain and B's time.

15. Divide $1660 into 2 sums, so that the true discount on one for 4 years at $3\frac{1}{2}\%$ per annum, may equal the true discount on the other for 6 years at $2\frac{1}{2}\%$ (simple interest).

16. What is the area of a right angled triangle, that has a hypotenuse 40 inches, and the difference of the sides 8 inches?

17. A dealer purchased on 6 months' credit goods to the amount of $520; after keeping them 3 months he sold them on credit for $575.96, and allowing money to be worth 8%, he found that he had made 10 per cent. on the transaction. On what term of credit did he sell the goods?

18. What rate of taxation must be levied on a corporation with $2800000 assessable property to secure money to build a $33950 school house, if the expenses for collecting amount to 3% of the total taxes?

19. Find the standard time of Picton, when it is 15 min. to 4 a.m. in Paris, if the longitude of Picton is 77° 10′ west and that of Paris 2° east.

20. A grocer buys a quantity of tea of a certain quality, and one-fourth as much of an inferior kind, the cost of the latter per pound being 80% of that of the former. He mixes them and sells the mixture at an advance of 10% on the cost per pound of the finer quality. Find his entire gain per cent.

18

1. A vessel containing 210 gallons of a mixture of coal-oil and water, of which 75% is coal oil, leaks at the bottom, from which a certain amount of water alone escapes. On an analysis of it there is found to be $87\frac{1}{2}$% of the remaining liquid coal-oil. Find the number of gallons of water which escaped.

2. A speculator holds $6840 stock, one-half in the 3 per cents at 72, and the rest in the 4 per cents at 95. If he sells out the first stock and with the proceeds buys more of the second stock, and also sells out the stock originally held in the second and invests in the first, find the change in his income.

3. A man removed from a town, where his income tax was 16 mills on the dollar, to another where the tax was only 7 mills on the dollar, and although in the latter place his salary was $500 more than in the former, his taxes were $6.40 less. Find his income tax in the latter place.

4. The ad valorem duty on 20% of the invoice price of a shipment of cigars was 40%, and on the remainder 35%; the whole duty being $810, what was the invoice price?

5. A commission merchant sold 2 lots for $800, on one of them he charged 2% commission for selling, and on the other 4%; his total commission being $26, find the selling price of each of the lot.

6. A grocer sells one kind of tea at 24 cents a pound and loses 20%, and another kind at $42\frac{1}{2}$ cents a pound and gains 25%. He mixes them in equal proportions and sells the mixture for $33\frac{1}{4}$ cents a pound. What is now his gain or loss per cent?

7. A sets off from M to N, and B at the same time from N to M, and they travel uniformly; A reaches N in 16 hours, and B reaches M 25 hours after they met on the road. Find in what time each performed the journey.

8. If a grocer's pound weight is $\cdot 0\dot{3}$ ounces too heavy, find his loss per cent. from its use.

9. Find the standard time of the City of Toronto, when it is 3 a.m. at Malta, if the longitude of the former place is 79° 25′ west, and that of the latter 14° 30′ east.

10. A grocer sells coffee for cash at a gain of $33\frac{1}{3}\%$ on cost. He also sells on credit, giving 8 pounds for what would buy 9 if paid in cash. How much per cent. above cost is his credit price?

11. My agent in New York sells a house for me on a commission of 2%, and remits to me the net proceeds, \$6000, by a draft purchased at $\frac{3}{4}\%$ premium. How much did the house sell for?

12. A merchant sold an article at 8% profit; if he had bought it for 10% less, and sold it for \$3 more, he would have gained $22\frac{2}{9}\%$. Find the cost.

13. A does $\frac{3}{8}$ of a piece of work in 15 days, and then gets B to help him. They work together for 5 days, when B leaves and A finishes it in 16 days more. How long would B take to do the whole work?

14. At what rate, simple interest, will a sum of money triple itself in 45 years?

15. A person pays \$87.50 for the insurance of a house at $4\frac{3}{8}\%$, and he finds that in case the house is destroyed, he will receive the value of the house, the premium of insurance and \$125 besides. What is the value of the house?

16. A bankrupt owes two creditors, A and B in the proportion of 3 to 5. A acts as assignee, and as such receives 3% of the assets, the remainder is divided in the proportion of the amounts owed to them. A receives in all \$1575. What were the bankrupt's assets?

17. If a train 88 yards long overtake a person, walking at the rate of 4 miles per hour along the railway, and pass him in 8 seconds, what is the rate of the train in miles per hour?

18. A merchant bought tea and marked it at an advance of 20% on cost, but afterwards mixed $\frac{1}{4}$ of its weight of inferior tea with it, which cost $\frac{2}{3}$ as much per lb. What per cent, does he now gain?

19. A merchant sends \$1956 in cash and butter to his agent with instructions to sell the butter and invest the proceeds, less his commissions, in tea. The agent charges 5% on the value of the goods he handles in each case. Find the value of the butter shipped, if his total commission amounts to $\$129.14\frac{2}{7}$.

1. Find the circumference of a circle whose area is equal to that of a square, having a side 3 inches in length.

2. The rates at which A and B walk are in the proportion of $2 \cdot \dot{4}$ to $3 \cdot \dot{7}$, and B walks $5 \cdot 2\dot{3}$ miles per hour. How many seconds start must B give A so as to just beat him in a ten mile race?

3. A speculator holds $10992 stock, one-half in the 5%'s at $95\frac{1}{2}$ and the other half in the 6%'s at 114. If he transfers his 5% stock to the 6%'s, and his original 6% stock to the 5%'s, find the alteration in his income, supposing he pays $\frac{1}{2}$% brokerage on each transaction.

4. Bought 3000 bushels of wheat at $1.15 per bushel payable at the end of a year. I immediately sell it at $1.04 per bushel, cash, and put the money at interest at 10%. How much will I gain or lose by the transaction at the end of the year?

5. A, B, and C invest capital in the proportion of 7, 8, and 12 at 4% per annum and realize $4050. If they invest it at $5\frac{1}{2}$%, find the yearly income of each.

6. The Merchants' Bank of New York having declared a dividend of 8%, a stockholder in Buffalo drew on the bank for the sum due him, and sold the draft at a premium of $1\frac{1}{2}$%, thus realizing $406 from the dividend. How many $100 shares did he own?

7. I owe $719.92, and give my note for 60 days. What must be the face of the note to pay the exact debt, when discounted at 8% per annum?

8. What sum of money, with its semi-annual dividends of 5% invested with it, will amount to $12750 in 2 years?

9. A merchant mixes 40 pounds of tea worth $37\frac{1}{2}$ cents a pound, with 64 pounds worth 45 cents a pound; he sells 24 pounds of the mixture at 50 cents a pound. At what price per pound must he sell the remainder so as to clear 25% on his whole outlay?

10. A merchant uses a false pound weight, which actually weighs $15\frac{3}{4}$ ounces. How much does he cheat a customer who buys goods to the amount of $50?

11. A hatter buys 325 hats at $2.75 each; he sells $\frac{3}{5}$ of them at an advance of 20% on cost, and the remainder for $2 each. Find his total gain.

12. What are the net proceeds from the sale of 1400 barrels of flour at $7 per barrel, freight and storage being 25 cents a barrel, commission for selling $2\frac{1}{4}\%$, and for guaranteeing payment $1\frac{1}{4}\%$, calculated on sales?

13. Assuming the earth a sphere, whose diameter is 7913 miles, and that 1000000 units are equal in length to one degree of longitude at the equator, find the length of the unit in inches. Given ($\pi = \frac{355}{113}$.)

14. An insurance company took a risk at $2\frac{1}{8}\%$, and re-insured $\frac{3}{4}$ of it at $2\frac{3}{8}\%$. The premium received exceeded the premium paid by $38.50. What was the amount of the risk?

15. A speculator increases his annual income $6 by transferring $4850 stock worth 80 to 5% stock worth 97. What was the per centage paid by the former stock?

16. A grocer buys goods for cash and receives a discount of 10% off the invoice price; in selling he uses scales which weigh light by $\frac{1}{2}$ oz. per lb., and sells for cash at an advance of $21\frac{7}{8}\%$ on the invoice price. Find his gain per cent.

17. Two men enter into partnership and agree to share all gains and losses in proportion to their investments. The former contributes $4683 and the latter $5978. Their net loss at the end of the year is $3198.30. How much of this loss should be sustained by each?

18. A bankrupt has book debts equal in amount to his liabilities, but on $24000 of them he realizes only $66\frac{2}{3}$ cents on the dollar, and the expenses of the bankruptcy are 5% of the book debts. He pays 65 cents on the dollar. Find the amount of his liabilities.

19. If stock at 7% discount will pay 6% interest on the investment, at what rate of discount would it have to be bought to pay 9% interest?

20. A can row a distance down a stream in 30 minutes and up in 40 minutes. What is the distance, if the rate of the stream is $\frac{1}{2}$ mile per hour?

21. A man employs 8 men and a certain number of boys, and pays on the average $97\frac{1}{2}$ cents a day to each. He pays each man $1.50 a day and each boy $62\frac{1}{2}$ cents a day. How many boys were employed?

1. A house and lot together are worth $1400; one-fifth of the value of the house is equal to one half the value of the lot. Find the value of each.

2. A merchant sold a piece of cloth for $24.60 and thereby lost 18%. What would have been his gain per cent. had he sold it for $36?

3. A drover bought a number of horses and cows for $10800. There were 5 times as many cows as horses, and a horse cost 4 times as much as a cow. If each cow cost $30, how many horses did he buy?

4. Eight men can do a piece of work in 40 days; after working 35 days they are joined by another man, and the whole work is completed in 39 days. What fraction of the work is performed by the last man in a day?

5. Divide 83 into two parts, so that the greater may exceed the less by 41.

6. A person lent a sum of money at 5% simple interest; in 16 years the interest amounted to $90 less than the sum lent. What was the sum lent?

7. A person invested $2304 in the 3 per cents at $95\frac{7}{8}$, and after receiving the half year's dividend, sold out at $94\frac{1}{2}$. Find his gain or loss, brokerage being $\frac{1}{8}$%.

8. Find the volume of a right circular cone, whose height is 4 feet, and the circumference of the base 9 feet.

9. A speculator sold a lot gaining 20% of the cost. What per cent. of the proceeds did he gain?

10. A dishonest merchant adds 5 gallons of water to every 20 gallons of coal-oil. How much does he cheat a customer who buys 15 gallons at 15 cents a gallon?

11. Three men start together to walk around a circular course 60 yards in circumference, their rates being $11\frac{1}{2}$, $14\frac{1}{2}$, and 16 yards per second, respectively. Find when and where they will first be together again.

12. The hour, minute, and second hands of a clock are on the same centre. When, first after 4 o'clock, will the second hand be midway between the other two?

13. A bankrupt's liabilities were $3000, his assets $540; the assignee charged 3% of the assets for his work. How many cents on the dollar is the bankrupt able to pay his creditors?

14. A sum of \$24 is to be contributed by 24 people; some contribute \$1.20 each, and the remainder $66\frac{2}{3}$ cents each. How many contributors were there of each kind?

15. The distance between Hamilton and Toronto is 39 miles. One traveller starts from Hamilton and travels at the rate of 4 miles an hour, while another starts from Toronto at the same time and travels at the rate of $2\frac{1}{2}$ miles an hour. It is required to know where and when they will meet.

16. A person expended \$3910 in $3\frac{1}{2}$% stock at $97\frac{3}{4}$, and after receiving the half-year's dividend, sold out at 95. Find his gain or loss on the whole transaction.

17. Find the difference between the interest on \$46 for $6\frac{1}{4}$ years at 9% per annum, (1) interest payable yearly, (2) half-yearly.

18. If a number be increased 20%, and that amount be increased 25%, the result will be 165. Find the number.

19. A grocer intended to gain 8% on a stock of tea, and fixed his price accordingly. When he had sold $\frac{2}{3}$ of the stock, he was compelled to reduce the price 10 cents a pound and so gained only half as much as he had intended. What was the original cost per pound of the tea?

20. If the 3 hands of a watch are on the same centre, when, first after 12 o'clock, will the minute hand be midway between the other two?

21. Two men start together to travel in the same direction around a circular track; their rates are as 11 to 14. Find where they will first meet.

22. A speculator buys stock that pays half-yearly dividends of 4%, at 20% discount. What rate of interest does he make on his money?

23. A can row a certain distance down a stream and back again in $5\frac{1}{3}$ hours. If the rate of the stream is one quarter of his rate in still water, find the time in going each way.

24. A and B were candidates for election in a constituency of 3000 voters. A received $37\frac{1}{2}$% of the total number of votes polled, and was defeated by a majority of 640. What percentage of voters did not vote?

1. In a 100 yard race, A can beat B by 5 yards and C by 10 yards. If B gives C 5 yards start in a 100 yard race, which wins and by how much?

2. A and B engage in trade; A puts into the business $400 for 6 months, and B a certain sum for 7 months. When they close up the business A receives $640 as his share of the stock and profits, while B receives $510 for his. What amount of capital did B put into the firm?

3. A man has a net income of $745.80 after paying an income tax of 12 mills on the dollar on all his income except $400 of an exemption. Find his total income.

4. A man exchanges $25000 of 6% stock at 70 for 8% stock at 120. Find the alteration in his income.

5. A boat can sail at the rate of 12 miles an hour in still water. How far may it go up a stream which is flowing at the rate of $3\frac{1}{2}$ miles an hour, so as just to occupy 7 hrs. 24 min. for the round trip?

6. A man has 50, 25, 10, and 5 cent pieces, the number of each being in the proportion of 1, 2, 3, and 4. If their total value is $10.50, how many has he of each?

7. A druggist gives an ounce troy instead of an ounce avoirdupois. How much per cent. does he lose?

8. A and B work $5\frac{1}{2}$ and $7\frac{1}{2}$ hours a day, respectively, for one day and receive the same wages; on the second day they work $6\frac{1}{4}$ and $8\frac{1}{4}$ hours a day, respectively; they receive $13.98 for their two days' work. How should it be divided?

9. A merchant lost 15% of certain perishable goods and sold the remainder at a price, such that he just cleared himself. Find at what advance on cost he sold them.

10. A bankrupt's liabilities are $4000, and after paying the assignee 4% of the total assets for winding up the business, he is able to pay only 21 cents on the dollar. Find his total assets.

11. A builder pays 4 times as much for material as for labor; had he paid 8% less for material and 13% more for labor, his contract would have cost him $220.78 less than it did. What was the contract price?

12. What must be the market value of 5% stock, so that, after deducting an income tax of 2%, it may yield 6% interest on the investment?

13. A man can row 5 miles an hour in still water. How far may he go down a stream which flows at the rate of 3 miles an hour, so that he may just take 2 hrs. 40 min. for the round trip?

14. A man insures a house worth $3200 for $\frac{3}{8}$ of its value, at $1\frac{3}{4}$% premium. If the house be destroyed, find the total loss sustained by the company.

15. A and B engage in trade, A invests $320 for a certain time, and B $415 for 4 months. In winding up the business it is found that A is entitled to $364 of the stock and profits, while B receives $156.50. Find the time A left his money in the business.

16. A mixture of black and green tea weighing 13 pounds is worth $4.92; if the proportions of each are interchanged, the mixture will be worth $4.44. The black tea is worth 30 cents a pound; find the price of the green tea per pound, and the number of pounds of each kind in the first mixture.

17. Two men working together can do a piece of work in 20 days. If the work is worth $120, and one man works 7 days less than the other, how should the money be divided?

18. A man sold a lot for $512 gaining 15% of the proceeds. What would he have sold it for, had he gained 15% of the cost?

19. A person holds $400 stock in the 4 per cents purchased at $104\frac{3}{4}$, and after receiving one-half year's dividend, he sells at an advance. If he gains $105 on the whole transaction, find the market value of the stock when he sold, brokerage $\frac{1}{8}$%.

20. The compound interest on $250 for 2 years is $50. Find the rate.

21. Two cisterns of equal dimensions are filled with water and the taps for both are opened at the same time. If the water in one will run out in 4 hours, and that in the other in 7 hours, find when one cistern will have 4 times as much water in it as the other.

22. Four-fifth of the selling price of goods is 25% less than cost. Find the gain or loss per cent. at which the goods were sold.

1. A person set out to walk from A to B, at the rate of 5 miles an hour. When he had travelled $1\frac{2}{3}$ miles he was overtaken by a coach from A, which was 10 minutes later in starting. At a distance of $11\frac{1}{6}$ miles from B he met the coach returning from B, where it had stopped 30 minutes. What is the distance from A to B?

2. A man has 8 hours at his disposal. How far may he ride in a coach which travels at the rate of 9 miles an hour, so as to return in time, walking at the rate of 4 miles an hour?

3. Divide $110 among A, B, C and D, so that B may have three times as much as A, C half as much again as A and B together, and D as much as A, B and C together.

4. If I exchange 75 railroad bonds of $500 each, at 36% below par, for bank stock at 5% premium, how many shares of $100 each will I receive?

5. Find the compound interest on $40 for $9\frac{1}{2}$ years at 8% per annum, interest payable half-yearly.

6. A cow is tied to the outside of a square enclosure (60 feet to the side) 20 feet from one end by a rope 100 feet long. Find how much ground she can procure grass from?

7. If the cost of an article had been 24% less, the gain per cent. (at the same selling price) would have been $3\frac{4}{5}$ times as great. What was the actual gain per cent.?

8. The length of a room is 25 feet, the height $12\frac{1}{2}$ feet, and the area of the floor is equal to $\frac{4}{9}$ of the area of the 4 walls. How much would it cost to carpet the floor at 80 cents per yard, if the carpet is $3\frac{1}{3}$ feet in width?

9. Two trains start at the same time, the one from A to B and the other from B to A. If they arrive in A and B respectively 4 hours and 9 hours after they pass each other, compare their rates of travelling.

10. A bankrupt owes $6000, his assets are $4500. If $2000 of the assets are secured to the creditors by a mortgage, find how many cents on the dollar he can pay the other creditors, the assignee's charges being 5% of the total assets.

11. Stock which sells at 12% premium, pays 8% dividend. What interest is made on the investment?

12. The compound interest on \$275 for 2 years is \$40. Find the rate per cent.

13. A merchant having bought a lot of goods sells $\frac{1}{3}$ of them at a loss of 9%. By what increase per cent. must he raise that selling price in order that by selling the remainder at the increased rate he may make 4% on the whole transaction?

14. A had \$150; he gave B $1\frac{2}{5}$ as much as B already possessed. After this transaction, B's money was $\frac{12}{25}$ of what A had left. How much had B at first?

15. If the cost of an article had been 10% less, the gain per cent. (at the same selling price) would have been $11\frac{2}{3}$ more. What was the actual gain per cent.?

16. Two trains start at the same time from A and B, and proceed toward each other at the rates of 30 and 50 miles per hour, respectively. When they meet it is found that one train has travelled 135 miles farther than the other. Find the distance between A and B.

17. A, B and C engage in trade. A and B together put in \$825, B and C \$650, and C and A \$725; they gain \$550. How should it be divided among them?

18. The difference between the simple interest and the true discount on a sum of money for 4 years, at a certain rate per cent., is \$$11\frac{2}{3}$, and for 8 years, at the same rate per cent., it is \$40. Find the sum and rate.

19. Stock selling at 7% discount yields 8% interest on the investment. What dividend does the company declare?

20. Find the compound interest on \$32 for $2\frac{1}{2}$ years at 12% per annum, interest payable every 2 months.

21. Two cisterns of equal dimensions are filled with water, and the taps of both are opened at the same time. If the water in one will run out in 5 hours and that in the other in 4 hours, find when one cistern will have twice as much water in it as the other.

22. The radius of the small wheel of a bicycle is $5\frac{1}{4}$ inches. A straight line drawn from the top of the small wheel to the top of the large one is $35\sqrt{3}$ inches in length, and makes with the vertical line an angle of 30 degrees. How many more revolutions does the small wheel make than the large one, in going a mile?

1. Find the ages of A, B and C, by knowing that C's age at A's birth was $5\frac{3}{5}$ times B's age, while now it is equal to the sum of A's and B's; also, if A were 1 year older his age would be $\frac{6}{7}$ of B's.

2. If the cost of an article had been 25% less, the gain per cent. (at the same selling price) would have been $5\frac{1}{3}$ times as great. What was the actual gain per cent.?

3. The circumference of the fore-wheel of a carriage is $7\frac{1}{2}$ feet, and that of the hind wheel 11 feet. In what distance will the fore-wheel make 250 revolutions more than the hind-wheel?

4. Which is the more profitable investment 5% stock at 120, or 3% stock at 75?

5. Find the compound interest on \$25 for 2 years at $1\frac{1}{2}$% per month, interest payable monthly.

6. Two cisterns of equal dimensions are filled with water, and the taps of both are opened at the same time. If the water in one will run out in 7 hours, and that in the other in 6 hours, find when one cistern will have 3 times as much water in it as the other.

7. A groccer bought sugar at 10 cents a pound; he sold it at a profit of 20%, as he thought, but by a mistake he used a pound weight $\frac{1}{2}$ ounce too heavy, and as a result he gained \$6 less than he intended. Find the number of pounds of sugar he bought.

8. If the cost price of an article be diminished 10 cents, and the selling price increased 10 cents, the gain would be increased from $16\frac{2}{3}$% to 60%. Find the cost.

9. A and B walk a race of 30 miles; A gives B 50 minutes start; A walks at the rate of $5\frac{1}{2}$ miles an hour, and catches B at the end of 25 miles. Find B's rate, and by what distance he is beaten.

10. A clock which gains 6 minutes in 24 hours is 4 minutes too slow at 7 a.m. Monday. What o'clock will it be when it indicates 14 minutes to 8 p.m. on the following Tuesday?

11. I sold through a broker, who charged $\frac{1}{8}$% brokerage, 136 (\$100) shares of stock at $68\frac{3}{4}$. Find what I realized.

12. The compound interest on \$320 for 2 years is \$60. Find the rate per cent.

13. A bankrupt can pay 27 cents on the dollar. If his assets were $480 more he could pay 30 cents. Find his liabilities and his assets.

14. A 90-day note is discounted by a banker at 10%. What rate of interest is he charging?

15. A and B dig a ditch 120 rods long. The soil at one end is clay and at the other end sand. If the whole ditch were sand A could dig it alone in 30 days and B in 24 days. If the whole length were clay A could dig it in 40 days and B in 60 days. A begins at the clay soil and B at the sandy; they together dig the ditch in 17 days. What length of the ditch was clay?

16. If the true discount off $25 for a certain time is $5, what would it be for $\frac{1}{3}$ of the time at the same rate?

17. Bought 80 barrels of flour, part of it at $10 a barrel and the rest at $6 a barrel; the whole cost was $724. How many barrels of each kind did I buy?

18. A train starts from A at 6 o'clock a.m. and reaches B at 2 o'clock p.m.; another train starts from B at 11 a.m. and reaches A at 5 p.m. At what time of the day do they meet?

19. A bankrupt's assets are $\frac{1}{3}$ of his liabilities, but $\frac{1}{4}$ of the assets prove to be worth only 40 cents on the dollar. How many cents on the dollar can he pay?

20. A man has $400 of his income exempt from taxation, and on the remainder he pays $12\frac{1}{2}$ mills on the dollar taxes. If his taxes amount to $7.50, find his total income.

21. Find 3 sums of money, the greatest exceeding the smallest by $47.50, so that the simple interest on the greatest at 6% per annum for 2 years shall be equal to that on the smallest at 4% for 4 years, and equal to that on the third at 5% for 3 years.

22. A man bought a house and barn, paying 4 times as much for the house as for the barn. If he had paid 25% more for the house and 20% less for the barn, they would have cost $248 more. What did he pay for each?

23. A can do a piece of work in 10 days, B in 12, C in 15. They all begin together, but only C continues till the work is finished, A leaving 3 days, and B $2\frac{1}{4}$ days before completion. In what time is the work done?

1. A man sold two horses for equal amounts, gaining 10% on one, and losing 10% on the other. If he lost \$4 on the whole transaction, find the cost of each horse.

2. Three pounds of tea and 8 pounds of sugar are worth \$2.90, but if the sugar increases 25% in value, and the tea depreciates 5%, they would cost \$2.90½. Find the price per pound of each.

3. A bill of \$745 was paid in twenty-five and ten cent pieces. If the whole number of coins was 49, how many coins were there of each kind?

4. A vessel can be emptied by 3 taps; by the first alone in 80 minutes, by the second alone in 200 minutes, and by the third alone in 5 hours. In what time will the vessel be emptied if the three taps are opened together?

5. A merchant imports goods, which cost him in England £720. He pays an ad valorem duty of 10%, and sells them through an agent for \$1200. If the agent charges 5% for selling them, find his gain (exchange being at par).

6. A person's net income, derived from the 4%'s at 98, after paying a 2% income tax, is \$600. If he sells out and invests the proceeds in 5% stock at 112, find the alteration in his net income.

7. In the preceding example, find the alteration, supposing the buying and selling of the stock were performed through a broker, who charged ½% brokerage on each transaction.

8. A commission merchant sold a certain amount of goods, from the proceeds of which he deducted his commission of 5%, and immediately remitted the balance to his employer; for his promptness he received ½%, which amounted to \$6.41¼. What did the agent realize on the transaction?

9. How much water will dilute 9 gal. 1 qt. 1 pt. of alcohol 96% strong to 81%?

10. A grain merchant spent a certain sum of money in the purchase of wheat, 3 times as much in barley, and \$1500 in oats. He sold the wheat at a loss of 6%, the barley at a gain of 9%, and the oats at a gain of 20%, receiving altogether \$9396. Find the sum laid out in wheat.

11. A mill valued at \$24000 is insured for $\frac{3}{4}$ of its value in two companies, the first taking $\frac{3}{4}$ of the risk at $\frac{7}{8}$%, the second the remainder at $\frac{3}{4}$%. What is the total amount of the premium?

12. Two men start together, and travel in the same direction; one is going at the rate of 4 miles an hour. Find the other man's rate, if he is always equi-distant from the starting point and the first man.

13. A and B can do a piece of work in 8 days; B and C in 10 days, C and A in 12 days. The work is done by the 3 together, with the exception of B not joining the others till after the first 3 days. In what time is the work done?

14. A watch, which is set accurately at 7 o'clock a.m., indicates 5 minutes past 8, at 8 p.m. of the same day. What is the exact time when it indicates 11.15 a.m. on the following day?

15. Which is the better to buy flour at \$5 a barrel on 6 months' credit, or \$4.87$\frac{1}{2}$ cash, money being worth 7%?

16. Divide \$100.75 between A and B, such that 21% of B's share may exceed 8% of A's by \$3.39$\frac{1}{4}$.

17. Property depreciates annually 8% for 4 years, and at the end of that time is worth \$4197.61$\frac{1}{2}$. What was its original value?

18. A merchant buys sugar at 5 cents a pound; in selling it he wastes 7 , and 20% of the sales are bad debts. At what price per pound must he sell the remainder so as to gain 24% on the whole transaction?

19. The capital stock of a railroad is \$895750; the passenger earnings in one year were \$74537.50, and the freight earnings \$94567.50; the disbursements were \$107963.00, and \$7397.00 was placed in a bank to the credit of the company. What rate of dividend can they declare if the remainder be distributed among the stockholders?

20. A man in building a house pays 3 times as much for material as for labor; had he paid 4$\frac{1}{2}$% more for material and 8% less for labor, his house would have cost \$66 more than it did. What was the cost of the house?

1. A man holds $15600 stock worth 60, which if he transfer to 4% stock at 78 he can increase his annual income $12; before he could effect the transfer, each stock increased 2 in price; find how his income is now altered.

2. A merchant buys goods for $304.50 on a credit of 3 months. At what price must he sell them, on a credit of 8 months, to make a ready gain of 25% (money being worth 6%)?

3. How may a farmer mix 4 kinds of wheat worth 75, 78, 93 and 100 cents a bushel, respectively, so as to form a mixture worth 80 cents a bushel, and have equal quantities of the first two kinds, as well as of the last two?

4. A merchant marks his goods at 10% above cost, but throws off 2% of this price for cash. What per cent. above cost is his cash selling price?

5. A has 54 five cent pieces and 1 dollar; B has 54 dollars and 1 five cent piece. What sum will have to change hands so that B will have exactly 10 times as much as A?

6. A merchant received an invoice of merchandise amounting to $618 on 4 months' credit, off which he was offered 4% discount for cash. If money is worth 9% per annum, how much cheaper can he get the goods by this offer?

7. A can beat B 3 yards in a hundred yard race, B can beat C 9 yards in a 300 yard race. Find how much A can beat C in a 500 yard race.

8. A man in England invests a certain sum of money in Canada at 6%. All but $100 of his income is taxed at 2%. If he pays $18.28 taxes, find the amount of English money invested, exchange at par.

9. A has $150, and 7% of his money exceeds 9% of B's by $6.09. Find B's money.

10. A, B, and C form a partnership, A contributing $3500, B $4000 and C $2500. They agree that $800 of the profits shall be placed to their account in the bank, and the remaining profits divided in proportion to the money invested. At the end of the year A's share of the divided profits is $2100. Find the percentage of profits realized on the entire capital.

11. What will I gain per cent. by purchasing goods on 9 months' credit, and selling them immediately for cash at the invoice price, money being worth 6%?

12. If $8\frac{1}{3}$% of the cost of an article be equal to $5\frac{5}{7}$% of its selling price, find the gain per cent. at which it was sold.

13. A man borrowed a sum of money at 6% per annum, and invested it in the 7 per cents at 120. At the end of the year he receives the dividend, and immediately sells out at 125; he finds that he has gained \$51.74$\frac{2}{5}$ after paying an income tax of 2%, reckoned on the difference between the total gain on the stock and the interest on the loan. What sum did he borrow?

14. A speculator transferred \$4200 stock from the 4 per cents to the 6 per cents at par, and thus increased his income 10%. At what price did he sell out the 4 per cents?

15. A grocer has 3 kinds of tea, costing 30, 45 and 60 cents a pound, respectively. What quantities of each must he take to form a mixture of 144 pounds, worth 40 cents a pound?

16. A man insures his house so that in case of loss he may recover the value of the house and the premium of insurance at $1\frac{1}{4}$%. The house is destroyed by fire and $\frac{4}{5}$ of the claim is allowed; he finds that he receives \$750 less than the value of the house. Find the value of the house and the premium.

17. Thomas Fraser can make as much progress with his studies in 9 days as Walter Hambly can in 10, but Fraser studies only 11 hours for every 12 Hambly does. On this basis, how should a total of 327 honor marks be awarded to each?

18. Thirty minutes after A has set out on a journey on horse back, at the rate of 9 miles per hour, B sets out from the same place and follows him on a bicycle at a uniform speed, overtaking him in 1 hour 30 minutes. At what rate per hour did B ride?

19. An importer bought $617\frac{1}{2}$ yards of silk at \$$1.87\frac{1}{2}$ per yard, and sold it so as to gain \$$260.62\frac{1}{2}$, after deducting \$$711.93\frac{3}{4}$ expenses. For what was the silk sold per yard?

1. A, B, and C run a 100 yard race; A beats B by 1 yard, and C by 2 yards. By how many yards can B beat C in a 100 yard race?

2. A man in England invests a certain sum of money in Canada at 6%, from which he receives an annual income of $870.36, after paying an income tax of 15 mills on the dollar on all his income except an exemption of $400. Exchange being at par, find the amount of English money invested.

3. A note drawn on August 3rd for 4 months is discounted on September 12th at 9% per annum. What rate of interest is charged?

4. Divide $338 between A and B, such that 3 times A's share may be $127 less than 4 times B's.

5. I owe $539.94 and give my note for 60 days. What must be the face of the note to pay the exact debt, when discounted at 8% per annum?

6. If I own a vessel valued at $7791 and wish to insure it at a premium of $4\frac{3}{5}$% so as to recover, in case of the destruction of the vessel, both the premium paid and the value of the vessel, for what sum must I insure?

7. A merchant buys 2590 yards of cloth at $1 a yard. He marks it to gain 25%, but in selling the first half he uses a 35-inch yard measure, and in selling the second half a 37-inch yard measure. Find the difference between his actual and his intended gain.

8. A and B form a partnership; they carry on business in different places. A alone makes $350 total profits, 25% of which goes to pay his expenses; B makes $375 total profits, 20% of which is used for his expenses. How will they settle, if they agree to divide the net profits equally?

9. A grocer mixes teas worth 32, 41 and 54 cents a pound, respectively, forming a mixture worth $13\frac{1}{2}$ cents a pound, which contains 12 pounds of the third kind and equal quantities of the first two kinds. How many pounds of each of these grades of tea must he use?

10. A speculator invests $9240 in the $3\frac{1}{2}$ per cents at $82\frac{1}{2}$, and, on their rising to 87, sells out and invests the proceeds in 5% stock at 112. Find the alteration in his income.

11. An owner of a house, worth \$4610.12½, insures it at 1¼%, so that in case of loss he will recover the value of the house, the premium of insurance and \$31.12½ besides. What was the amount of insurance he placed on the house?

12. A can do as much work in 10 hours as B can do in 11, but he does not work the same time. If A earns \$29.70 and B earns \$30.00, compare the times they each worked.

13. If a company takes an accident risk of \$3000 at 1½%, and reinsures ⅔ of it in another company at $2\frac{1}{5}$%, what will the first company gain by the transaction, if no accident occurs?

14. A grocer sold tea at 32 cents a pound, and coffee at 24 cents a pound. Having one day sold 2 pounds more of coffee than of tea, he found the amount of his sales equal. What were the quantities of each sold that day?

15. A person bought a Swiss watch bearing a duty of 25%, and sold it at a loss of 10%, but had he sold it for \$26 more, he would have cleared 3% on his bargain. What did he receive for the watch?

16. In a 400 yard race A wins, B being 10 yards, and C 40 yards behind. How much would B beat C in a mile race?

17. I bought goods for \$30.40 on 4 months' credit, and sold them immediately for \$37.20, with such allowance of credit as made my immediate gain 20%. How long credit did I give, money being worth 4% per annum?

18. Divide \$40.50 between A and B, such that 12% of A's share is 83 cents more than 9% of B's.

19. A man sold two lots for \$208, gaining 20% on one and losing 20% on the other. Find the cost of each, if he gained \$8 on the whole transaction.

20. If 15 men, 18 women, and 26 boys receive \$64.40 for a day's work, and 2 men receive as much as 3 women or 10 boys, what does each man, woman, and boy, respectively, receive for a day's work?

21. I shipped my agent 6000 bushels of wheat, which he sold at 85 cents a bushel, and deducted 3% commission for selling it. He invested the net proceeds, less a commission of 2% calculated on the amount invested, in real estate. Find his total commission.

1. If salaries above $1000 are assessed for the full amount, and those below have $700 exempt, what is the total salary of a man whose net salary is $2.78 less than that of a man whose total salary is $995 (the rate of assessment being 18 mills on the dollar)?

2. A sold goods to B at 12% profit, B sold them to C at 8% loss, and C sold them at 5% profit, realizing $81.92 more than their first cost. Find the average rate of profit on the goods and their first cost.

3. A grocer mixes sugars worth 6, $7\frac{1}{2}$, 8, and 10 cents a pound, respectively, forming a mixture, worth 9 cents a pound, which contains equal quantities of the first three kinds. In what quantities may he mix them?

4. A merchant bought a quantity of goods for $340, on 3 months' credit; how must he sell them to gain 25% ready money, after giving 9 months' credit (money being worth 8% per annum)?

5. A man owns $29680 stock in the 3 per cents, which he sells at 82; he invests one-half of the proceeds in the 4 per cents at 106, and the other half in the $4\frac{1}{2}$ per cents at 112. Find the alteration in his income.

6. A banker discounts a 143 day note at 10% per annum. What rate of interest is he charging for the money advanced?

7. Bought 600 barrels of flour at $7 per barrel; $\frac{1}{2}$ of which was to be paid in 6 months, and the balance in 9 months. If I paid cash in full on the day of purchase, find what I must pay, money being worth 6% per annum?

8. The wages of A and B together for $14\frac{1}{2}$ days amount to the same sum as the wages of A alone for 25 days. For how many days will this sum pay the wages of B alone?

9. What must I ask for cloth which cost me $1.20 a yard, so that I may fall 10%, and still make 20%, after deducting 5% of the sales for bad debts?

10. A grain dealer bought wheat at $1.40, barley at $0.50, and oats at $0.39 per bushel. In what proportion may he mix them so as to sell the remainder at $76\frac{1}{2}$ cents per bushel on a credit of 6 months, and make a cash gain of 20%, money being worth 4%?

11. Divide \$118 among A, B and C, so that B may have \$12 more than A, and C \$25 more than B.

12. Owing to a depression in trade, a merchant marks his goods at 7% loss, but upon business assuming a more favorable turn, he increases this price 8% and clears at the latter price \$22. What was the cost of the goods?

13. A huckster bought bananas at the rate of 3 for 4 cents, and retailed them at the rate of 20 cents a dozen for the first 12 dozen, on which he made his original outlay, the remainder he sold at 3 cents each. What did he gain per cent. on the whole transaction?

14. I sold an article so as to gain 8%. If I had bought it for 8% less and sold it for 48 cents less I would have gained 16%. Find the cost price.

15. A, B, and C form a joint partnership with a capital of \$1550; A's stock continues in trade 8 months, B's 6 months, and C's 5 months. A's gain is \$72, B's 90, and C's \$112.50. Find the stock which each put into the business.

16. A teacher has a net salary of \$891, after paying an income rate of 18 mills on the dollar on all his salary, except \$400. What was his salary?

17. By buying 3% stock at a certain price, I find that, after paying an income tax of $18\frac{3}{4}$ mills on the dollar, my net income is \$83.12½. Find the amount invested if I make $3\frac{1}{4}$% interest on my money.

18. A and B invest capital in a joint business in the ratio of 2 to 3; at the end of 4 months A decreases his capital 30%. How should a gain of \$1858.50 be divided between them at the end of the year?

19. A commission merchant charged \$348 for selling grain on a commission of 3% and for investing the proceeds, less this commission, and a commission of 2% (reckoned on the amount invested) in bank stock at 87. Find the amount of bank stock purchased.

20. Three men are employed on a work, working 8, 9, and 10 hours a day respectively, and receiving equal daily wages. After 3 days, each works 1 hour a day more, and the work is finished in 3 days more. If the sum paid for wages is \$14.25$\frac{5}{8}$, how should it be divided?

1. A merchant sold $\frac{5}{9}$ of a lot of goods at a gain of 25%, and the remainder for 85% of cost; his whole profit was $260. Had he sold $\frac{5}{9}$ at a gain of 15%, and the remainder for $\frac{3}{4}$ of cost, what would have been his gain or loss?

2. A clock which was 15 minutes fast at 11 a.m., on Wednesday, is exactly right at 4.30 p.m. on Friday. How many minutes will it be slow at 10 minutes to 7 p.m. on Saturday?

3. A man who owned a flock of sheep sold 20 more than 15% of them to A, 70 more than 13% of the remainder to B, 72 more than 16% of what then remained to C, and had 50% of his original flock left. What number had he at first?

4. I bought tea at 55 cents a pound cash, and sold it immediately for 68 cents a pound on 3 months' credit. If money be worth 8%, find my immediate gain.

5. A garrison of 1500 men was victualled for 70 days. After 11 days, it was reinforced by 1500 men. How long will the provisions last?

6. Divide $381 between A and B, such that 29% of A's share may exceed by $7.59, 41% of B's.

7. The length of a rectangular field is 3 times its width; another field which is 20 yards longer and 10 yards broader than the former, contains 1700 square yards more. Find the size of the former field.

8. In a city with $2000000 assessable property, the high school rate is $\frac{1}{2}$ mill on the dollar; the fees collected from the pupils amount to $3500, the Government grant to $1760, the expenses of running the school, not including teachers' salaries, to $1150. The trustees reserve $1000 of the net receipts, and the remainder exactly pays the teachers, $\frac{3}{5}$ of whose salaries are taxed at the rate of 2 cents on the dollar. Find the net amount expended by the city in salaries.

9. A grocer sells sugar at the rate of 49 ounces for 50 cents, which he bought at the rate of 50 ounces for 49 cents. Find his gain per cent.

10. A man paid $165 to 55 laborers, consisting of men, women and boys; each man received $5, each woman $1, and each boy 50 cents. How many were there of each?

11. I invested \$4690 in the $3\frac{1}{4}$ per cents at $87\frac{1}{2}$, and on the price of the stock rising to 90 I sold out and invested the proceeds in the 3 per cents, whereby my income was decreased \$1.40. At what price did I buy the 3 per cents?

12. A man insures a house worth \$4900 at 2%, so that in case of loss he may receive back both the value of the house and the premium of insurance. What amount did the policy call for?

13. A tailor buys cloth at \$1.90 a yard, which in sponging shrinks 5%. At what price per yard must he sell it to gain 10% on his outlay?

14. If 5 acres of grass, together with what grows during the time of grazing, keep 20 oxen 10 weeks, and 8 acres keep 29 oxen 16 weeks, how many weeks will 15 acres keep 70 oxen?

15. What sum of money is such that if $\frac{5}{8}$ of it be distributed among a number of persons, and the remainder be improved at 7%, simple interest, for 10 years, and then $\frac{5}{6}$ of the amount be again distributed, there will be \$170 left?

16. The sum of the ages of A and B is now 75 years, and their ages 18 years ago were as 2 to 1. Find their present ages.

17. Three men do $.5\dot{3}$ of a piece of work in $2 \cdot \dot{6}$ days. How long will it take 8 boys to finish it, 4 men and 3 boys having done a similar piece of work in 3 days?

18. An article which cost 19 guineas per cwt. is retailed at $4\frac{1}{2}$ shillings per pound. What is the rate per cent. profit, there being a waste of 5%? (112 lbs = 1 cwt.)

19. If 4 men or 7 boys can do a piece of work in 42 days, in what time can 7 men and 4 boys do it?

20. Milk is worth 28 cents a gallon, but by watering it the price is reduced to 5 cents a quart. Find the proportion of water to milk in the mixture.

21. A, B, and C engage in trade with a capital of \$500; A's money was in trade for 5 months, B's for 6 months, and C's for 9 months. The profits were divided equally; what stock did each invest?

22. A lot of land in the form of a rectangle contains 6 ac., 132 sq. rods; its length is to its width as 21 to 13. Find the number of rods of fence required to enclose it.

1. A man embarks his whole fortune in 3 successive speculations. In the first he gains 80% and in each of the others he loses 20%. What is his gain per cent. on the whole transaction?

2. A man who holds $3750 stock in the 5 per cents sells out $\frac{3}{4}$ of it at 117, and invests the proceeds in the 3 per cents at 75. Find the alteration in his income.

3. An agent sells 450 reapers for $125 each. He is to be responsible for the bad debts, which amount to 10% of the entire sales, and is to receive 14% of the good debts for his commission. What are his net earnings?

4. A drover sold oxen at $28 each, cows at $17, and sheep at $7.50, and received $749 for the lot. There were twice as many cows as oxen, and three times as many sheep as cows. How many were there of each kind?

5. A carter bought a horse for $60, a harness for $25, and a cart for $33; he earned on the average $1.50 a day throughout the entire year; he payed $11 a month for horse keep, $17.50 for repairs during the year, and $2 a month for a license. At the end of the year he sold his whole outfit for $100. Find his assets at that time.

6. Water expands 10% in freezing. Find the weight of water in a solid piece of ice 10 yards long, 5 feet wide, and 10 inches thick. (A cubic foot of water weighs 1000 oz.).

7. If 10 acres of grass keep 48 oxen 15 weeks, and 7 acres keep 34 oxen 14 weeks, how many acres will keep 38 oxen 16 weeks, the grass growing uniformly all the time?

8. A sold a lot of goods to B, B disposed of them to C, and C sold them to D for $1247.40. A gained 5%, B 8%, and C 10%. What did the goods cost A?

9. What rate of trade discount taken off 3 times in succession is equivalent to $57\frac{13}{16}$% off?

10. A barrel of sugar, containing 150 pounds, costs 6 cents a pound, and is retailed at 8 cents a pound. Supposing that there is 7% of the sugar wasted, and that the merchant is charged 25 cents for the barrel, which is worthless when empty, find his gain per cent.

11. In a shooting match, a bull's eye counts 4, a centre 3, and an outer 2. If a company of marksmen, consisting of 20, fire one round and score 40, 6 misses and 3 bull's eyes being made, find the number of centres and outers.

12. Divide $2567.50 among A, B, C and D, so that A's share may be to B's as 4 to 5, B's to C's as 6 to 7, and C's to D's as 8 to 9.

13. Two men and 5 boys can do a piece of work in 20 days, 1 man and 8 boys can do it in 18 days. In what time can a man or a boy do it?

14. What is the difference in the expense of fencing 2 fields of 25 acres each, one square and the other in a form of a rectangle, whose length is twice its breadth, the fence costing $62\frac{1}{2}$ cents a rod?

15. A commission merchant's charges, together with the money he invests, amount to $1700. If he charges a commission of 2% for investing the money, and his expenses amount to $17\frac{1}{2}$% of his total commission, what clear profit does he make?

16. If hay costs $12 a ton, and oats 40 cents a bushel, what will it cost to keep a horse from March 10th to Oct. 25th of the same year, supposing that the horse eats 24 lbs. of hay and 3 gallons of oats daily?

17. A merchant in Hamilton buys 25 tons of a cheap grade of coal for 75 cents a ton. He pays $1.75 a ton for freight; it contains 25% of bituminous coal, on which he pays a specific duty of 60 cents a ton, and he also pays an ad valorem duty of 20% on the whole quantity, reckoned on the prime cost. Find his gain by selling it at $3.50 a ton.

18. A farmer has 14 bushels of wheat worth $1.50 a bushel, 19 bushels of barley worth $.48 a bushel. How much oats worth 34 cents a bushel must he mix with the wheat and barley, to make a mixture worth $65\frac{1}{2}$ cents a bushel?

19. A Toronto firm is owed $4020 by a merchant in Vancouver. A draft for the amount on Vancouver is at $\frac{1}{2}$% discount in Toronto, and a draft on Toronto bought in Vancouver is at $\frac{1}{2}$% premium. Which is the better way to cancel the debt, and by how much?

1. A person has a sum of money to invest which will buy $1200 more 4% than $4\frac{1}{2}$% stock, when the former is selling at 87 and the latter at 99. Find the difference in the incomes from investing the sum in these securities.

2. A can do a piece of work in 8 days, B in 10 days, C in 12 days. The work is done by the 3 together, with the exception of B, who quits work 2 days before it is completed. In what time is the work done?

3. What rate of trade discount deducted twice is equivalent to $33\frac{63}{64}$% off.

4. Water expands 10% in freezing. Find the weight of the ice stored in a building 40 feet long, 32 feet wide, and 17 feet high, supposing there is no ice above the plate, and $\frac{1}{8}$ of the space is filled with sawdust.

5. A grocer sold a quantity of sugar at 12% advance on cost, and gained $40. Had he sold it at 15% advance he would have gained $50; find the cost.

6. Divide 635.75 into two parts, such that the simple interest on one part for 6 years at 5 may be $29.37 more than the simple interest on the other part for 4 years at 8%.

7. In a constituency, $26\frac{1}{4}$ of the voters refused to vote: one candidate polled 70 of those promised him, the other polled 75% of those promised him, and was elected by 150 votes. Had they each polled all promised, the successful candidate would have been defeated by 200 votes; how many voters were there in the constituency?

8. A can earn $10 in the same time that it takes B to earn $8, but at a piece of work on which they are both engaged A works only $\frac{2}{3}$ as long as B. How should $55 which was paid to both be divided?

9. A building society borrow $30000, and with this sum build 12 houses at an equal cost. They sell 4 of them at once for $2800 each, and rent 6 of them for $25 a month, and the remaining 2 are idle. If the houses which they hold are assessed for their full value at 15 mills on the dollar, and they pay 5 per annum for their borrowed capital, find their gain or loss at the end of the year.

10. Find the amount of money a person invests in the 3 per cents at 84, so that after paying an income tax of 2 cents on the dollar, his net income is $102.90.

11. 14% of a shipment of goods were admitted free of duty on account of damage received, and on the invoice price of the remainder a duty of 26% was charged; the duty amounted to $111.80 What was the invoice price of the shipment?

12. A quantity of goods invoiced at $1278, cost me in store $1452.38, after paying the duty and $14.63 for freight. What was the rate of duty?

13. A wine merchant has 35 gallons of wine worth $3.25 a gallon. What quantity of water must he add to lower the price to $2.60 a gallon?

14. For what price should a man sell a lot, which costs $550, so as to gain 12% of the proceeds on the sale?

15. A farmer has oxen worth $45 each, and sheep worth $6.25 each. The number of oxen and sheep being 35, and their value $645, find the number he had of each.

16. The compound interest on $5000 for 3 years is $1298.56. Find the rate charged.

17. One-fifth part of some goods was destroyed by fire, one-third of the remainder was sold at a loss of 5%. At what increase per cent. on cost must the balance be sold so that no loss may be sustained?

18. A person sold a house for $700, gaining 12% of the proceeds. What would he have sold it for, had he gained 12% of the cost?

19. Max Pierce rents a farm for $300 a year. If he does not pay the rent for 5 years, what will be the amount due, interest at 7%?

20. A tailor bought 50 yards of broadcloth, $1\frac{1}{2}$ yards wide, but on sponging, it shrunk 5% in width and 5% in length. He bought flannel $1\frac{3}{4}$ yards wide to line it, which shrunk 1 yard for every 16 yards in length, and $2\frac{2}{5}$ inches in width; how many yards of flannel are required?

21. What is the difference between 60% discount, and 20% taken off 3 times?

22. A, B and C engaged in trade with a joint capital of $2128; A's capital was in 5 months, B's 8 months, and C's 12 months. A's share of the profit was $228, B's $226.40, and C's $330. What was the capital of each?

1. The Commercial Company issued a policy of insurance on a vessel for $\frac{3}{5}$ of the value of the vessel and cargo at $4\frac{1}{2}\%$, and immediately re-insured $\frac{1}{2}$ the risk in the Manhattan Company at 5%. During the voyage the ship was lost, and the Commercial Company lost $120 more than the Manhattan Company. What did the owners lose?

2. A man invests his whole capital in 4 successive speculations. In the first he gains 60%, in the second he loses 25%, in the third he loses 20%, and in the fourth he gains 10%. What is his gain or loss per cent. on the whole transaction?

3. A and B run a hundred yard race; A takes 7 steps while B takes 9, but 11 of A's steps are equal in length to 13 of B's. Which will win the race, and by how much?

4. A person having a certain sum of money to invest, finds that by investing in the $3\frac{1}{2}$ per cents at 78, his annual income will be $7.50 greater than if he invested in the 5 per cents at $112\frac{1}{2}$. Find the sum he had to invest.

5. Two persons travelling together agree to pay expenses in the ratio of 8 to 5. The first (who contributes the greater sum) pays away on the whole $79.65, the second $24.35. What must one pay the other to settle the bill according to agreement?

6. Sold goods at $3.15 and gained $\frac{1}{4}$ of the cost price. What part would I gain if I sold for $3.33?

7. A vessel had two taps running into and one running out of it; the taps running into it can fill it in 5 and 7 hours respectively, and the third can empty it in 4 hours. If the vessel be empty and the three taps started simultaneously, how long before the vessel will be filled.

8. The compound interest on $15 for 3 years is $4.96\frac{1}{2}$. Find the rate per cent. charged.

9. What per cent. does a fruiterer gain, who buys lemons at the rate of 21 for 50 cents, and sells them at the rate of 40 cents a dozen?

10. A and B start at the same time to walk in the same direction. A is 8 miles behind B and travels at the rate of 5 miles an hour, while B's rate is 4 miles. Where will C have to start from, and what will be his rate, provided he wants to remain equi-distant from them while travelling?

11. A merchant marked his goods so as to gain 20%, but sold them for 8% less than his asking price. He gained altogether $130.65, what was the cost price of the goods?

12. What is the final value of an annuity of $750 for 9 years at 7% per annum?

13. A merchant gains 20% of his capital in each year for 3 successive years, and at the end of that time he is worth $10800. Find his original capital.

14. The law requires that a teacher's salary shall be paid quarterly. Find the value of assessable property in a school section which pays the teacher $500, and is compelled to give 3 notes for 9, 6 and 3 months, respectively, at 8% per annum, to pay the salary as it comes due, if a $4\frac{1}{2}$ mill rate leaves it a surplus of $124.81 after paying the teacher.

15. A grocer mixes teas worth 36, 48 and 60 cents a pound, respectively, forming a mixture worth $49\frac{1}{2}$ cents a pound, having equal quantities of the first two kinds and 10 pounds of the third kind. How many pounds of each of the first two kinds does he use?

16. A man buys 4% stock at $86\frac{1}{4}$, and after receiving one dividend sells out at $93\frac{1}{8}$, and clears altogether $108.75. How much did he invest at first?

17. A person bought a house for $3000. He insures it at $1\frac{1}{4}$% for $\frac{4}{5}$ of its value, pays $1.25 a month water rates, and the assessor assesses it at $\frac{5}{6}$ of its value (the rate of taxation being 19 mills on the dollar). If the property depreciates in value 5% during the year, what rate of interest does the owner make on his money, supposing that he receives $35 a month rental for the property?

18. A wine merchant buys 1240 gallons of wine at $3 a gallon. After adding 1 gallon of water to every 5 of wine, he bottles it in pint-and-half bottles, each of which costs 5 cents. At what price per bottle must he sell it to clear 25% on his whole outlay?

19. A mixture of black and green tea, weighing 16 pounds, costs $6.15\frac{1}{2}$. If the proportions are interchanged the mixture is worth $5.85; the black tea is worth 35 cents a pound. Find the number of pounds of each kind in the first mixture.

1. After paying an income tax of 18 mills on the dollar, a man has $245.50 left. What had he at first?

2. Find the compound interest on $17 for $8\frac{1}{2}$ years, at 6% per annum.

3. A merchant buys 1440 yards of cloth. He sells $\frac{1}{3}$ of it at a gain of 8%, $\frac{1}{6}$ at a gain of 12%, $\frac{1}{8}$ at a gain of 14%, and the remainder at 9% loss. Had he sold the whole at a gain of 5% he would have received $23.50 more than he did. What was the cost price per yard?

4. A merchant sold $\frac{1}{4}$ of a lot of goods at a profit of 12%, $\frac{1}{3}$ at a profit of 21%, $\frac{1}{6}$ at a profit of 25%, and the remainder at a loss of 45%. How much was his average gain per cent.?

5. A man rented a house for $25 a month for $2\frac{1}{2}$ years. What sum would pay the entire rent in advance, interest being calculated monthly at $\frac{1}{2}$% a month?

6. Twelve men engage to do a piece of work in 9 days. How long may 5 men remain away, and the work be finished in the same time by their bringing 10 men more with them?

7. A man, whose net income is $1156.10, secures it from the following sources, a fixed salary, and the rent of a house. What is his fixed salary, if on the house which rents for $25 a month, there is a mortgage of $1000 at 6% per annum, a $2000 insurance at $1\frac{1}{4}$%, taxes (calculated at the rate of 19 mills on the dollar) on an assessment of $2500, and on his salary (with $400 exempt) the same rate of taxation?

8. A, B, and C engage in business; A puts in $400 at first, and $100 more at the end of 6 months; B puts in $900 at first, and withdraws $\frac{1}{3}$ of his capital at the end of 6 months; C puts in $200 at the end of every 6 months; at the end of 2 years they gained $6700. What share of the profits should C receive in addition to 25% of the total profits for managing the business?

9. A man borrows a sum of money at 6% per annum, and invests it in 5% stock at 105; he receives the dividend at the end of the year, which is subject to an income tax of 2%, and immediately sells his stock at $112\frac{1}{2}$. If his net gain is $18.80, what sum did he borrow?

10. How much wine at $1.15 a gallon must be mixed with 16 gallons at $1.80 a gallon, and 46 gallons at 90 cents a gallon, to make a mixture worth 1.14\frac{1}{12}$ per gallon?

11. A trader bought merchandise as follows:—July 3rd, $35.26; July 4th, $48.65, on 30 days; August 17th, $6.48; September 12th, $50. What is due on account on October 12th (interest at 9%)?

12. A merchant mixes 9 pounds of one quality of tea with 3 pounds of another quality, and the mixture is worth 45 cents a pound, but if the quantities were interchanged in a mixture it would be worth 35 cents a pound. Find the price of each kind in the mixture.

13. A man has 7 hrs. 48 min. 45 sec. at his disposal. How far may he walk at the rate of 4 miles an hour, so as to return in time riding at the rate of 11 miles an hour?

14. Find the compound interest on $23 for 10 years, at 6% per annum.

15. A person bought cloth at the rate of 48 yards for $50, and sold it at the rate of 50 yards for $48. Find his loss per cent.

16. Received an invoice of crockery, 12% of which was broken. At what advance per cent. on cost must the remainder be sold to clear 25% on the invoice?

17. A man spends $25 a year in tobacco for 30 years. What would this amount to if placed in a bank, at 4% per annum?

18. A commission merchant sold flour for his principal at a loss of 10%, but if the flour had cost $1 a barrel less he would have gained 5% on the sale. What was the cost of the flour per barrel?

19. A and B start from different places to walk around a circular track; their rates are 6 and 4 miles per hour, respectively. Where would C have to start from, and what must his rate of walking be, so as to be alongside B every time that A is?

20. A can do as much work in 9 days as B can do in 10, but A works only 11 hours for every 12 B works. If their total earnings on a job be $109, how should the money be divided?

1. A man worth $10000 was offered a salary of $2000 per annum, and a chance to invest his $10000 on a first-class mortgage at 6% per annum. He, however, decided to go into business, and invested his $10000 in goods, on which he paid a duty of 24%, except on 12% which were damaged. He sold the undamaged goods at 65% advance on prime cost, and those which were damaged at 20% of the cost. He lost 5% of the sales in bad debts, and paid 1% on the remainder for collecting. Did he gain or lose by not accepting the first offer, and how much? (Interest reckoned on mortgage only.)

2. A man increases his capital yearly 20% of what it was at the beginning of each year What per cent. does he increase it in 4 years?

3. Three persons were to share $10000 in the proportion of 3, 4, and 5, but the first dying it is required to divide the whole sum equitably between the other two.

4. A man's income from the 3% consols is $720. If he sells out $\frac{1}{3}$ of it at 84, and invests the proceeds in other stock at 120, he will increase his income by $40. What per cent. does the latter stock pay?

5. Ten per cent. of an army was slain on the field of battle, 8% of the remainder was mortally wounded: the difference between the killed and the mortally wounded was 504. How many men went into battle?

6. A vessel A contains 2 gallons of wine and 3 of water; another vessel B contains 3 gallons of wine and 1 of water. How many gallons must be drawn from each cask so as to produce by their mixture 1 gallon of wine and 1 of water?

7. A square box, whose depth is 11 inches, has a cubical content of 4 cubic feet 113 cubic inches. Find the length of a side.

8. A man buys a farm for $8000, which he agrees to pay for in 12 years by equal annual instalments, without interest; the owner being pressed for money offers to take $6000 cash. Which is the better for the purchaser, money being worth 5% per annum?

9. Find the compound interest on $27 for $4\frac{1}{4}$ years, at 8% per annum, interest payable quarterly

10. A merchant in London (Eng.), owed another in Petersburg 9842 roubles, which he remitted through Paris, when the exchange was 25·35 francs for £1, and between Paris and Petersburg 3·39 francs for 1 rouble. Shortly after, the exchange between London and Paris was 25·625 francs for £1, and between Paris and Petersburg 3·37 francs for 1 rouble. How much would he have gained by the delay?

11. A merchant, who puts 5 gallons of water into every 25 gallons of coal oil, sells it for 2 cents a gallon less than it can be procured for elsewhere. If good coal oil is worth 18 cents a gallon, how much does he cheat a customer who buys 40 gallons?

12. The capital stock of a railway is $1750000, and its debt is $675000; its gross earnings for a year are $565-000, and expenses $384500. After paying the interest on the debt at 6%, and $52500 of the debt, what rate of dividend are they able to declare?

13. A grocer intended to gain 10% on a stock of tea, and fixed his price accordingly. When he had sold $\frac{3}{4}$ of the lot, he was compelled to lower his price 8 cents a pound, and so gained only two-thirds as much as he had intended. What was the original cost per pound of the tea?

14. If stock bought at 5% premium will pay 6% on the investment, what per cent. will it pay if bought at 15% discount?

15. I sold a consignment of goods through a factor, who charged me $1\frac{1}{4}$%; I was allowed $2\frac{1}{4}$% commission and $3\frac{1}{4}$% on the sales for insuring payment, and I cleared $51. Find the sum remitted to my employer.

16. A farmer rents a farm of 450 acres on the following terms: he pays a fixed rent of $1.50 an acre, and a corn rent of 100 bushels of wheat, 40 bushels of barley, and 75 bushels of oats. The price of the wheat, barley, and oats being 75, 48, 35 cents a bushel respectively. Find the entire rent paid.

17. How many railway shares ($100 each) at 10% discount must be sold in order that the proceeds, invested in bank stock, which is 4% below par, and pays a dividend of 7%, may yield an income of $1680?

1. Find the decimal of a foot, which differs from an inch by less than the millionth part of a yard.

2. A certain article of consumption is subject to a duty of 12 cents per pound: in consequence of a reduction in the duty, the consumption is doubled, and the revenue is increased one-quarter. Find the duty per pound after reduction.

3. Find the compound interest on \$25 for $8\frac{1}{2}$ years at 4% per annum, interest payable half-yearly.

4. A merchant in New York wishes to transmit 4500 marcs banco to Hamburg, and the exchange between New York and Hamburg is 35 cents for 1 marc banco. He finds, however, that the exchange between New York and Lisbon is \$1.08 for 1 milree, that between Lisbon and Paris is 6 milrees for 38 francs, and that between Paris and Hamburg is 19 francs for 10 marcs banco. How much will he gain or lose by the circuitous exchange?

5. What is the value of a perpetuity of \$450 a year to be given at the end of 15 years, money being worth 8% per annum?

6. Three men invest capital in business, in the proportion of 4, 5 and 6, on the understanding that the last is to receive 10% of the total profits for managing the business, and the remainder is to be divided in the proportion of the capital invested by each. The manager receives in all \$1600, find the total profits.

7. 14 oxen eat 2 acres of grass in 3 weeks, and 16 oxen eat 6 acres of grass in 9 weeks. How many oxen will eat 24 acres of grass in 6 weeks, the grass on each acre being equal at first and growing uniformly?

8. A merchant buys 12 dozen of port at \$18 a dozen, and 48 dozen at \$10 a dozen; he mixes them, and sells the mixture at \$14.50 a dozen. What profit per cent. does he make?

9. A bankrupt whose total assets amount to \$6000, owes A \$5000 and B \$1600; the assignee charges a certain rate per cent. of the assets for winding up the business, and other expenses amount to \$132. A receives for his share of the estate \$2931.25. Find the rate charged by the assignee.

10. Coffee, costing 40 cents a pound, is mixed with chicory worth 10 cents a pound, in the proportion of 7 to 3, and the mixture is sold for 35 cents a pound. Find the gain per cent. at which the mixture is sold.

11. One man sells stock and another buys it; if the broker who charges $\frac{1}{8}$% on each transaction makes \$56, find the amount of stock handled.

12. In building a house the owner paid twice as much for material as for labor; had he paid 5% more for material and 7% more for labor, the house would have cost \$10144. What was its cost?

13. The imperial gallon contains 277·274 cubic inches, and a cubic foot of water weighs 1000 ounces. Find the weight of a quart of water correct to three places of decimals.

14. A man divided a farm among 3 sons; to the first he gave 40 acres, to the second $\frac{1}{5}$ of the whole, and to the third $\frac{3}{4}$ as much as to both the others. How many acres did the farm contain?

15. Two casks contain equal quantities of liquid; from the first 27 quarts are drawn, and from the second 15 gallons; the quantity remaining in the first cask is double of that remaining in the second. How much did each cask originally contain?

16. What will it cost to paint a cistern without a cover, inside and out, at 12 cents a square yard, if the cistern is 30 feet long, 21 feet wide, and $8\frac{1}{2}$ feet deep?

17. The compound interest on \$430 for 3 years is \$90. Find the rate charged.

18. A merchant in Boston wishes to pay £4000 in Liverpool. Exchange on Liverpool is at par; on Paris, 5 francs 25 centimes for \$1; and on Hamburg 40 cents to a guilder. The exchange between France and England at the same time is 25 francs for £1, that of Hamburg on England $12\frac{1}{4}$ guilders for £1. Which is the most advantageous, to transmit direct, through Paris, or through Hamburg?

19. Find the present value of an annuity of \$475, to begin at the end of 7 years and to run for 9 years, money being worth 5%.

1. Find the measure of the altitude and the area of a triangle, the measures of whose sides are a b and c, respectively.

2. A farmer rents a piece of land for $120 a year. He lays out $625 on 75 sheep. At the end of a year he sells them, having expended $12.50 in labor. How much per head must he gain on them, in order to realise his rent and expenses and 20% gain on his original outlay for the sheep?

3. A merchant imported a quantity of goods, paying 15% for freight and insurance and 10% for duty (reckoned on the prime cost). He sold them at a loss of 10%, but had he sold them for $600 more than he actually did he would have made a profit of 2%. Find the invoice price of the goods.

4. A person invested $8001 in the 4 per cents at $95\frac{1}{4}$, and when they rise to 98 sells out and invests in the 3 per cents at 84. What amount of the latter stock does he obtain?

5. What per cent. of the first loss is the difference between 8% loss on the cost and 8% loss on that selling price?

6. A speculator bought 368 acres of land at $57.50 an acre, borrowing the money at 4%. At the end of the year he sells $\frac{3}{4}$ of it at $63 an acre, and the remainder at $50 an acre. How much does he lose by the transaction?

7. A merchant in London owes another in Petersburg a debt of 160 roubles, which must be remitted through Paris. He pays the requisite sum to his broker, at a time when exchange between London and Paris is 23 francs for £1, and between Paris and Petersburg 2 francs for 1 rouble. The remittance is delayed until the rates of exchange are 24 francs for £1, and 3 francs for 2 roubles. What does the broker gain or lose by the transaction?

8. A grocer by selling 10 pounds of tea for a certain price, gained 15%; afterwards he increased the price, giving only 8 pounds for the same money. What per cent. did he make at the increased price?

9. Having received a stock dividend of 6%, I find that I own $291\frac{1}{2}$ ($100 shares). How many had I at first?

10. A person leaves $12670 to be divided among 5 children and 3 brothers, so that after the legacy duty has been paid, each child's share shall be twice as great as each brother's. The legacy duty on a child's share being one per cent., and on a brother's share three per cent., find what amount they respectively receive?

11. A horse dealer sold two horses for $160 each, gaining the same per cent. on one as he lost on the other, and on the whole he lost 13\frac{1}{3}$. Find the per cent.

12. An English mile is ·2136 of a German mile. What time will a train, which travels 20 English miles an hour, take to travel 3$\frac{1}{4}$ German miles?

13. The compound interest on $435 for 6 years is $205. Find the rate per cent.

14. A merchant wishes to transmit 5600 marcs banco to Hamburg. He finds exchange between Montreal and Hamburg to be 36 cents for 1 marc. The exchange between Montreal and London (Eng.) is $4.83 for £1; that between London and Paris is 26 francs for £1; and that between Paris and Hamburg is 47 francs for 25 marcs. By which way should the merchant transmit?

15. A, B, and C engage in manufacturing shoes. A puts in $1920 for 6 months; B a sum not specified for 12 months; and C $1280 for a time not specified. A receives $2400 for his share of the stock and profits, B $4800, and C $2080. Required B's stock and C's time.

16. A speculator bought 35 ($100) shares of stock at 20% premium, and gave in payment a draft on New York for $4000. What was the rate of premium of the draft?

17. The net taxes raised by a corporation were $38600 from an assessment rate of 2%, subject to a commission of 3$\frac{1}{2}$% for collecting. Find the value of assessable property.

18. A sells goods to B at a loss of 4%, B sells them to C at a loss of 6$\frac{1}{4}$%, and C sells them to D for $130.20, gaining 8$\frac{1}{2}$%. Find the prime cost of the goods.

19. A baker's outlay is 70% of his gross receipts, and other trade expenses are 20%. The price of flour rises 50%, and trade expenses are thereby increased 25%. What advance must be made in the price of a 7$\frac{1}{2}$ cent loaf, that he may still realize the same amount of profit from it?

SOME PROPERTIES OF NUMBERS.

1. A number is divisible by 2 when its last digit is divisible by 2, or is zero.

2. A number is divisible by 3 when the sum of its digits is divisible by 3.

3. A number is divisible by 4 when its last two digits taken in order form a number divisible by 4, or are zeros.

4. A number is divisible by 5 when its last digit is 5 or zero.

5. A number is divisible by 6 when its last digit is divisible by 2 or is zero, and the sum of its digits is divisible by 3.

6. A number is divisible by 7 when the sum of *once* the units, or first digit, 3 times the second, 2 times the third, 6 times the fourth, 4 times the fifth, 5 times the sixth, once the seventh, 3 times the eighth, etc., is divisible by 7.

7. A number is divisible by 8 when its last three digits taken in order form a number divisible by 8, or are zeros.

8. A number is divisible by 9 when the sum of its digits is divisible by 9.

9. A number is divisible by 10 when its last digit is zero.

10. A number is divisible by 11 when the difference between the sum of the digits in the odd places and the sum of those in the even places, is divisible by 11 or is zero.

11. A number is divisible by 12 when its last two digits taken in order, form a number divisible by 4, or are zeros, and the sum of the digits is divisible by 3.

12. A number of three digits is divisible by 13, when 4 times the hundreds digits, increased by 3 times the tens digit, and then diminished by the units digit, gives a result divisible by 13.

13. A number is divided into periods of three figures each, beginning at the right. If the sum of the odd periods differs from the sum of the even periods by a multiple of 13, the number is divisible by 13.

14. The theorem of example 13 is true of divisors 7 and 11.

15. An even number is divisible by 14 when the test of example 6 is satisfied.

16. A number of three digits is divisible by 14, when twice the hundreds digit, increased by the units digit, and diminished by four times the tens digit, gives a result divisible by 14.

17. A number is divisible by 15, when its last digit is 5 or zero, and the sum of its digits is divisible by 3.

18. A number of four digits is divisible by 16, when the result of eight times the thousands digit, less 4 times the hundreds digit, increased by 6 times the tens digit, less the units digit gives a result divisible by 16.

19. A number of 4 digits is divisible by 17 when the sum of 3 times the thousands digit, twice the hundreds digit, and 7 times the tens digit, diminished by the units digit, gives a result divisible by 17.

20. An even number is divisible by 18, when the sum of its digits is divisible by 9.

21. A number of four digits is divisible by 19, when the result of 7 times the thousands digit, less 5 times the hundreds digit, increased by 9 times the tens digit, diminished by the units digit, gives a result divisible by 19.

22. A number is divisible by 25 when its last two digits, taken in order, form a number divisible by 25, or are zeros.

23. A number is divisible by 125 when its last three digits, taken in order, form a number divisible by 125, or are zeros.

24. Any number formed by writing down an odd number of digits, and then repeating them in order, is divisible by 11.

25. A number, formed by writing down three digits and then repeating them in order, is divisible by 7 and 13.

26. A number, formed by writing down four digits and repeating them in order, is divisible by 73 and 137.

27. Show that the difference between two numbers consisting of the same digits, arranged in different order, is divisible by 9.

28. If the sum of the digits of a number be subtracted from the number, the remainder is divisible by 9.

29. A number is divisible by 8 when 4 times the hundreds digit, twice the tens digit, and the units digit form a sum divisible by 8.

30. A number of three digits is divisible by 19, when 5 times the hundreds digit, added to the number formed by the other two digits, gives a sum divisible by 19.

31. A number of three digits is divisible by 9, if eight times the units digit equals the number formed by the other two digits.

32. A number of three digits is divisible by 13, if 9 times the units digit equals the number formed by the other two digits.

33. A number formed by writing any units digit and placing before it double the units digit is always divisible by 3 and 7.

34. A number of 2 or 3 digits is divisible by 7, when the units digit is one-ninth of the part on the left.

35. A number of 3 digits is divisible by 7 when twice the hundreds digit, added to the number formed by the tens and units digits is divisible by 7.

36. If a number be divisible by 11, the number formed by writing its digits in reverse order is also divisible by 11.

37. A reversible number, consisting of 2n—1 digits is divisible by 11, if the remainder on dividing the first n digits by 11 is one-half of the nth digit.

38. The difference between the square of a number of two digits and that of the number formed by reversing these digits' is divisible by 99.

39. To multiply a number by 25 affix two zeros at the right, and divide by 4.

40. To multiply a number by 125, affix three zeros at the right, and divide by 8.

41. To multiply a number by 125, affix two zeros at the right, and increase this result by ¼ of itself.

42. To multiply a number by 328, multiply first by 8; multiply this product 40, and take the sum of the two products.

43. Show how to multiply a number by 121728144, using three lines of partial products.

44. The product of any three consecutive numbers, increased by the middle number, is a perfect cube.

45. The sum of any fraction and its reciprocal is greater than two.

46. The difference of the squares of two consecutive numbers is equal to the sum of the numbers.

47. The square of any whole number ending in 5 may be found thus :—Strike off the 5, find the product of the resulting number and the next consecutive whole number, and affix 25 to this product.

48. The square of any mixed number, of which the fractional part is $\frac{1}{2}$, may be found as follows :—Find the product of the whole number and the next consecutive number, and affix the fraction $\frac{1}{4}$ to this product.

49. Show that $\frac{1}{99}$ produces a recurring decimal with a two-digit period.

50. Show that $\frac{1}{999}$ produces a recurring decimal with a three-digit period.

51. Show that $\frac{1}{99999}$ produces a recurring decimal with a five-digit period.

52. Show that $\frac{12347}{99999}$ produces a recurring decimal with a five-digit period.

53. Show that $\frac{4}{27}$ $(=\frac{148}{999})$ produces a recurring decimal with a three-digit period.

54. Show that $\frac{5}{271}$ produces a recurring decimal with a five-digit period.

55. Show that $\frac{5}{4649}$ produces a pure repetend with a seven-digit period.

56. Show that $\frac{3}{1010101}$ produces a pure repetend with an eight-digit period.

57. Show that $\frac{37}{333667}$ produces a pure repetend with a nine-digit period.

58. Show that $\frac{123456789}{265371653}$ produces a recurring decimal with a thirteen-digit period.

59. Show that the number of places in a repetend, when the denominator of the common fraction producing it is a prime, is always equal to the number of units in the denominator, less 1, or to some factor of this number.

60. A number, whose digits are all 9's, is either a multiple of 11, or is 2 less than a multiple.

61. Show that a fraction whose denominator is of the form 10^n+1, and whose numerator is unity, produces a repetend with a 2n-digit period.

62. The number of figures in a repetend cannot exceed the number of units in the denominator of the common fraction producing it, less 1.

63. Reduce $\frac{1}{7}$ to a repeating decimal, and hence show that a number consisting entirely of nines must contain at least six digits before it is exactly divisible by 7. Also, show that a number consisting of six one's is divisible by 7.

64. How many digits must there be at least in a number consisting entirely of 4's, so that it may be exactly divisible by 7?

65. What is the least number of digits greater than six, which must be in a number consisting entirely of 4's, so that it may be exactly divisible by 7?

66. Prove that a number consisting entirely of 9's must contain 16 digits to be exactly divisible by 17.

67. Prove that a number consisting entirely of 1's must contain 28 digits to be exactly divisible by 29.

68. Why does a proper fraction whose denominator is 37, a prime number, repeat after three decimal places?

69. In reducing $\frac{1}{17}$ to a repetend, when we obtain the remainder 16 we have found one-half of the repetend, and the remaining half may be found by subtracting the terms of the first half respectively from 9.

70. When 10^{16} is divided by 17 the remainder is 1.

71. Without dividing, show that 37 is a factor of 71824963214.

72. Given $\frac{1}{17} = .\dot{0}58823529411764\dot{7}$, show how to find $\frac{5}{17}$ by a short method.

73. Show that $\frac{5}{269}$ produces a recurring decimal, with not less than a 67-digit period. (It gives a 268-digit period).

74. Show that $\frac{5}{47}$ will produce a recurring decimal of not less than 23 digits.

75. Why does the square root of 87 produce a non-terminating, non-repeating decimal?

76. In reducing a proper fraction to a recurring decimal, the denominator being prime to 10, show that when in the process of division a remainder is produced which is equal to the difference between the numerator and denominator of the fraction, one-half of the period has been found; and the remaining half of the period may be found by subtracting in succession each digit found from 9.

Solution.—Let $\frac{a}{b}$ represent the fraction n the number denoted by the partial quotient when the remainder $b - a$ has been found, r the number of digits in n.

$$\text{Then } \frac{a}{b} = \frac{n + \frac{b-a}{b}}{10^r} = \frac{n + 1 - \frac{a}{b}}{10^r}$$

$$= \frac{n + 1 - \dfrac{n + \frac{b-a}{b}}{10^r}}{10^r}$$

$$= \frac{n}{10^r} + \frac{(10^r - 1) - n + \frac{a}{b}}{10^{2r}}$$

77. Change $\frac{8}{13}$ into a recurring decimal, obtaining only one digit by dividing by 13.

78. Change $\frac{12}{17}$ into a recurring decimal, obtaining only one digit by dividing by 17.

79. Change $\frac{16}{17}$ into a recurring decimal, obtaining only 2 digits by dividing by 17.

80. Change $\frac{7}{23}$ into a recurring decimal, obtaining only 1 digit by dividing by 23.

81. Find the quotient of 12 millions by 17, obtaining only 1 digit by dividing by 17.

82. Any power of an even number is even, and, conversely, the root of an even number, which is a complete power, is even.

83. A number consists of two digits; another number is formed by reversing the digits. One digit of the difference between the two numbers being 4, find the other.

84. Any power of an odd number is odd, and, conversely, the root of an odd number, which is a complete power, is odd.

85. In finding the product of two numbers, why are the partial products placed in a diagonal column?

86. In multiplying one number by another we can obtain the partial products in any order we please, and arrive at the true result.

87. Multiply 46987 by 4967, beginning with the 4 of the multiplier; then using the 9, and so on.

88. Explain clearly the reasons for the different steps in a problem in long division.

89. Prove that any number containing 7 digits is greater than the square of any number containing 3 digits.

Hence show that the square of the number represented by the 3 right-hand digits of any number containing 7 digits is less than the number of units represented by the remaining 4 digits.

90. Every time a remainder is obtained in the process of extracting the square root of a number, what kind of a number has been subtracted from the original number?

91. Between what numbers do the successive complete remainders indicate the differences, in the process of finding the square root of a number?

92. Between what numbers do the successive complete remainders indicate the differences, in the process of finding the cube root of a number?

93. In long division, if we neglect to introduce into the work at the proper time the successive digits of the dividend, we can, by neglecting in order the successive digits of the divisor, beginning at the right, obtain a certain number of digits in the quotient.

Show that this principle will hold in a case in square root, after a number of digits in the square root has been obtained by the ordinary method.

Example.—Extract the square root of 11 to 10 decimal places: first find 7 digits in the answer by the ordinary method, and then use contracted division.

94. A prime number cannot divide the product of two factors without dividing at least one of the factors.

95. A number cannot be resolved into prime factors in more ways than one.

96. If any number divides two other numbers, it will divide the sum or difference of any multiples of those numbers.

97. Multiply 46·493278 by 3·2946, correct to four decimal places, and explain why the following method will generally give the correct result:

Place down the partial products as far as the fourth decimal places, thus: 46·493 × ·2, is all of the one partial product that we keep. In *carrying*, find the product in the fifth place; if this product is less than 5 carry 0 to the fourth place, if between 15 and 24 inclusive, carry 2, and so on. Add the partial product thus obtained, and the result will generally be correct.

98. Divide 59·4687 by 47·32548, correct to 5 decimal places.

Use 47·3254 for first divisor instead of placing a 0 after the 7 in the dividend, and go back one place in the divisor to see what should be carried to the product of 4 and the first digit in the quotient.

Use 47·325 for second divisor instead of bringing down a 0 from the dividend and back to the 4 in the divisor to see what should be carried to the product of 5 and the second digit in the quotient.

Continue this operation as far as possible, employing the same rule for carrying as you have used in example 97. Explain the process.

99. Show without dividing that $\frac{1}{81}$, when reduced to a repetend, gives the digits in order with the omission of 8.

100. In our ordinary system of notation the value of any digit is increased ten-fold by removing one place to the left. How many units would be indicated by 23, if the value of a digit be increased eight-fold by removing one place to the left?

NOTE.—A number in which the value of each digit increases eight-fold by removing one place to the left, is said to be expressed in the scale of *radix* eight.

101. How many units are expressed by 123 in the scale of radix 4?

102. How many units are expressed by 10101 in the scale of radix 5?

103. Express in ordinary scale 12345 which is in the scale of 6.

104. Divide 125 by 4; divide the quotient by 4; divide the second quotent by 4. Show that $125 = 1 \times 4^3 + 3 \times 4^2 + 3 \times 4 + 1$.

105. Express 125 in the scale of radix 4.

106. Express 12345 in the scale of radix 5.

107. Express 52361 in the scale of radix twelve.

108. Express, in the ordinary scale, the number 4321, which is in the scale of 7.

109. Add 4678, 5236, 7134, 8216, which are in the scale of 9, and express the result in the scale of ten.

110. Subtract 4136 from 7014, each number being in the scale of 9; express the remainder in the scale of ten.

111. Multiply 1687432 by 6, each number being in the scale of 9; give the result in the scale of ten.

112. Multiply 41625 by 254, each number being in the scale of 7; give the result in the scale of 7.

113. Square the number 425, in the scale of 6; give the result in the scale of 6.

114. Divide 1738 by 144, in the scale of 9; give the quotient in the scale of 9.

115. Divide 1000002 by 110, in the scale of 4; give the quotient in the scale of 4.

116. Form the multiplication table up to six times in the scale of radix 7.

117. Write down the first ten numbers in the scale of 2.

118. Any number in the scale of 7, when divided by 6 gives the same remainder as the sum of its digits divided by 6.

119. How can a grocer in one weighing with a balance obtain $3\frac{1}{2}$ pounds of sugar, if the only weights he has at his disposal are a $\frac{1}{2}$, 1, and 4 pound weight respectively?

120. How can a grocer, by using not more than one of each of the weights, 1, 2, 4, 8, 16, 32 pounds respectively, weigh at one time 51 pounds of sugar?

121. By using not more than one of each of the following weights: 1, 3, 3^2, 3^3, etc., pounds, how can a grocer weigh 427 pounds at one weighing?

122. Every power of 6 necessarily ends with 6.

123. The product of 2 numbers, differing by 2, is one less than the square of the intermediate number.

124. Show that if any number which is a perfect square be divided by 3, it can never leave 2 for a remainder.

125. Show that the square root of 3 differs from $1\frac{3}{4}$ by less than $\frac{1}{50}$.

126. The square of 12345 is 152399025. Find the square of 12344, without going through the ordinary operation of multiplication.

127. Explain the following method of determining in which hand a person has the even number of coins, knowing that the person has an even number in one hand and an odd number in the other.

"Desire the person to multiply the number in the right hand by any even number whatever, and that in the left by any odd number; then bid him add together the two products, and if the whole sum be odd, the even number will be in the right hand, if the sum be even, in the left."

128. If we take any two numbers, then either one of them, or their sum, or their difference, is divisible by 3.

129. If we take any term of the series, $1\frac{1}{3}$, $2\frac{2}{5}$, $3\frac{3}{7}$, $4\frac{4}{9}$, etc., in which the whole numbers and the numerators are the successive natural numbers, and the denominators the successive odd numbers, beginning with 3, and convert it into an improper fraction, the numerator and denominator will represent two sides of a right-angled triangle, of which the hypothenuse would be represented by the number greater than the numerator by 1.

Thus $4\frac{4}{9} = \frac{40}{9}$, gives the sides 9, 40, 41. Explain.

130. Investigate the reason for the following rule for dividing a number by 365:—

Move the decimal point in the number two places to the left and divide by 4; then divide this quotient by 11, then this last quotient by 20; then again by 11 and 20 in the same way, and so on by 11 and 20 continually until the required degree of accuracy is reached: then the sum of all these quotients will be the quotient required.

131. In dividing by 73000 it is advantageous to do so by the following method :—

Having written down the number to be divided we write under it one-third of itself, then one-tenth of this second number, neglecting remainders, and lastly one-tenth of this third number. The sum of these four numbers with the last five figures, reckoned as decimals, will be the quotient required.

Establish the correctness of the method.

To what extent can its accuracy be depended upon?

Indicate a slight extension of the method which will enable any required degree of accuracy to be attained.

132. Prove the correctness of the following approximate method of dividing a number by 73 ;

Multiply the number by 137 ; subtract $\frac{1}{10000}$ of the product, and point off four decimal places.

133. To divide a number by 137 ;

Multiply the number by 73 ; subtract $\frac{1}{10000}$ of the product, and point off four decimal places.

134. If a is a little less than the cube root of a number N, then $\frac{2N + a^3}{N + 2a^3}a$, is a closer approximation.

135. By successive applications of the principle in example 134, find the cube root of 5, true to 3 decimal places.

136. The square of the sum of a series of the natural numbers, beginning with 1, is equal to the sum of the cubes of the same numbers.

PROBLEMS SELECTED FROM TORONTO UNIVERSITY MATRICULATION PAPERS.

1. The value of the old Spanish dollar (which was the unit of exchange between England and America), was 4s. 6d. sterling, but gold became the standard of the U. S. currency by the acts of 1834-7, which made the gold eagle weigh 258 grains, being nine-tenths fine. The English coinage is of metal 22 carats fine, 40 pounds being coined into 1869 sovereigns. With these data explain why the bank par of exchange between New York and London is said to be $109\frac{1}{2}$.

2. Prove the following rule for computing interest at six per cent. per annum for a period of months and days, the substance of which was given in the *Leader* of March 11, 1865.

Multiply the number of months by 5, and add one-sixth the number of days; multiply this sum by the principal expressed in dollars; the result will be the interest expressed in mills.

3. A tradesman who gives six months' credit abates 5 per cent. for cash. Find the rate of interest in order that this may be the true discount.

4. "We are advised by telegraph that, on Friday, United States Five-Twenties were sold at the London Stock Exchange for $68\frac{1}{2}$ gold. The same class of securities brought here $110\frac{1}{4}$ currency. Taking the average gold premium at fifty per cent., and adding the par of exchange, which is about eight per cent., the reader can calculate for himself that $68\frac{1}{2}$ gold in London is pretty near $110\frac{1}{4}$ currency in New York, or, more exactly speaking $110\frac{3}{4}$." *New York Times*, Tuesday Aug. 21, 1866.

Give a full explanation of the preceding extract.

5. A grocer buys 150 pounds of coffee at 14 cents per pound, and 39 pounds of chicory at 6 cents per pound; he pays an import duty of $12\frac{1}{2}$ per cent. ad valorem, and mixes and sells them at 25 cents per pound, but by the use of a false balance gains $\frac{1}{4}$ ounce on every apparent pound sold. Find the profit per cent. made on his outlay.

6. By the Canadian Statute it is provided that the silver coins of the Canadian currency shall bear the same relation to the pound currency that the sterling silver coins bear to the pound sterling, being also of the same standard of fineness. Sterling silver is 92.5 per cent. fine, and from one pound troy of this metal are coined 66 shillings. The pound sterling is said to be equal to £1 4s. 4d. currency, or \$4.86$\frac{2}{3}$, the pound currency being \$4. In Martin & Trubner's "Currency," the Canadian ten cent piece is said to weigh 38·42 grains and to be $\frac{9}{10}$ fine, but an analysis by Prof. Croft shows that the fineness is that of sterling. The American mint asserts the value of this piece to be about 9$\frac{7}{8}$ cents, their dollar containing 345·6 grains pure silver.

Examine the consistency of these statements.

7. Two persons, A and B, borrow \$300 on joint mortgage from a building society, A taking 200 and B 100, the amount being repayable, principal and interest, by equal monthly instalments, to which A and B contribute proportionally. After a few payments have been made, they desire to borrow each \$200 additional and propose to merge the old debt and the new into a single mortgage for \$700, whereupon the account stands thus;

Amount required to pay off old mortgage..	\$245
" of new loan........................	400
Surplus..................................	55
	\$700

Discuss the interest which each has in this surplus of \$55, and the proportion in which they should contribute to the instalments payable in future.

8. A quantity of pulp, filling a trough 3 feet deep, 10 feet 7 inches long, and 11 inches wide, is made into paper of the same width and of such a thickness that 12 sheets would measure a quarter-inch. Find the length of the paper made, the pulp losing $\frac{2}{5}$ of its bulk in manufacture.

9. Explain what is meant by interest and discount. Find the time for which the discount on a certain sum of money will be equal to the interest on the same sum for a year; the rate of interest in both cases being 5 per cent.

10. Two persons, A and B, start from the same corner of a rectangular field to race around it in opposite directions, A giving B $31\sqrt{2}$ yards' start. A's speed is greater than B's by one-tenth, and he has not reached the opposite corner by $11\sqrt{2}$ yards when he meets B. Prove that the area of the field cannot be greater than 6050 yards, nor its diagonal less than 110 yards.

11. The Sovereign weighs 123·275 grains, being of 22 carats, and the Napoleon (20 francs) weighs 99·561 grains, being nine-tenths fine. It is proposed to coin a piece of twenty-five francs, and to make this and the sovereign interchangeable. Supposing that the relative alteration in the standards of the two countries is to be the same, how can this be done with the least disturbance of the coinages, and what will be the weight of the pure gold in each?

The Chancellor of the Exchequer is reported (*Times*, June 7), to have said that by retaining one grain out of each sovereign (that is—diminishing its weight by one grain) by way of seignorage, it would be identical with the 25-franc piece. Is this so? If not, what did the Chancellor probably mean?

12. What is meant by "the funds?" Explain why the English funds rose on the birth of the Prince Imperial of France.

A person holds stock in the English $3\frac{1}{2}$ per cents., which are at 98, to the amount of £1500 sterling. This he transfers to Canadian Government 6 per cents., which are at 105. Find the alteration in his income in dollars, if one pound sterling is worth $4.87.

13. Explain the distinction between *simple* and *compound* interest, and between interest and discount.

What rate per cent. per annum interest is discount on a note for one year at 7 per cent.?

What rate per cent. per annum interest (compound) is discount on a note for half a year at $3\frac{1}{2}$ per cent.?

14. Find the present value of a $1000 Government debenture with coupons for semi-annual payments of interest at 7 per cent. attached, the debentures to mature at the end of 3 years, when I want 8 per cent. interest for money, interest also payable semi-annually.

15. The following is the rule given in the *Lilavati* for finding the square of a number :—

"Place the square of the last digit over the number; and the rest of the digits doubled and multiplied by the last are to be placed above them respectively; then repeating the number with the omission of the last digit, perform the same operation (and take the sum of these numbers); thus to find the square of 297,

```
4
36
 81
 28
 126
   49
-------
  297
88209."
```

—*Peacock's Arithmetic.*

Investigate the reason for this rule.

16. A grocer buys a stock of tea, and sells $\frac{5}{6}$ of its nominal amount at 84 cents per pound, thus clearing $190; he now calculates that if he sells the remainder at 87 cents per pound he will, on the whole, make 30 per cent. on his outlay. But he has forgotten to take into account a loss of weight of two per cent. by waste in handling. How much less cash will he receive than he expected?

17. A person buys a quantity of tea in New York at 90 cents per pound, when gold is at 250, and pays a duty on it of 20 per cent. in bringing it into Canada. He sells it when greenbacks have risen to 60, for silver on which there is 4 per cent. discount in buying Canadian bills (gold in the purchase of them being at one per cent. premium in New York). Find the rate at which he sells, so as to make ten per cent. on the outlay.

18. A person pays in British gold coin for an English acre of land, at the rate of ten francs per metre. What weight in gold does he give, supposing an English sovereign to be $4.86, and the alloy to be worth one-tenth part of the gold, a French metre being 39·38 inches and a franc 17 cents, the gold in the sovereign weighing 110 grs. and the alloy 10 grs.?

19. A person purchases a quantity of goods in Liverpool for £37 10s. sterling, and sells in Montreal for £65 Canadian currency; he pays in Montreal an ad valorem duty of $4\frac{1}{2}$ per cent. Neglecting other incidental expenses, what is the gain per cent., supposing a pound sterling to be worth $4.87?

20. A, B and C engage in trade; A contributes £150, B £200 and C £250. At the end of two years A draws £100, and one year after B draws £150. When the partnership is wound up at the end of four years, it is found that there is to the credit of the firm the sum of £1000. Are the data sufficient to enable us to make an equitable distribution? Give reasons for your answer. If sufficient, determine the amount to which each is entitled.

21. Explain the common system of numerical notation. If the decimal point in any number be moved one place to the left, and then again, and so on, and the numbers thus formed be added together, the sum is the result of dividing the original number by nine.

22. Sterling gold is 22 carats fine, and from 40 pounds of it are coined 1869 sovereigns. Jewellers' gold is 18 carats fine. An ornament made of the latter, and weighing 22 ounces, was sold at an advance of two-thirds of its value by weight, and the jeweller's profit was equivalent to £$1\frac{409}{660}$ per ounce on the pure gold contained in it. What was the charge for workmanship, disregarding the value of the alloys?

23. To divide a number by a divisor 111 . . . . . (where 1 is repeated any number of times), remove the decimal point in the number as many places to the left as there are one's in the divisor; repeat this operation successively, and add together the numbers thus formed, then subtract the number which is their sum from the same number with the decimal point removed one place to the right.

Prove the above rule.

24. A grocer has two kinds of tea which cost him seventy and eighty-five cents a pound respectively, and mixes them in the proportion of two pounds of the former to one of the latter. At what price should he sell the mixture in order to realize thirty per cent. profit?

25. Find the square root of 3 to twelve places of decimals and deduce the values of $\frac{1}{\sqrt{3}}, \frac{1+\sqrt{3}}{1-\sqrt{3}}, \sqrt{0 \cdot 3\dot{3}}$.

26. The value of a diamond (brilliant cut) varies as the square of its weight, and a diamond of one carat is worth £16; find the value of the Koh-i-nor, which weighs $102\frac{1}{2}$ carats.

27. The Yorkshire coal field extends over 940 square miles, with an average depth of seam 70 feet; coal is about $1\frac{1}{4}$ times as heavy as water, of which a cubic foot weighs 1000 ounces. The annual consumption of coal in Great Britain being 80 millions of tons, how many years will this field furnish the supply at that rate?

28. Find the present worth of a promissory note of $728, due 2 years hence (a) at simple interest, (b) at compound interest, money being at 6 per cent.

29. A person holds $1000 of Bank of Toronto stock, and directs a broker to dispose of it, and invest the proceeds in Ontario Bank stock, after deducting his commission. The broker sells at 179 and buys at 107, and charges one-third per cent. commission for each transaction. How is the income of the owner affected, supposing the Bank of Toronto to pay twelve and the Ontario Bank eight per cent. per annum?

30. A dealer sells at a profit of 25 per cent. His purchaser fails, paying 75 cents on the dollar. How much per cent. does the dealer gain or lose?

31. A person bought an amount of tea at the rate of 8 pounds for 5 dollars, and sold half of it at the rate of 5 pounds for 3 dollars, and found he was losing money. He then sold the remainder at 3 pounds for 2 dollars, and on the whole transaction gained $1. What was the loss per cent. on the first sale, and the gain per cent. on the last?

32. A person has three notes, one for a certain amount, and each of the others for half as much, and all three due in twelve months; he gets the note for the largest sum discounted at 10 per cent., and the two others at 8 per cent. and 12 per cent., respectively, and finds that the sum of these two discounts is less than the discount on the other note by $\$3\frac{58}{189}$. Find the amounts of the notes.

33. A ladder, 100 feet long, stands in a vertical position against a tower. How much will the top of the ladder be lowered by drawing out its foot 10 feet in a horizontal plane?

34. Investigate the method of extracting the square root of any number, giving reasons for the successive steps in the process. Obtain the square roots of

1746238 ; 59810432 ; 894651027 ;

to four decimal places.

35. Gold is purchased in New York to the amount of 3300 dollars at the quotation $112\frac{1}{3}$. It is then brought to Toronto, where it is all expended in the purchase of United States currency notes at the quotation $88\frac{1}{2}$. These notes are then expended in the purchase of gold at the quotation $111\frac{1}{2}$. How much was gained by the transaction, and what percentage was made on the amount originally laid out?

36. Prove the rule for pointing in the extraction of the cube root of a number.

There is a metal cubical box of 96 feet surface and $1\frac{1}{2}$ ft. thickness; also, three solid cubes of another kind of metal, whose surfaces are as the numbers 1, 4 and 9, and whose combined weight equals that of the box. Find the lengths of the edges of the cubes, the weight of the latter metal being to that of an equal bulk of the former as 3 is to 4.

37. $500 is offered by a building society, to be repaid in two annual instalments of $285 dollars each, so that the debt is liquidated at the end of two years from the present. Find the society's rate of interest.

38. A bank wishes to realize 4 per cent. interest on its discounting operations. Form for it a table of the rates at which it must discount notes payable in 30, 60 and 90 days, respectively.

39. A person who holds stock in the bank of Montreal, which is at 185, to the value of $20,000, sells out and invests in Royal Canadian bank stock, which is at 99. If the Montreal bank pays a dividend of 10 per cent., and the Royal Canadian a dividend of 6 per cent., find whether his income is increased or decreased, and to what extent.

40. State the advantages arising from the employment of Bills of Exchange. Define "Par of Exchange" and "Course of Exchange;" mention any causes that influence the latter.

41. Gold is quoted at 113 in Boston, and Nova Scotia bank notes at 110; American gold is at a premium of 3 per cent. in Nova Scotia. If a Halifax merchant wishes to pay a Boston creditor $6000 American currency, determine whether it would be more advantageous to remit gold than Nova Scotia currency, and how much (in Nova Scotia currency) he will save by following the more profitable course.

42. A grain dealer laid out a certain sum in oats, 25 per cent. more in barley, and in wheat 20 per cent. more than in oats and barley together; he sold the oats at a profit of 8 per cent., the wheat at a profit of 14 per cent., and the barley at a loss of 5 per cent, receiving altogether $12829.20. Find the amount invested in each kind of grain.

43. A man holds three notes, the first for $1000 due April 1st; the second $1600, due July 1st; the third $1200, due Sept. 1st. He has them exchanged for two others, one of which is for $2000 payable May 1st. Find when the second note matures.

44. A person borrows $100 at 10 per cent. simple interest, and repays the same, principal and interest, in four equal quarterly payments. Find the amount of each payment.

45. A merchant becoming embarrassed compounds with his creditors and gives them forty cents on the dollar and has $5000 left for himself. Had he made an assignment and given up everything to the creditors they would have received forty-five cents on the dollar, after paying the expenses of the insolvency, which would amount to one-tenth of the whole estate. Find the assets and liabilities of the insolvent.

46. Between 2 and 3 o'clock the minute hand was between 3 and 4, and about an hour afterwards the hour and minute hands had changed places. What was the exact time at the first observation?

47. Define interest and discount.

What rate of discount is equivalent to seven per cent. per annum interest?

A merchant discounts with a bank at the rate of seven per cent. per annum a note for $950 at three months, bearing interest at eight per cent. per annum. With what amount is he credited?

48. A person sells $12000 of Dominion bank stock at 112 and invests the proceeds in New York Central railway stock selling at $98\frac{3}{4}$. What yearly dividend should the latter stock pay in order that his income may be unchanged, the Dominion bank paying half-yearly dividends of four per cent., and gold being quoted at $112\frac{3}{4}$?

49. A at his death leaves $35,000 to be divided between his two sons, in proportion to their ages, on the elder coming of age. Had he left it to be divided in proportion to their ages on the younger coming of age, the latter would have received $1000 more than he does. Find the difference in their ages.

50. A is engaged to do a piece of work and is to receive $3 for every day he works, but is to forfeit one dollar for the first day he is absent, two for the second, three for the third, and so on. Sixteen days elapse before he finishes the work and he received $26. Find the number of days he is absent.

51. Having found a certain number of figures in the extraction of the square root of a number, show how we may obtain as many more, less one, by a simple process of division.

Example :—Find the square root of 265·32 to within a unit of the fifth decimal place.

52. There are three towns, A, B and C; the road from B to A forming a right angle with that from B to C. A person travels a certain distance from B towards A, and then crosses by the nearest way to the road leading from C to A, and finds himself three miles from A and seven from C. Arriving at A he finds he has gone further by one-fourth of the distance from B to C than he would have done had he not left the direct road. Required the distance of B from A and C.

53. On a railway are two parallel tracks; on one of these, trains pass a certain point every $49\frac{1}{2}$ minutes; on the other, the same point, every $52\frac{1}{4}$ minutes. A train on the former track has just passed this point, and in $27\frac{1}{2}$ minutes one on the latter will do so. Will trains on these tracks ever pass this point at the same instant? If so, in what time from the passage just mentioned?

54. Prove the following rule for finding approximately the interest on $1 for any number of days, at six per cent.: Divide the number of days by six and call the quotient mills.

The interest on a certain sum of money for 73 days obtained by this method, differed from the true interest by 5 cents. What was the sum of money?

55. A and B are to race from M to N and back; A moves at the rate of 10 miles an hour, and gets a start of 20 minutes. On A's returning from N he meets B moving towards it, and one mile from it; but A is overtaken by B when one mile from M. Find the distance from M to N.

56. A mortgage dated 1st January, 1872, payable in three equal annual payments of $200 each, with interest on the whole, payable half-yearly at six per cent., is sold on the 1st July, 1872. What sum must the purchaser pay so that the investment may be worth 8 per cent.?

57. The captain of a privateer seeing a trading vessel 10 miles a head, sailed 15 miles in direct pursuit of her, and then observing that the trader steered at right angles to her former course, changed his own course so as to overtake her without changing the direction of his ship's motion, the privateer running 10 and the trader 7 miles an hour. Find the whole distance travelled by the privateer before overtaking the other, and the time occupied.

58. A person has a certain amount of bank stock which he sells at $110\frac{1}{8}$, and invests the proceeds in 5 per cents at $79\frac{7}{8}$. When this has risen $5\frac{1}{4}$ per cent., he purchases the same amount of the original stock as he held at first at $109\frac{7}{8}$, which now pays 8 per cent., and finds that while $110 remain in cash his income has fallen $8. Find the percentage originally paid by the bank stock, allowing $\frac{1}{8}$ per cent. brokerage on each transaction.

59. Three laborers, whose rates of working are equal, can complete a certain piece of work in three hours, working together. They work together an hour when one stops, and each of the others works half as hard again; at the end of the second hour another stops, and the remaining one works half as hard again as he did during the second hour. If in each successive hour he increases his rate of working in the same ratio, in what time will the work be completed.

60. A and B are a certain distance apart. A person to whom this distance is known has two foot-rules, which have expanded uniformly at different rates, the difference in their lengths being $\frac{1}{8}$ of an inch; he lays off a certain distance from A towards B, and with the same rule lays off from B towards A, the nominal difference between the whole distance and the distance first laid out, and notes the distance between the two points thus found. He performs the same operation with the second rule, and finds the distance between the two points in the latter case to be one inch less than in the former. What is the distance from A to B?

61. Explain the carat system of expressing the quality of gold.

What is the quantity of standard metal contained in 21 lbs. 10 oz. 18 dwt. 12 grs. of gold reported worse 1 carat $3\frac{1}{4}$ grs.?

62. An oarsman finds that during the first half of the time of rowing over any course he rows at the rate of five miles an hour, and during the second half at the rate of four and a half miles. His course is up and down a stream which flows at the rate of three miles an hour, and he finds that by going down the stream first and up afterwards it takes him one hour longer to go over the course than by going first up and then down. Find the length of the course.

63. A man has real estate from which he receives an income at the rate of ten per cent., without allowing for taxes. On both income and property he is taxed at the rate of $19\frac{1}{2}$ mills on the dollar. At what rate is his property taxed altogether?

64. Find the smallest sum that can be exactly paid with both shillings sterling and currency, one pound sterling being worth \$4.84.

65. Bank of Commerce stock is worth 120 and pays a dividend of 8 per cent. per annum. Find the income from 100 shares, and the amount obtained by the sale of them, allowing the broker a commission of $\frac{1}{8}$ per cent.

66. A grocer mixes 40 gallons of whiskey at 75 cents, 40 at \$1.50, and a certain number of gallons at one dollar. After keeping the mixture a year, by selling it at \$1.35 a gallon, he would have gained twenty per cent. profit and six per cent. interest on his capital; but owing to a leakage he gains his interest and $16\frac{1}{2}$ per cent. profit. Find the number of gallons that leaked out.

67. A man is to row over a certain course in a certain time. By rowing at the rate of four miles an hour he would arrive 5 minutes too late; and by rowing at the rate of five miles an hour he would arrive 10 minutes too early. Find the length of the course and the time of rowing.

68. A room is 20 feet long, 16 ft. wide and 12 ft. high, with openings of area 94 sq. feet. It takes as much plastering as another room which is as long as broad, 10 ft. high, and whose openings have an area of $70\frac{1}{4}$ square ft. Find the length of the second room.

69. \$1200 is to be distributed among A, B and C. From part of it they receive equal amounts, and of the rest B's share is 10 per cent. more than A's share, and C's is 10 per cent. more than B's. Altogether B's share is $8\frac{16}{123}$ per cent. more than A's, and $7\frac{23}{36}$ per cent. less than C's. Find the part of the \$1200 that was divided equally.

70. A certain amount of six per cent. stock at 95 is sold out; and, being invested in the $7\frac{1}{2}$ per cents at a certain price, it is found that the resulting income, after deducting an income tax of one per cent. is two per cent. more than the previous one, after deducting an income tax of 2 per cent. Find at what the second stock is quoted.

71. There are four cannon balls of diameters 3, 4, 5 and 6 inches, respectively. Show that the weight of the largest is equal to the combined weight of all the others.

72. A person invests $5445 in stock paying six per cent. when at $90\frac{3}{4}$, and on the stock rising to 91 transfers to another stock paying 7 per cent., which is selling at $97\frac{1}{2}$. How much is his income increased?

73. Find the amount accumulated at the end of three years by a person who invests $5000 now, and does the same at the beginning of each succeeding year, at 8 per cent. compound interest on the whole sum invested.

74. Find the difference between the simple and compound interest on $5000 for 5 years, at 8 per cent.

75. The price of diamonds per carat varies as their weight. If a diamond of three carats is worth $342, what is the value of a diamond of 4 carats?

76. A person has $20000 invested in stock paying six per cent., which he sells and invests in stock paying seven per cent. at $87\frac{1}{2}$. If the increase in his income is $40, what is the price of the first named stock?

77. Divide 111 into three parts so that the products of each pair may be as 4:5:6.

78. Distinguish between interest and discount and show that if P, I, D be respectively the principal sum, and the interest and discount upon it for any given time,

$$\frac{1}{D} = \frac{1}{I} + \frac{1}{P}.$$

79. A person has an income derived from £3360, which was originally invested in the 4 per cents. at 96. If he now sells out at 94, and invests one-half of the proceeds in railway stock at $82\frac{1}{4}$, which pays a dividend of 3 per cent., and the other half in bank stock at $164\frac{1}{2}$, paying $8\frac{1}{2}$ per cent. dividend, what difference will he find in his income?

80. A started from Ottawa at 9 a.m. to walk to Chelsea. After he had walked $1\frac{1}{8}$ miles, B started and overtook A half way there. A then increased his pace one-fifth and B decreased his one-ninth, and they reached Chelsea together at $11{:}28\frac{1}{2}$ a.m. Find the distance to Chelsea.

81. A merchant receives $300 for sales in one day. On $200 he gains 40 per cent. What does he gain or lose per cent. on the remaining $100, in order that his profit for the day may be $50?

82. A county borrows $150,000, to be paid off, principal and interest, in twenty equal annual instalments. Find the annual payment, interest at six per cent.

83. A proprietor of three per cent. consols receives his half-yearly dividend, and lays it out in the purchase of more consols at 90. His next half-year's dividend is £457 10s.; how much does this dividend exceed the former?

84. Find the value of

$$\left(\frac{2}{\cdot\dot{3}} - \frac{4}{\cdot\dot{7}}\right) \text{ of } \cdot 6\dot{3}5\dot{5} \text{ of } £13\ 16s.\ 6d.$$

85. A merchant buys cotton (27 inches wide) at 5 cents per square yard. He pays a duty of 2 cents per square yard and 15 per cent. ad valorem. For what price per yard should he sell it in order to gain 25 per cent. on his outlay?

86. A person bought a lot of land for $10000. He sold one-half of it at a gain of 50 per cent., two-fifths of it at $40 an acre, and the remainder at a loss of 40 per cent. He gained 45 per cent. on the whole. Find the number of acres in the lot.

87. A and B can do a piece of work in three-fifths of the time in which A can do it alone. B can do three-fourths of the work by working five days longer than it takes A to do the whole work. Find in what time each can do the work.

88. Sulphuric acid contains $\frac{100}{49}$ per cent. of hydrogen. When zinc is put into sulphuric acid all the hydrogen is set free and zinc sulphate is left, which contains 40·32 per cent. of zinc. If 50 cubic inches of hydrogen weigh one grain, how much zinc would be necessary for the preparation of enough hydrogen to fill a balloon ten feet in diameter?

89. Extract the cube root of 2007 to two decimal places. Show how the cube root of any perfect cube less than a million may be determined by inspection.

90. If $300 be laid out at simple interest for a certain number of years it will amount to $360. If the same be allowed to remain two years longer, and at a rate of interest one per cent. higher, it will amount to $405. Find the rate and the number of years.

91. A Canadian company borrows in Paris 294000 francs, for which it pays an annual interest of $2920. This loan is transmitted through London, when exchange on London is quoted at 25·30 francs, and sterling exchange is 109$\frac{3}{8}$. Find what rate of interest the company pays on the money actually received.

92. Assuming a metre to be 39·37 inches, express a yard in terms of the metre, and an inch in terms of the centimetre.

The areas of Ontario and Canada are respectively 121260 and 3500000 square miles. Express these in square kilometres.

93. A quantity of sugar, valued at $42134 Spanish gold, was entered for duty at 30%. In consequence of Spanish gold having been taken at par, whereas it was only worth 92$\frac{1}{2}$ cents on the dollar, a refund of duty was afterwards claimed. Calculate the amount.

94. How much will $1000 amount to in 2$\frac{1}{2}$ years, compound interest at 4 per cent. per annum, payable half yearly?

95. A person pays $292.50 for $300 due in three months hence. What rate per cent. interest does he receive?

96. What is meant by the expression, "Sterling exchange 9$\frac{1}{2}$ p.c. premium?"

A person pays $181.50 for £37 10s. stg. What per cent. premium is sterling exchange?

97. A waterman rows a given distance a and back again in b hours, and finds that he can row c miles with the stream in the same time as d miles against it. Find the time each way and the rate of the stream.

98. Four points, moving each at a uniform speed, take 198, 495, 891 and 1155 seconds, respectively, to describe the length of a given straight line. Supposing them to be together at any instant at the same end of the line, and to move in it from end to end continually, what interval of time will elapse before they are together at the same point again?

99. A person invests $4700 in shares which are at 98, paying 3$\frac{1}{2}$ per cent., and the same in 3 per cent. consols at 94. Find the income from each.

100. Two trains, 92 and 84 feet long, respectively, moving with uniform velocity in opposite directions on parallel rails, pass each other in $1\frac{1}{2}$ seconds; when moving in the same direction the faster passes the other in 6 seconds. Find the rate at which each train moves.

101. A cistern is 2 metres long, 5 decimetres broad and 8 centimetres deep. What is the quantity and weight of water it will contain?

102. A crystal weighs 1·53 ounces in water and 1·73 in naphtha of specific gravity ·85. Find its actual weight, bulk and specific gravity.

103. Given one pound sterling equals $\$4.86\frac{2}{3}$. Obtain short methods for the conversion of sterling into currency and currency into sterling, and illustrate by examples.

104. A man bought \$500 three per cent. stock at $93\frac{1}{2}$, and after receiving one dividend sold at $96\frac{3}{4}$. What did he gain?

105. Explain the metric system of weights and measures, and give the equivalent of each metric unit.

106. The issue price of certain railway shares was \$50, to be paid in 5 instalments of \$10 each, the first on application. After a "call" or second payment of \$10, the shares stood at one dollar a share premium. A person then invested \$756, and after paying a further call of \$10, a dividend was declared of $8\frac{1}{2}$ per cent. per annum on the paid-up capital. What is the amount of his dividend, and what interest has he got for his money?

107. Three boats started at the same moment at intervals of 100 yards apart; in 6 minutes the third overtook the second and in two minutes more it overtook the first. How soon will the second overtake the first?

108. The interest on a sum of money for two years is \$349.58, and the discount on the same sum for the same time is \$310.74; simple interest in both cases. Find the rate per cent. and the amount.

109. A person shooting at a target, at a distance of 515 yards, hears the bullet strike the target 4 seconds after he fired. A spectator, equally distant from the target and the shooting point, hears the shot strike $2\frac{1}{2}$ seconds after he heard the report. Find the velocity of sound.

110. A in Toronto pays B in Paris 1000 francs by a bill of exchange on London, exchange at Paris being 25·25 francs for £1 sterling. Find the amount of the bill and its value in currency (£1 = \$4.86$\frac{2}{3}$). When the bill reaches Paris exchange is at 25·23. Find the amount in francs for which the bill sells.

111. If the minute hand of a clock be 4 inches long and the hour hand 3 inches, find the times between 4 and 5 o'clock, when their ends are 5 inches apart.

112. Lead weighs 11·324 times as heavy as water: cork weighs $\frac{6}{25}$, and fir $\frac{9}{20}$, respectively, of the weight of an equal volume of water. How much cork and lead must be combined together so that the mass may be equal to 80 pounds, the weight of a beam of fir timber of the same magnitude?

113. Two casks, A and B, contain each four gallons of two different kinds of fluid. A gallon is taken out of A and put into B, and then a gallon of the mixture is taken out of B and put into A. This double operation is repeated 12 times. How much of the original fluid is there then in A?

114. What must be the gross produce of an estate in order that, after paying a 10 per cent. income tax, and a rate of 12$\frac{1}{2}$ cents on the dollar on the residue, there may remain \$1612?

115. A grocer sold 60 pounds of coffee and 80 pounds of sugar for \$25, but he sold 24 pounds more of sugar for \$8 than he did of coffee for \$10. What was the price of a pound of each?

116. Give a short method of approximating to any root of decimals nearly equal to unity.

117. How many square feet of inch and a half plank will be required in the construction of a closed box, whose external dimensions are to to be 3, 4, and 6 feet?

118. Find the number of square feet of inch-and-a-half plank necessary for the construction of a closed box, whose internal dimensions must be 3′ 6″, 3′ 3″, and 3′ 1$\frac{1}{2}$.″

119. It is between two and three o'clock, and the hour and minute hands of a clock are inclined to each other at an angle of 60 degrees. Find the time.

120. Multiply 359,999,999,999 by 799,999 and divide the result by 599,999.

121. Simplify

$$\frac{2+\sqrt[4]{3}}{2-\sqrt[4]{2}} \div \frac{3+\sqrt[4]{2}}{3-\sqrt[4]{2}} \times \frac{5+\sqrt[4]{6}}{5-\sqrt[4]{6}}.$$

122. The crew of a boat row 6 miles down a river and half way back again in two hours. Supposing the stream to have a current of $2\frac{1}{2}$ miles an hour, find at what rate they would row in still water.

123. Show how a piece of paper 8 inches square may be cut so as to form a parallelogram, whose sides are very nearly 13 and 5, and whose angles are very nearly right angles.

124. A vessel 5 ft. 10 inches long, 4 ft. 2 in. wide, and 3 ft. 4 in. deep is filled with water. Find the weight in pounds of the contents of the vessel; a cubic foot of water weighing 1000 ounces.

125. A Glasgow merchant ships to his Montreal agents for sale goods for which he pays £116 sterling in Glasgow. He pays an ad valorem duty of 12 per cent. upon the goods and a commission of 7 per cent, to his agents for their services. The goods realize in Montreal $780. Find the merchant's net gain, a pound sterling being equal to $4.86.

126. A person ordered $150 to be distributed among some poor people, but before the distribution had taken place two more unexpectedly appeared, in consequence of which the former received $2.50 each less than they otherwise would have done. What was their number at first?

127. A grocer can sell coffee at 30 cents a pound, and realize a profit of 25 per cent. He, however, mixes the coffee with chicory, which costs him 6 cents a pound, and selling the mixture at 25 cents a pound realizes a profit of 40 per cent. How many per cent. of coffee does the adulterated mixture contain?

128. The area of the base of a cylinder is 2 square ft. and its height 30 inches. Find the height of a cylinder the solid content of which is 3 times as great, but whose diameter is only two-thirds that of the given one.

129. Find accurately to three places of decimals the cube root of 78836421.

130. If 10 men and 3 women working 8 hours a day perform a piece of work in 12 days, how many days would be required for 8 men and 5 women working 9 hours a day to perform the same work, if 3 women do as much work in a given length of time as 2 men?

131. A person borrows $540, which he agrees to pay in yearly payments of $90 each, together with interest at the rate of 8 per cent., payable annually, the borrower having the privilege of paying a greater sum than $90 on account of principal annually, if he choose to do so, the lender agreeing to allow him interest at 7 per cent. upon all principal money paid in excess of $90 a year. The borrower makes three annual payments of $150 each and pays the balance of his indebtedness at the end of the fourth year. Find the amount of the last payment.

132. The plate of a mirror is 36 inches by 24 and is to be framed with a frame of uniform width, whose area is to be equal to $\frac{16}{27}$ that of the glass. Find the width of the frame.

133. The diameter of the hind wheel of a carriage is one foot greater than that of the fore wheel, and in a journey of seven miles the fore wheel makes 210 revolutions more than the hind wheel. Find the diameter of each wheel, assuming the ratio of the diameter to the circumference of a circle to be as 7 to 22.

134. A goldsmith mixes gold of 15, 19, 23 and 24 carats so that the compound may be 20 carats fine. What quantity of each must he use?

Any consistent answer will be sufficient.

135. A market woman buys eggs at the rate of 5 for twopencé, sells half of them at 3 for a penny, and the remaining half at the rate of two for a penny, and finds that she has gained one shilling. How many eggs did she buy?

136. A snail crawls up a wall c ft. high, going up b feet per day and slipping back a feet at night. When will it reach the top?

Point out the conditions of possibility.

137. A room 24 ft. long, $13\frac{1}{2}$ ft. wide, and 11 ft. high has two doors and two windows, each 8 ft. high and 4 ft. wide. How many square yards of plastering will it require, including the ceiling, and how many yards of carpeting, 27 inches wide for the floor, allowing no waste?

138. Find the sixth root of

$$2,565,726,409.$$

139. A square number cannot be of the form $12n+5$.

140. The product of three consecutive numbers cannot be a perfect square.

141. Define the terms circulating decimal and perfect repetend.

If $\dfrac{1}{2n+1}$ is a perfect repetend, and n figures of the repetend are known, show how the remaining n figures can be found by simple inspection.

142. Given $\frac{1}{17}=0\cdot0588\frac{1}{17}$, write down four more figures by simple multiplication, and the remaining eight by inspection.

143. The scale of a map is $\frac{1}{150000}$ and the distance on the map between the positions of two places is $2\frac{2}{3}$ inches. Find their actual distance in miles.

144. Find the radius of a sphere whose volume is equal to the sum of the volumes of three spheres whose radii are 7, 8 and 9 feet, respectively.

145. The present income of a railway company would justify a dividend of six per cent. if there were no preference shares; but as there is £400000 of the stock consisting of such shares which are guaranteed $7\frac{1}{2}$ per cent. per annum, the ordinary shareholders receive only 5 per cent. Find the amount of ordinary stock and the company's income.

146. A man has a piece of land 109 rods one yard $\frac{3}{10}$ ft. long, and 66 rods 1 yard $1\frac{9}{20}$ ft. wide, which he wishes to lay out in the largest possible square lots of equal size. How many lots will there be?

147. A person annually increases his capital 20 per cent., less a yearly expenditure of \$500. At the end of four years his capital amounts to \$18,052. Find his original capital.

148. A person left \$6000 to be divided among his three sons, whose ages were 6 years 2 months, 9 years 8 months, 13 years and 2 months, respectively, in such proportions that the share of each at simple interest at six per cent. should amount to the same sum when they should arrive at the age of 21 years. What was each one's share?

149. A merchant bought a certain quantity of corn, for which he paid a certain sum of money, but on measuring it he found only $\frac{39}{40}$ of the quantity he expected; he sold it, gaining $\frac{1}{8}$ of the cost, and received \$2160, which was at the rate of $12\frac{4}{13}$ cents a bushel more than he would have paid had he received the quantity he expected. How many bushels did he suppose he had bought, and at what price?

150. A railroad train travels for $\frac{1}{4}$ of the distance at a rate of 30 miles an hour, the next $\frac{1}{4}$ of the distance at the rate of 35 miles an hour, and the remaining distance at the rate of 40 miles an hour. What is the average rate in miles per hour?

151. The recent $3\frac{1}{2}$ per cent. Canadian loan of £5000000 was placed in London at the rate of £91 2s. 6d. for each £100. If the proceeds were placed in Canada, by drafts on London, drawn at the rate of $8\frac{7}{8}$% premium (old par of exchange), what was the total amount in Canadian currency realized from the loan?

152. What sum should be paid for a \$100 debenture to run for 20 years at 4 per cent. per annum, in order that the investor may realize 5 per cent. per annum on his outlay?

153. A contractor engaged to complete 1000 yards of railway in 50 days, and employed 100 men working 9 hours a day, but at the end of 30 days he found only 450 yards finished. How many additional men must he hire in order that all working 10 hours a day may finish the work in the given time?

154. Standard gold is \$18.94$\frac{23}{24}$ an ounce. Find the least number of ounces that can be coined (1) into an exact number of \$5 pieces, (2) into an exact number of sovereigns, and find the number of coins in each case.

155. The value of diamonds varies as the square of their weights, and the square of the value of rubies varies as the cube of their weights; a diamond of a carats is worth m times a ruby of b carats, and both together are worth £c. Find the value of a diamond and ruby, each weighing x carats.

156. A watch which is 10 minutes too fast at 12 o'clock noon on Monday loses 4 minutes and 12 seconds per day. What will be the true time on the following Saturday morning, when the watch shows 8 o'clock?

157. The L.C.M. of two numbers is 100793; the G.C.M. is 17; the difference of the numbers is 1224. Find the numbers.

158. At what advance on cost must a merchant mark his goods so that after allowing 10 per cent. of his sales for bad debts, 8 per cent. of the costs for expenses, and an average credit of 9 months (money being worth 4 per cent.), he may make a clear gain of 20 per cent. on the first cost of the goods?

159. What will be the true interest on $1000 for 6 months, it being supposed that if this interest is invested for the next six months that the whole interest for the year shall be exactly six per cent.?

160. A merchant in London remits to Amsterdam £1000, at the rate of 18d. per guilder, directing his Amsterdam agent to remit the same to Paris at 2 francs 10 centimes per guilder, less $\frac{1}{2}$ per cent. for commission, but the exchange between Amsterdam and Paris happened to be at the time the order was received at 2 francs 20 centimes per guilder. The merchant at London, not apprised of this, drew upon Paris at 25 francs per pound sterling. Did he gain or lose, and how much per cent?

161. Calculate to within $\frac{1}{1000}$ of one per cent. the rate per cent. that paid quarterly is equivalent to 8 per cent. paid annually.

162. Two boats start to row a race at 3 o'clock. The race is over at $6\frac{2}{3}$ minutes past 3, the losing boat being 40 yards behind at the finish. At 4 minutes past 3 this boat was 700 yards from the winning post. Find the speed of each boat in miles per hour.

163. A sent his agent 5000 bushels of wheat, which he sold at $1.20 per bushel on a certain commission. After deducting this commission and also a second one at the rate of 4 per cent., he invested the remainder in silks for A. The two commissions amounted to $500. At what rate was the first one charged?

164. Three persons, A, B and C, are concerned in a cotton mill; A puts in $4800 for 8 months; B, a sum unknown, for 10 months; and C, $6000 for a time not known. When the accounts were settled, A, B and C received, respectively, $6000, $7000 and $7800 for stock and profits. Find B's stock and C's time.

165. A person having to pay $1085 at the end of two years, invested a certain sum of money in the three per cent. consols, allowing the dividends to accumulate until the payment of the debt, and also an equal sum the next year. Supposing the investments to be made and the debt to be paid when the consols are at 73, what must be the sum invested on each occasion, that there may be just sufficient to pay the debt at the proper time?

166. A ladder 30 feet long just reaches a window in a house on one side of a street 42 feet wide, it is then turned about its lowest point and just reaches a window on the other side. If the two positions of the ladder be at right angles, find the height of the windows.

167. A crew rows three miles down stream and back again in 1 hour and 20 minutes. If the rate of the current be three miles per hour, find their rate in still water

168. A merchant annually increases his capital by 50 per cent. of itself except an expenditure of $2000 a year, and at the end of five years finds he is worth $34337.50. Find his capital at first.

169. A, B and C engage in business. A puts in $400 at first and $400 more at the end of six months. B puts in $900 at first and withdraws one-third of his capital at the end of six months. C puts in $200 at the end of every six months. At the end of two years they have gained $6700. What share of the profit should C receive in addition to 25 per cent. of the total profits for managing the business?

170. Lumber worth \$20 per thousand feet is exported from Canada to the United States, to be manufactured at a cost of \$25 per thousand feet for re-importation into Canada. If the export duty on lumber be 10 per cent., and the import duty on the manufactured article be 30 per cent., find the cost of manufacture in Canada that the Canadian manufacturer, after making a profit of 20 per cent., may sell his goods at a price which would leave the American manufacturer no profit.

171. A ship carrying a certain number of passengers is becalmed with provisions for six weeks, but it is found that one person dies at the end of each week and the provisions last for 8 weeks. How many were on board at first?

172. Find the time between m and $m+1$ o'clock, when the hands of a watch are exactly n minute spaces apart.

173. A ladder when placed upright is 4 feet higher than a wall; when the foot of the ladder is removed 16 ft. from the wall it just reaches the top. Find the height of the wall.

174. A rope 512 ft. long has one end fastened to a corner of a house 25 ft. square, around which it is then completely wrapped. How far will a man holding the other end of the rope, and keeping it stretched, walk before the rope is re-wrapped round the house in the contrary direction?

175. Find the effect of adding the same quantity to both terms of a fraction.

Employ your result to compare the values of the fractions

$$\frac{397}{483} \text{ and } \frac{433}{519};\ \frac{3731}{4568} \text{ and } \frac{3798}{4628};\ \frac{796}{799} \text{ and } \frac{799}{801}.$$

176. A cubic inch of gold weighs $\frac{19300}{1728}$ oz. (avoir.), one of silver $\frac{10500}{1728}$ oz., and of copper $\frac{8900}{1728}$ oz. A coin whose volume is ·01 cubic inch is composed of gold, silver and copper in the ratio (by volume) of 12:2:1. How many grains of each kind of metal does the coin contain?

177. A mortgage for \$1000, bearing interest at the rate of 8 per cent. per annum payable half yearly, has two years to run. Find its present value if money is worth six per cent. per annum.

178. Which is the better investment:

(1) Railway shares at 70 paying yearly dividends of 4 per cent., brokerage $\frac{1}{8}$ per cent., or

(2) Bank stock at 140 paying half yearly dividends of 4 per cent., subject to an income tax of 17 mills on the dollar?

179. A person borrows $1000 for two years, and discharges the debt by paying $600 at the end of one year and $600 at the end of two years. What rate per cent. per annum (compound) interest did he pay?

180. Out of a cask containing 440 quarts of pure alcohol a quantity is drawn off and replaced by water. Of the mixture a second quantity, $91\frac{2}{3}$ quarts more than the first is drawn off and replaced by water. The cask now contains as much water as alcohol. Find how many quarts were taken out the first time?

181. A man insures his life for $\$a$, at a premium of $\$d$ per annum; he died after n years, and the insurance office neither gained nor lost in the transaction. Find n, reckoning compound interest at the rate of q per cent. per annum.

182. If B and C, working together, take p days to do a piece of work, for which C and A together took q days, and A and B together r days, find how long each would take by himself.

183. Define exchange and par of exchange. Find the par of exchange between England and the United States, having given that the United States' $10 gold piece weighs 258 grs. and is $\frac{9}{10}$ fine, and that 40 lbs. troy of English standard gold, $\frac{11}{12}$ fine, are coined into 1869 sovereigns.

184. If a person invests in the three per cents so as to have three per cent. clear on his investment, after paying an income tax of 9d. in the pound, what per cent. clear does he receive, when the income tax is reduced to 6d. in the pound?

185. A person who bought a house and lot sometime ago for $4000, now finds them to be worth $4700, the value of the lot having doubled while that of the house has fallen off 10 per cent. What will it cost him to insure the house at $\frac{3}{4}$ per cent. on $\frac{4}{5}$ of its value?

186. A note for \$1056, dated August 30, for three months, and bearing interest at the rate of 6 per cent. per annum, is discounted at a bank on October 5 at 8 per cent. What are the proceeds of the note?

187. Which is the better investment, 3 per cent. consols at $96\frac{13}{16}$, the dividend being payable 5 months hence; or, U. S. 4 per cents at $130\frac{3}{4}$, dividend payable in 2 months, money being worth three per cent. per annum interest?

188. Two men, A and B, enter into business; A is to receive \$25 a month and 10 per cent. of the sales for managing the business, the remaining profits to be divided equally. The goods are sold at an average profit of 30 per cent. Find the total sales, in order that B may receive 25 per cent. of the profits.

189. A gallon of fresh water weighs 10 lbs., and has a volume of 277·274 cubic inches. How many cubic feet of salt water (which is three per cent. heavier than fresh water), will a vessel weighing 500 tons displace, the weight of the water displaced being equal to the weight of the vessel?

190. A and B are travelling on the same road towards Toronto, A at the rate of a miles and B at the rate of b miles per hour. At noon A was m miles, and at 6 p.m. B was n miles from Toronto. Find how many hours from noon A passed B, a being greater than b. Interpret the result when $m = 40$, $a = 5$, $b = 3$ and $n = 26$; also, when $n = 18$.

191. A man divides \$1300 into two sums, and lends them at different rates of interest. He finds the incomes from them to be equal. If he had loaned the first at the rate of the second he would have received \$36, and the second at the rate of the first he would have obtained \$49. Find the rates of interest.

192. A circular plate of lead 2 inches thick and 8 inches in diameter, is converted without loss into spherical shot of the same density and each ·05 inch radius. Find the number of shot.

193. If 10 square chains make an acre, find the length of a link in inches. Find the distance in links between the opposite corners of a square field containing 10 acres.

194. A man's income consists of a salary of $510 per annum, of dividends on shares, which pay 5 per cent. per annum, and of rents. If his dividends form $\frac{1}{6}$ of his total income and his rents $\frac{1}{8}$, find the amount of capital which he has invested in shares.

195. A man invests $5000 equally in the shares of two banks. The shares of the one are at 3 per cent. discount, and of the other at 5 per cent. premium; the price of stock in the former rises 7 per cent. and that in the latter falls 6 per cent. lower than when the purchase was made. If the man now sells out what will he gain or lose?

196. A house is let for $200 a year, payable quarterly. What present sum will pay the rent for three years, if money is worth 6 per cent. per annum, payable half yearly?

197. State clearly the difference between simple and compound interest. Show that at 10 per cent. per annum compound interest a sum of money will double itself in a little more than seven years.

198. A in Toronto owes B in Liverpool £2500; in what different ways may payment be made? Explain the method usually adopted. If exchange is at par, what sum in Canadian money will discharge the debt?

199. Define the terms, Stock, Annuities, Consols, Brokerage. A person invests $3840 in the 3 per cents at 84, and when they have risen to 86 transfers $\frac{3}{4}$ of his capital to the 4 per cents at 98. Is his income increased or diminished, and by how much?

200. Water is discharged at the rate of 500 gallons a minute from a reservoir of 2700 square feet surface into one of 1800 square feet surface. Find how long it will take to diminish the difference between their levels by 5 inches, a cubic foot of water weighing 1000 ounces, and a gallon 10 pounds.

201. To finish a certain piece of work 16 masons were employed for 30 days, 20 carpenters for 32 days, and 15 painters for 16 days. If a mason receive 5 per cent. less, and a painter 5 per cent. more per day than a carpenter, and the total cost of the work is $2965.60, find how much each workman receives per day.

202. A workman in the city of A finds that he has to pay 50 per cent. of his wages for food, 15 per cent. for clothes, and 12 per cent. for rent, and saves the remainder. On moving to B he finds that food costs him four-fifths as much as at A, clothes 3 times, and rent two-and-a-half times as much, but wages are 50 per cent. higher. How long will his savings for a year support him in each of the two places, respectively.

203. An importer enters for duty (at 30 per cent. ad valorem) an article which he values at $260. The value of the article is however found to be $330, and the customs regulation in such cases is that there shall be levied on such goods (in addition to the duty payable when properly entered) a sum equal to the same percentage of the proper duty as the percentage of undervaluation of the original entry. Find the full amount payable in this case.

EXAMINATION PAPERS.

UNIVERSITIES OF QUEEN'S, TRINITY AND VICTORIA.

MATRICULATION, 1888.

1. A man owns $\cdot\dot{2}\dot{7}$ of a certain patent. He sells $\cdot 41\dot{6}$ of his share for \$3240. What is the value of the patent?

2. Find the value of

$$1\cdot 26 \div \cdot 57 \times 2\cdot\dot{3} \div 1\cdot\dot{0}\dot{1} + \frac{15\cdot 79\dot{3}}{7\cdot 107} \div 1\frac{1}{9}.$$

3. Divide \$12.46 among A, B, C and D, so that A's share may be $\frac{3}{10}$ of D's, C's share $\frac{3}{10}$ of A's, and B's share the sum of A's and C's.

4. If 125 men dig a trench 100 yds. long, 20 yds. wide and 4 ft. deep, in 4 days, working 12 hours a day; how many men will be needed to dig one 500 yds. long, 8 yds. wide and 6 ft. deep, in 3 days, working $7\frac{1}{2}$ hours a day?

5. A man left $\frac{5}{12}$ of his estate to one of his sons, and 50 per cent. of the remainder to another, and the rest of the estate to his widow; the difference of his sons' legacies was \$783. How much did his widow receive?

6. Find the true discount on a note of \$259.05, due 4 years' hence at $2\frac{1}{2}$ per cent. per annum.

7. The discount on a certain sum due 2 years hence is \$300, and the interest on the same sum for 2 years is \$342. Find the sum and the rate per cent. per annum.

8. The area of a rhombus is 990 sq. yds., and the length of one of its diagonals is 55 yds. Find the other diagonal.

9. The cost of fencing a circular plot of ground at \$1.25 a yd. was \$715.00. Find the length of a straight path running from side to side through the centre.

10. The pressure of wind on a plane surface varies jointly as the area of the surface and the square of the wind's velocity. If the pressure on a square foot is 1 lb. when the wind has a velocity of 15 miles an hour, find the pressure on a square yard when the wind is moving with the velocity of 75 miles an hour.

Q., T., V. MATRICULATION, 1889.

1. How do we know by inspection when a number is divisible by 2? by 3? by 4? by 5? by 6? by 8? by 9? by 11?

2. Work out to a single final result each of the following:

$$(a)\quad \frac{1}{2+\frac{1}{3\frac{1}{4}}}-\frac{1}{4+\frac{1}{3\frac{1}{2}}}.$$

$$(b)\quad \frac{\sqrt{2}+2}{\sqrt{2}+1} \text{ to 4 decimals.}$$

$$(c)\quad 0{\cdot}1\dot{9}\dot{6}\times\frac{0{\cdot}1\dot{2}}{0.\dot{3}} \text{ expressed in vulgar fractions.}$$

3. A and B run a mile race which B can accomplish in $6\frac{1}{3}$ minutes. A gets 10 seconds the start and comes in 16 rods ahead. How would the race finish if B got 9 seconds the start?

4. If I buy $\frac{2}{3}$ of $4\frac{1}{2}$ yards of cloth for \$0.65 and sell $\frac{1}{5}$ of $1\frac{1}{2}$ yards for 24 cents, at what rate per yard must I sell the remainder so as to gain 20 per cent. on the whole transaction?

5. A, B and C form a partnership for a year. On Jan. 1st, A puts in \$300, B \$500, and C \$600. B draws out \$300 on May 1st, and C draws out \$300 on Sept. 1st. For the first six months the gain is 10 per cent. on the mean stock, and for the second six months it is \$112. Find the total profits, and show how they are to be divided.

6. A person sells \$6250 of 3 per cent. consols, when they are worth 96, and invests the proceeds in railway stock at 75 and paying $2\frac{1}{4}$ per cent. How does this transfer affect his income?

7. A floor 16 by 20 is to be completely covered with carpet 27 inches wide, and having a pattern which matches every 3 feet. If the carpet is to be matched, in which direction must the strips run so as to give the least waste, and how much is the waste?

8. Of how many gallons capacity is a box 3 feet long, 2 feet wide and one foot deep?

9. A silver plate of uniform thickness in the form of a square is worth $3.20. What is the value of the largest circular disc which can be cut from it?

Q., T., V. MATRICULATION, 1890.

1. Simplify $\frac{\frac{10}{12}}{1-\frac{25}{144}}$ and $\frac{\frac{1}{70}-\frac{1}{99}}{1+\frac{1}{70}\cdot\frac{1}{99}}$, and divide their difference by their product increased by unity.

2. Reduce $0{\cdot}\dot{4}8\dot{1}$ of $0{\cdot}\dot{5}1\dot{8}$ of $0{\cdot}0\dot{8}\dot{1}$ of $0{\cdot}06875$ of 405 miles to yards.

A railway train moves at the rate of $27{\cdot}3\dot{7}$ yards per second. How many miles an hour is this?

3. A horse is bought for $360. At what price must he be sold to make $12\frac{1}{2}$ per cent. on the transaction?

4. A piece of cloth, five times as long as it is broad, costs £38. Find its dimensions if the price is 9s. 6d. per square yard.

5. $200 is distributed among 15 men, 30 women and 35 children, each woman receiving $3 and each man as much as a woman and 2 children. How much does each receive?

6. There are 50 coins consisting of 25-cent and 20-cent pieces worth $11.75. If each 25-cent piece were a 20-cent piece, and each 20-cent piece a 25-cent piece, the value of the coins would be $10.75. How many coins are there of each kind?

7. A goldsmith pays half as much per ounce avoirdupois as he sells for per ounce troy. What is his gain per cent.?

8. What two numbers, each of four figures, have 101 for their greatest common measure, and 27573 for their least common multiple?

9. The discount on a certain sum for $2\frac{1}{2}$ years is $\frac{40}{47}$ of the interest for the same time, simple interest being reckoned in each case. Find the rate per cent., and if the interest and discount together are $365.10; find the principal sum.

10. The external length, breadth, and height of a closed rectangular wooden box are 18, 10 and 6 inches, respectively, and the thickness of the wood is half an inch. When the box is empty it weighs 15 pounds, and when filled with sand 100 pounds. Compare the weights of equal bulks of wood and sand.

EDUCATION DEPARTMENT.

THIRD CLASS, 1883.

1. Add together $\frac{3}{7}$ of £13, $\frac{1}{3}$ of $\frac{1}{2\frac{4}{5}}$ of $\frac{3}{5}$ of £2 12s., and $\frac{4}{7}$ of 9d.

Reduce 13s. $4\frac{1}{2}$d. to the decimal of 19s. 6d.

2. Find by Practice the value of ·8596 lbs. at £10 18s. $7\frac{1}{2}$d. each.

3. A person borrows \$500 on April 10th, and on June 22nd pays his debt with \$510.20. At what rate per cent. per annum was he charged interest?

4. A man having a certain sum of money to invest has an opportunity of purchasing 7 per cent. stock at 95, but delays until it has risen to 110. What per cent. is his income less than if he had purchased at the first price?

5. At an international exhibition one country was awarded 5 gold, 9 silver and 11 bronze medals; and another 4 gold, 15 silver and 10 bronze. Find a ratio of values for such medals that these countries may be regarded as equally fortunate.

6. In a box there is a certain number of sovereigns, three times as many guineas, and twice as many marks (13s. 4d.) as guineas. The entire amount in the box is £815. How many coins of each kind are there?

7. Find, when first after 2 o'clock, the hour and minute hands of a clock make an angle of 60 degrees with each other.

8. For each of three succeeding months the population of a North-West town rose 50 per cent.; and at the end of the third month was 2700. What was the population at the beginning of the time?

9. Leap year is omitted once in every century, except those centuries whose number is divisible by 4. What is the average length of a year?

10. A cube is formed of a certain number of pounds avoirdupois of a substance, and the same number of pounds troy of the same substance. What proportion will a side of the cube bear to a side of a cube formed of the same number of pounds as before, but all avoirdupois (175 lbs. troy = 144 lbs. avoirdupois)?

THIRD CLASS, 1884.

1. Simplify $\dfrac{(4\frac{1}{5} - 3\frac{1}{8}) \times (1\frac{1}{3} - \frac{5}{6})}{(1\frac{1}{7} - \frac{3}{14}) \div (\frac{2}{11} - \frac{1}{7})} \div \dfrac{\frac{15}{13}}{\frac{17}{31}}$.

2. Find the cost of ·0625 of 112 lbs. of sugar, where one pound costs ·0703125 of 17s. $9\frac{1}{3}$d.

3. A and B were employed to do a piece of work for $60. They were to be paid in proportion to their ability to work, which was 4 to 5, and to the time each worked, which was 3 to 4. How much did each receive?

4. A quantity of silk was sold at a loss of 1 per cent.; had it been sold for 4s. $2\frac{1}{2}$d. per yard there would have been a gain of 1 per cent. Find the actual selling price.

5. A person rides to town at the rate of $8\frac{1}{4}$ miles an hour, and after resting 35 minutes walks back at the rate of $2\frac{3}{4}$ miles an hour. The whole time occupied was 7 hours $20\frac{5}{11}$ minutes. Find the distance.

6. Instead of a yard measure a draper uses a stick which is 36.35 inches long. What does he lose per cent. by so doing?

7. When the course of exchange between London and New York is quoted at 4·96, London exchange (*i.e.* English money) is said to be at 2 per cent. premium. From this calculate the par of exchange.

8. If silver is worth $1.10 per ounce, and gold $17 per ounce, find the weight of a ten-dollar coin containing 37 parts in 40 of gold, and the rest silver.

9. Equal volumes of iron and copper are found to weigh 77 oz. and 89 oz. respectively. Find the weight of $10\frac{1}{2}$ feet of circular copper rod, when 9 inches of iron rod of equal diameter weigh $31\frac{9}{10}$ ounces.

10. The expense of carpeting a room 15 feet wide was \$52.80; but if the length had been a yard less, the expense would have been \$46.20. Find length of the room.

11. A rectangular solid $4\frac{1}{2}$ feet long, $3\frac{1}{2}$ feet broad, and $1\frac{1}{3}$ feet thick, is increased 11 inches in thickness. By how much must the breadth be diminished, so that the solid may retain the same bulk as before?

THIRD CLASS, 1885.

1. Define prime number, factor, common multiple, discount, exchange.

Draw a diagram showing that there must be $30\frac{1}{4}$ sq. yds. in a sq. rod, if the linear rod contains $5\frac{1}{2}$ yds.

2. A merchant bought 124 yds. of cloth at \$3.62½ per yd., and 87½ yds. at \$4.12½ per yd. At what price per yd. must he sell the whole to realize a profit of 20 per cent.?

3. Simplify the following and give the result in £ s. and d. :—

$$\frac{3}{5}(3\cdot\dot{3} + 1\cdot25) \text{ of } £1 + \tfrac{1}{4} \text{ of } \frac{1\cdot125 - \frac{1}{3} \text{ of } 1\frac{5}{8}}{\frac{1}{10} \text{ of } 3\frac{1}{3} + \frac{1}{12}} \text{ of } 9\text{s.} + \frac{2\cdot1\dot{6}}{2\cdot\dot{0}\dot{9}}\text{d.}$$

4. A farmer sold two loads of wheat, in all 110 bushels, for \$91.95. One load was sold at 97 cts. per bushel and the other at 72 cts. per bushel. How many bushels were there in each load?

5. A and B engage in trade; A invests \$6000, and at the end of 5 months withdraws a certain sum, B invests \$4000 and at the end of 7 months \$6000 more. At the end of the year A's gain is \$5800 and B's is \$7800. Find the amount A withdrew.

6. A merchant bought cloth at \$2 per yd., and sold the whole at a profit of \$120; had he sold it at 20 per cent. less he would have lost \$96. How many yds. did he buy?

7. What will be the cost of insuring a property worth \$47580, at the rate of $\frac{7}{8}$ of 1 per cent., so that in case of loss the owner may recover both the value of the property and the premium paid?

8. Divide \$4841 among A, B and C, so that 9 months' interest on A's share at $8\frac{1}{2}$ per cent. per annum, 9 months' interest on B's share at $3\frac{3}{4}$ per cent., and 9 months' interest on C's share at $4\frac{1}{2}$ per cent. may all be equal.

9. I owe a man \$850, and give him my note at 90 days. What must be the face of the note to pay the exact sum, if discounted at $1\frac{1}{4}$ per cent. a month (bank discount)?

10. If a brick 8 in. long, 4 in. wide and 2 in. thick weighs 5 lb., what will be the weight of a brick of the same material 16 in. long, 8 in. wide and 4 in thick?

11. The top of a ladder reaches to the top of a wall when its foot is at a distance of 10 ft. from the bottom of the wall, but if the foot of the ladder be drawn 4 ft. farther from the wall the top of the ladder will reach a point 2 ft. below the top of the wall. Find the length of the ladder.

THIRD CLASS, 1886.

1. A had \$7 less than B had, and B had \$10 less than C had. A gave \$5 to B and \$12 to C. How many dollars had C more than A then?

2. One-quarter of the time which a man spent on a journey from M to T, he travelled by steamboat, at an average rate of 14 miles an hour; two-thirds of the time he travelled by railway train at an average rate of 25 miles an hour; and the remaining hour of the time he rode the remaining 7 miles of his journey. Find the distance from M to T.

3. At what time between 4 and 5 p.m. is the minute hand exactly 2 minute-spaces ahead of the hour hand of a watch marking correct time?

4. A man, assisted part of the time by a boy, completed a job in 15 hours. The man received five-sixths of the pay and the boy received one-sixth, but the man was paid at double the rate the boy was, in proportion to the amount of work each did. How long would the man unassisted have taken to accomplish the job?

5. How much water must be added to a mixture of 15 gallons of vinegar, costing 52 cents a gallon, and 13 gallons costing 40 cents a gallon, that \$5 may be gained by selling the whole at 15 cents a quart?

6. A total of 250 marks is to be allowed to a paper of 10 questions. To the first 7 questions the average is given. Divide the remaining marks so as to allow 7 marks to the tenth question and 5 marks to the ninth for every 3 marks allowed to the eighth.

7. A bookseller charges on certain books 35 cents on the shilling of the published price and gives a discount of 35 per cent. What is the actual rate he charges on the shilling?

8. A bill for \$253.03, dated 7th October, and payable at London in 3 months from date, was discounted in Toronto on 20th October, the discount being at the rate of 9 per cent. per annum, and 45 cents being charged for exchange. Find the proceeds of the bill.

9. A cubic foot of water weighs 62·426 pounds and a gallon of water weighs 10 pounds. How many gallons will a cylindrical cistern of 5 feet diameter by 4 feet deep hold?

THIRD CLASS, 1887.

1. Prove the rule for the multiplication of two fractions.

Simplify $\dfrac{(7\frac{1}{4}-3\frac{1}{2})\times\{4\frac{1}{5}-(2\frac{1}{3}-1\frac{7}{10})\}}{(7\frac{1}{4}+3\frac{1}{2})\div(1\frac{1}{2}-9\frac{1}{2}\times\frac{9}{77})}$.

2. A, B, C, rent a pasture for \$102; A puts in 6 horses for 8 weeks, B 12 oxen for 10 weeks, C 50 cows for 12 weeks. If 5 cows are reckoned as 3 oxen, and 4 oxen as 3 horses, what shall each pay?

3. A does a work in 10 days, B in 9 days, C in 12 days; all begin together, but A leaves in $3\frac{2}{5}$ days before the completion, B in $2\frac{3}{5}$ days before the completion. In what time was the work done?

4. Prove the rule for division of decimals. Divide to 6 decimal places, ·0078539 by ·9921461.

5. On March 23rd a bank gives me \$845 for a note of \$860. When is the note due, interest 8%?

6. Find the cost, in sterling, of 184 tons 17 cwt. 3 qrs. 14 lbs. of copper, invoiced to a Toronto importer at £87 17s. 11d. per ton? (qr. = 28 lbs.).

7. I bought certain 4 per cent. stock at 75 and after a number of years sold out at $94\frac{1}{2}$, and found that I had made $7\frac{1}{2}$% per annum simple interest. How long did I hold the stock?

8. There is a mixture of vinegar and water in the proportion of 93 parts vinegar to 7 parts water; how much water must be added so that in 25 parts of the mixture there may be 2 parts water?

9. I invested \$10,000, but sold out at 20% discount. How much must I borrow at 4% so that by investing all at 8% I may just retrieve my loss?

10. A square field containing $27\frac{1}{2}$ acres has a diagonal path across it. What is the length of the path in yards?

11. When the temperature of a cube of zinc is raised from 32°F to 212°F each dimension is thereby increased ·3 per cent. Find the percentage of increase in the bulk.

12. Water is flowing at the rate of 10 miles per hour through a pipe 14 in. in diameter, into a rectangular reservoir 187 yds. by 96 yds. In what time will the surface be raised 1 inch?

THIRD CLASS, 1888.

1. Simplify

$$(a)\ \frac{\frac{3}{4} \text{ of } \frac{7}{9} \text{ of } \frac{15}{28} - 2\frac{1}{4} \text{ of } 3\frac{2}{3} \text{ of } \frac{1}{72}}{4\frac{1}{2} - (3\frac{1}{3} + 4\frac{2}{7}) + 3\frac{7}{8} + \frac{3}{56}}$$

(b) What fraction of $365\frac{1}{4}$ days is 349 days 8 hrs. 52 min. $\frac{19}{23}$ sec.?

2. A can do a work in one-half the time that B requires, B can do it in two-thirds of the time that C takes. All working together do it in 18 days. How long would it take each one separately?

3. A man got a 90 days' note for $1360 for a lot which cost him $1200 cash just a year before; money 6 per cent. Find his net gain at time of sale (bank discount, 360 days to a year, no days of grace).

4. Bought 78ac. 3r. 15per. 7ds. 1ft. 9in. of land at $80 per acre; sold $\frac{2}{5}$ of it at $120 per acre, and the rest at $.005 per sq. foot. Find gain.

5. A number of men and women earned $93 a day, each man getting $2.25 and each woman $1.50. Had there been 6 more men and 7 more women the whole number of women would have earned the same as the whole number of men. Find the actual number of each.

6. A commission merchant receives 125 bbls. of flour from A, 150 bbls. from B, 225 bbls. from C; he finds on inspection that A's is 10 per cent. better than B's, and C's is $5\frac{5}{11}$ per cent. better than A's. He sells the whole lot at $7.00 per barrel, charging 4 per cent. commission. What sum must he remit to each?

7. A compound of tin and lead weighs 10·43 times as much as an equal bulk of water, while tin weighs 7·44 times, and lead 11·35 times as much as equal bulks of water. Find the number of pounds of each metal in 765 lbs. of the compound.

8. A bankrupt had goods worth $7950, which, if sold at their full value, would give his creditors $81\frac{1}{4}$ per cent. of their claims. But $\frac{3}{5}$ of them were sold at $17\frac{1}{2}$ per cent. below their value, and the remainder at $23\frac{3}{4}$ per cent. below their value. How many cents on the dollar did his creditors realize?

9. A begins business with a capital of $3200; after 3 months B is admitted as partner with $2400; after 3 months more C is admitted with $1600. What fraction of the year's gain should each have?

10. What is the cost of polishing a cylindrical marble pillar, 2 ft. 6 in. in diameter and 12 ft. long, at $1.25 a square foot?

11. If it cost $11.20 for paper for a room 25 ft. 3 in. long, 19 ft. 9 in. wide, and 12 ft. high, when the paper is $\frac{3}{4}$ yd. wide, find cost of the paper per linear yard. (No allowance for doors and windows).

12. A square field, containing 16 ac. 401 sq. yds., has a walk around it outside 12 ft. in width. Find the area of the walk in yards.

THIRD CLASS, 1889.

1. *(a)* Simplify $\dfrac{\cdot 5 \times \cdot 006}{\frac{9}{15} \text{ of } \frac{4}{5} \times (\frac{1}{4})^2} + \dfrac{\frac{1}{5} \text{ of } \frac{15}{16} \times (\frac{2}{3})^2}{1\cdot 6 \times \cdot 625}$ (Answer in fractional form.)

(b) Find the average, correct to 4 places of decimals, of $12\frac{14}{25}$, 21, $7\frac{3}{4}$, ·034, 3·125, 0, 24·58 and $12\frac{9}{20}$.

NOTE.—No marks will be allowed for either *(a)* or *(b)* except the answer be perfectly correct.

2. In what time will $30441 gain $2210·10 if, at the same rate, the gain on $24944·10 for 1 year and 15 days is $2596·92? What is the rate per cent. per annum (365 days to a year)?

3. A house that cost $15500 rents for $155 a month. It is insured for $10850 @ $\frac{4}{5}$% yearly; the taxes are 15 mills on an assessment of $12450, and $346.45 is spent each year on repairs. What rate of interest does the investment pay?

4. A rectangular field, whose width is $\frac{3}{4}$ of its length, contains 15 acres, 123 per. In going from one corner to the opposite how much shorter is it to take the diagonal than to go around the two sides?

5. A note of $2450, dated Halifax, June 1st, 1886, for 4 mos., bearing interest @ 6%, is discounted at a bank on Aug. 15th @ 8%. Find the proceeds.

6. A farm cost $3\frac{3}{4}$ times as much as a house; by selling the house @ 10% *loss* and the farm @ $7\frac{1}{7}$% *gain*, $3993.30 is received. Find cost of each.

7. Bought 64 yards of cloth @ $5.70 a yard. If it shrank 5% in length, find the selling price per yard to gain 20%.

8. A and B are partners, A's capital being $\frac{3}{5}$ of B's. At the end of 5 months A withdraws $\frac{1}{4}$ of his capital, and at the end of 9 months B withdraws $\frac{1}{3}$ of his. How should they divide a gain of $4222.33 at the end of the year?

9. A man sold his 5 per cents @ 78 and invested the proceeds in 6 per cents @ 104. His change in income being $385, find how much 5 per cent. stock he had.

10. A dealer shipped 400 bushels wheat @ $1.40, 800 bushels @ $1.62½, and 300 bushels @ $1.20, to his agent, who sold the first at 20 per cent. gain, the second at 15 per cent. gain, and the third at $4\frac{1}{6}$ per cent. loss; the agent's commission was 3 per cent., and other charges were $83.44. Find the dealer's gain per cent.

11. What is the cost of boards, at $1 for 50 sq. ft., to make a closed box 7 ft. 10 in. long, 3 ft. 8 in. wide, 2 ft. 6 in. high (outside dimensions), the boards being 1 inch thick?

12. Reckoning a pint to be 30 cub. in.; if 462 gals. are taken out of a cylindrical cistern 7 ft. in diameter, how many inches will the surface of the water be lowered? ($\pi = 3\frac{1}{7}$.)

THIRD CLASS, 1890.

1. (*a*) Show how to find the L.C.M. of two or more numbers.

(*b*) Find the L.C.M, of 24, 105, 180, 96, 336, 84, and of

(*c*) 4410, 7350, 7875.

2. (*a*) Prove the rule for finding the product of two fractions.

(*b*) Simplify

$$\tfrac{4}{5}(3\tfrac{1}{3} + 1\tfrac{1}{4})£ + \frac{1\tfrac{1}{8} - \tfrac{1}{3} \text{ of } 1\tfrac{5}{6}}{\tfrac{1}{10} \text{ of } 3\cdot\dot{3} + \tfrac{13}{72}} \text{ of } \cdot 95 \text{ of } 5s + \frac{8\cdot 4}{\cdot 012}\, d.$$

3. If the Avoirdupois lb. is equal to 7000 grains Troy, and if 6144 sovereigns weigh 133 lbs. 4 oz. Troy, how many sovereigns will weigh an oz. Avoirdupois?

4. How many bricks, 9 inches long, $4\frac{1}{2}$ inches broad and 4 inches thick, will be required to build a wall 45 ft. long, 17 ft. high and 4 ft. thick, supposing the mortar to increase the volume of each brick $6\frac{1}{4}$ per cent.?

5. A man engages a sufficient number of men to do a piece of work in 84 days, if each man does an average day's work. It turns out that three of the men do respectively $\frac{1}{6}$, $\frac{1}{7}$. and $\frac{1}{9}$ less than an average day's work, and two others $\frac{1}{8}$ and $\frac{1}{10}$ more; and in order to complete the work in the 84 days, he procures the help of 17 additional men for the 84th day. How much less or more than an average day's work on the part of these 17 men is required?

6. A circular race-course is 22 yds. wide and has an area of 12 acres. Find the diameter of the inner circle.

7. The area of each of the longer walls of a room is 330 square feet; the area of each of the other walls is 220 square feet; the area of the floor is 384 square feet. Allowing $\frac{1}{25}$ of area of walls for doors and windows, how many yards of paper, 18 inches wide, are required to cover the walls?

8. The pressure of compressed air varies inversely as its volume. If the pressure on the inner surface of a cylinder fitted with a piston be 20 lbs. on the square inch, and when the piston is forced in 2 inches, the pressure becomes 30 lbs. on the square inch; what is the length of the cylinder?

9. A man has $20000 Bank Stock which is at 170 and pays a half-yearly dividend of 5%; he sells out and invests in Stocks at 108, which pays $3\frac{1}{2}$% half-yearly. Find the change in his half-yearly income.

10. Bought goods at $5.70 on 4 months' credit and sold them immediately at $6.12 on such a term of credit as made my immediate gain $6\frac{2}{3}$%. Reckoning interest at 4% per annum, how long credit did I give?

11. (*a*) What is meant by averaging accounts?

(*b*) Find the equated time for the payment of the following account

Dr. JOHN SMITH. *Cr.*

1888.			1888.		
June 10	To mdse. @ 30 days,	$950	July 10	By Cash -	$450
July 15	" " " 45 "	300	Aug. 15	" " -	350
Aug. 20	" " " 60 "	250	Sept. 5	" " -	200
Sept. 1	" " " 30 "	150			

12. A merchant in Montreal drew on Hamburg for 10000 guilders at \$·415; how much more would he have received if he had ordered remittance through London to Montreal, exchange at Hamburg on London being $11\frac{1}{4}$ guilders for £1, and at London on Montreal $9\frac{1}{4}$%, brokerage being $1\frac{1}{4}$% for remittance from London?

SECOND CLASS, 1883.

1. *Prove* that $\frac{1}{4}$ of $\frac{3}{7} = \frac{3}{28}$.
Simplify

$$(2\tfrac{4}{7} \text{ of } 3\tfrac{1}{18}) + \tfrac{4}{9} - (1\tfrac{1}{3} \text{ of } 1\tfrac{5}{18}) - (1\tfrac{3}{4} \text{ of } 4\tfrac{4}{7} \text{ of } \tfrac{3}{14}).$$

2. The pendulum of one clock makes 24 beats in 26″; that of another 36 beats in 40″. If they start at the same time, when first will the beats occur together?

3. A can do as much work in 4 hours as B in 6; and B in $3\frac{1}{2}$ as C in 5. A does half a certain piece of work in 12 hours; in what time can it be finished by B and C, working separately equal times, and C succeeding B?

4. A note for \$500, made March 9th at three months, is discounted April 11th, at 8 per cent. What is received for the note? (True discount.)

5. The unclaimed dividends on a certain amount of stock which pays 6 per cent. per annum amounted in 3 years to \$1152. The stock was sold at a discount of $12\frac{1}{2}$ per cent. on its par value. What sum was realized?

6. Teas at 3s. 6d., 4s. and 6s. a pound are mixed to produce a tea worth 5s. a pound. What is the least integral number of pounds that the mixture can contain?

7. A man buys 150 lbs. of sugar, and after selling 100 lbs. finds he has been parting with it at a loss of 5 per cent. At what rate per cent. advance on the cost must he sell the remaining 50 lbs. that he may gain 10 per cent. on the entire transaction?

8. The hour, minute and second hands of a watch are on concentric axes. When first, after 12 o'clock, will the direction of the second hand produced backwards bisect the angle between the hour and the minute hands?

9. Each member of a pedestrian club walks as many miles as there are members in the club, and the expense of the trip is for each member as many pence per mile as there are members in the club. The total expense is £50 13s. 11d. How many members are there?

SECOND CLASS, 1884.

1. Simplify—

$$\frac{(1\frac{1}{4} - 1\cdot002) \div (\frac{3}{4} - \cdot006)}{\cdot002 \div \cdot06} \times \cdot2\dot{9}\dot{9} \times 3\cdot6$$

2. A man mixes 28 lbs. black tea with 36 lbs. of an inferior quality, which costs 20 cents a pound less, and by selling the mixture at $58\frac{1}{2}$ cents a pound, gained 20 per cent. Find the cost of each kind of tea.

3. When the temperature of a cube of zinc is raised from 32° F. to 212° F., each dimension is increased ·3 per cent. Find the percentage of increase in the bulk.

4. On a quantity of tea a grocer fixed a price to make a gain of 25 per cent., but $\frac{1}{4}$ of the quantity was found to have been damaged, and he had to reduce the price on this 25 cents a pound, and so his whole gain was $48\frac{1}{3}$ per cent. less than the sum he had expected to gain. What price did he pay for the tea?

5. In a mile race between a bicycle and a tricycle their rates were as 5 to 4; the latter had half a minute's start and was beaten by 176 yards. Find the actual rate of each.

6. If 8000 metres be equal to 5 miles, and if a cubic fathom of water weigh 13440 lbs., and a cubic metre of water 1000 kilogrammes, find the ratio of a kilogramme to a pound avoirdupois.

7. A tradesman marks his goods at two prices, one for ready money and the other at a credit of six months. What is the ratio of these prices, if money is worth 10 per cent.?

8. What amount of American currency is equal to £500 14s. 6d., if gold is quoted at $15\frac{1}{2}$, and the course of exchange is 489?

9. The external dimensions of a rectangular iron chest are 2 ft. 3 in., 1 ft. 8 in., 1 ft. $2\frac{1}{2}$ in., and the sides, lid and bottom are one inch thick. Of how many cubic inches of iron is it formed?

10. A dealer has three prices for his goods—a year's credit price, a six months' credit price, and a cash price. The year's credit price is 35 per cent. in advance of cost, his six months' price is 6 per cent. off his year's credit price, and his cash price 10 per cent. off his year's credit price. At what advance on cost must he mark a six months' credit price on an article whose cash price is $12.15?

SECOND CLASS 1885.

1. A man bought a house, which cost him 4 per cent. on the outlay to put it in repair; it remained empty for a year, during which time he reckoned he was losing 5 per cent. on his total outlay. He then sold it for $1192, which paid for repairs and loss and also gave a profit of 10 per cent. on the cost price of the house. Find the cost price.

2. A railway train moving with uniform speed is met and passed in 5 seconds by an engine and tender $30\frac{2}{3}$ feet long and running 30 miles an hour; the engine and tender return shortly afterwards and pass the train in 25 seconds after overtaking it. Find the length of the train.

3. A person invested $8420 in 8 per cent. stock on the 7th day of January, at $109\frac{1}{2}$, and on the 12th day of February, of the same year, sold it out at $117\frac{1}{2}$, paying $\frac{1}{4}$ per cent. brokerage on each transaction. Find his gain per cent. on what the stock cost him—money being worth 8 per cent. per annum (360 days).

4. I bought French goods for 7490 francs, and paid an import ad valorem duty of 15 per cent. I sold the goods for £420. Find my gain or loss in dollars and cents, if the £ = 25.22 fr. = $4.87.

5. A triangle, altitude 60 feet, is bisected by a line drawn parallel to the base. Find the perpendicular distance between the base and the dividing line.

6. A merchant bought 3885 yds. of cloth and marked it at an advance of $33\frac{1}{3}$ per cent. on cost; in selling the first half of it he gave only 35 ins. for a yd., but in selling the remainder he gave 37 ins. for a yd. He gained on the whole transaction $3897. What did the cloth cost him per yd.?

7. I invested in 7 per cent. stock at $78\frac{1}{8}$, and having received a half-year's dividend I sold out at $79\frac{3}{8}$, paying $\frac{1}{8}$ per cent. brokerage on each transaction, and increased my capital altogether by $292.50. How much did I invest?

8. In an election 15 per cent. of the constituency refused to vote; of two candidates, one received 45 per cent. of the votes in the constituency and was elected by a majority of 150. Find the number of votes cast for each.

9. A person bought a quantity of goods for $224, payable in 2 months, and sold them at once for $274, payable in 4 months. Find the gain in ready money, allowing trade discount at 6 per cent. per annum.

10. A, B, and C walk from P to Q, each at a uniform rate, A's rate being equal to $\frac{4}{5}$ of C's, and B's rate was 4 miles an hour. B started 45 minutes after A, and C started 27 minutes after B. They all arrived at Q at the same time. Find the distance from P to Q.

11. The areas of the several faces of a rectangular solid are 57, 27, and 19 square feet. Find its dimensions.

SECOND CLASS, 1886.

1. A sold $\frac{1}{3}$ of his goods at cost and the remainder at a loss of 25 per cent. on cost. Had he received $25 more for them than he did he would have gained 25 per cent. on the whole cost. Find that cost.

2. One-half of a ball of lead, 3 inches in diameter, is melted down and cast in the form of a right circular cone 3 inches in height. Find the diameter of the base of the cone.

3. The men employed in a certain factory numbered three less than twice the number of women employed in it. The men received $1.55 per day, the women 85 cts. per day, and the total weekly wages amounted to $469.80. How many men were employed in the factory?

4. A and B agree to share the profits of a certain transaction in the proportion of $11 to A for every $7 to B. In connection with the transaction, A has received $960 and paid out $470, and B has received $1370 and paid out $330. How much must B pay to A to settle the accounts of the transaction?

5. M and N starting at the same moment from the same place, and in the same direction, walk around a circular track, M at the rate of $8\frac{1}{2}$ yds. to every $5\frac{1}{2}$ yds. by N. At what point of the track will M first overtake N, and how many rounds will each have then made?

6. At an election the successful candidate received $\frac{5}{8}$ of the total number of votes cast, and had a majority of 832 over his rival. Of the total number of electors in the constituency $\frac{3}{16}$ did not vote. How many electors were there in the constituency?

7. Between 1871 and 1881 the county of A lost 24·73 per cent. of its population by deaths and removals, but during the same time it gained 42·41 per cent. by births, etc., the percentages being reckoned on the population in 1871, In 1881 the population was found to be 26478. What was it in 1871?

8. Find the difference between the discount on 10th Sept., at 8 per cent., on a bill for $128 drawn on 3rd Sept. at 3 months, and the interest at 8 per cent. for the same time on the proceeds. (In reckoning the discount include the 3 days of grace, but no other charges.)

9. The length of the sides taken in order of a quadrilateral field are 20 rd., 21 rd., 21rd., and 22 rd., and the angle between the first and second of these sides is a right angle. Find the area of the field to the nearest square rod.

SECOND CLASS, 1887.

1. (*a*) In reducing a vulgar fraction to a decimal, explain how you determine whether it will be a finite or a circulating decimal, pure or mixed. What is the limit as to the number of repeating digits?

(*b*) Express as a decimal $\frac{1}{820}$ of $6\cdot\dot{3}0769\dot{2} \times 1\cdot\dot{4}2857\dot{1}$.

2. Bought goods at 4 months' credit, and after 7 months sold them for $1500, $2\frac{1}{2}$ per cent. off for cash, and gained 15 per cent. Money being worth 6 per cent., what did the goods cost?

3. An alloy of gold is mixed with an alloy of silver in the proportion of 11·4 to 2·6. The percentage of dross in the silver is 13·5 and in the gold 17·35; what is the percentage of dross in the mixture?

4. (*a*) At 10 per cent. for 4 years, what fraction of the simple interest is gained by charging compound instead of simple interest?

(*b*) The compound interest on $500 for 3 years is $95·508, find the rate.

5. What is meant by the *par* of exchange? The *course* of exchange?

When the course of exchange between London and New York is quoted at 2 per cent. premium, what will be obtained in New York money for a bill of £240 12s. 8d.?

6. A 60-day note was discounted at the bank at 1 per cent. a month, and $4.80 more than True Discount was charged. Allowing days of grace, find the face of the note.

7. Bought 16 cows and 120 sheep for $465, the animals of the same kind costing a uniform price. Sold for $496.50, gaining $7\frac{1}{2}$ per cent. on the cows, and 6 per cent. on the sheep. Find the cost of each a head.

8. Bought a 6 per cent. mortgage for $2500 at 5 per cent. discount, with two years to run. What rate of interest is obtained if the mortgage is satisfied at maturity?

9. A man invested $5500, a part in the 4 per cents at $83\frac{3}{4}$, and the rest in the 5 per cents at $102\frac{1}{4}$, brokerage $\frac{1}{4}$ per cent. in each case. His total income being $\$266\frac{2}{3}$, find the sum invested in each stock.

10. A garden whose width is 9 rods and length 15 rods is to have a wall $3\frac{3}{4}$ ft. thick around it outside. What will be the cost of digging a trench for it, $2\frac{1}{2}$ feet deep, at $1\frac{3}{5}$c. per cubic foot?

11. A circular race course 22 yds. wide covers 12 acres, find the diameter of the inner circle. ($\pi = 3\frac{1}{7}$).

12. A conical tent whose slant height is 12 ft. requires 132 sq. ft. of canvas to make it; how much ground does the floor of the tent cover?

SECOND CLASS AND JUNIOR MATRICULATION, 1888.

1. A person sells out 3 per cent. consols at $94\frac{1}{2}$, and invests the proceeds in bank stock which sells at 225, and pays yearly dividends of $8\frac{1}{2}$ per cent. If his income is changed to the extent of \$57, how much money had he invested?

2. The profits of a loan company for a year were sufficient to enable the directors to add \$20,000 to a reserve fund, to pay \$5965 for cost of management, to pay two half yearly dividends of $3\frac{1}{2}$ per cent. on a paid-up capital stock of \$309056, and to have still on hand \$4236. Find the profits for the year.

3. A and B enter into partnership for 3 years. A puts in \$20000 and B \$5000; B is to manage the business, and the profits are to be equally divided; but at the end of the first year A increases his stock to \$36000. How shall they divide a gain of \$28500 at the end of the three years?

4. Find, most readily, to six places of decimals,

$$\frac{1}{\sqrt{5}}, \ \frac{1}{\sqrt{5}-1}, \ \sqrt{6+2\sqrt{5}}, \ \text{where } \sqrt{5} = 2 \cdot 236+$$

5. A train leaves Toronto for Hamilton at 5.55 p.m., running at the rate of 26 miles an hour. Another leaves Hamilton for Toronto at 6.35 p.m., running 40 miles an hour. Before they meet the first loses 5 minutes and the second 10 minutes by stoppages. At what time will the trains meet, Toronto and Hamilton being 39 miles apart?

6. How may we know, without dividing, when a number is exactly divisible by 9? State and explain how we may find the quotient in such a case without actual division.

7. A note bearing interest at 8 per cent. per annum, having two years to run, is offered for sale. What per cent. advance on its face value can a purchaser offer for it so as to receive six per cent. interest for his money?

8. In reducing to a decimal any proper fraction with 7 for denominator the same digits $\dot{1}4285\dot{7}$ are produced in circular order. Explain this.

What is the shortest method of finding $\frac{2}{17}$ when it is known that $\frac{16}{17} = \cdot\dot{9}41176470588235\dot{2}$?

9. A merchant buys a quantity of goods and sells $\frac{2}{3}$ of it at an advance of 15 per cent., and $\frac{1}{4}$ of it at an advance of 20 per cent. He now discovers that 10 per cent. of his goods are quite unsalable. What per cent. profit must he obtain on the remainder that he may gain 15 per cent. on the whole transaction?

10. The French 20-franc piece, or Napoleon, weighs 6·45161 grammes (a gramme = 15·43235 grains), and is $\frac{9}{10}$ pure gold. The sovereign is $\frac{11}{12}$ fine, weighs 123·274 grains, and is worth \$4·8665. How much is the Napoleon worth?

11. A grocer mixed together two kinds of tea and sold the mixture, 144 lbs., at an advance of 20 per cent. on cost, receiving for it \$62.10. Had he sold each kind of tea at the same price per pound as he sold the mixture he would have gained 15 per cent. on the one and 25 per cent. on the other. How many pounds of each kind were there in the mixture, and what was the cost of each per pound?

12. The money deposited in a savings bank during the year 1885 was 5 per cent. greater than that deposited in 1884. In 1886 the deposits were $33\frac{1}{3}$ per cent. greater than in 1885, while the amount deposited in 1887 exceeded the average of the three previous years by 20 per cent. The aggregate for the four years was \$150937.50. Find the amount deposited in each year.

13. If 76 men and 59 boys can do as much work in 299 days as 40 men and 33 boys can do in 557 days; how many men will do as much work in a day as 15 boys?

SECOND CLASS AND JUNIOR MATRICULATION, 1889.

1. A note for $876, dated May 17, for 90 days, and bearing interest at the rate of 8 per cent. per annum, is discounted at a bank on July 3, at 6 per cent. What are the proceeds of the note?

2. Explain the terms Exchange, Bill of Exchange, Par of Exchange.

What is meant by saying "the rate of sterling exchange is $4·87 for 60-day bills"?

How is the par of exchange between two countries arrived at?

3. What capital should be invested in 6 per cent. stock at 104 to produce an income one-third greater than that derived from $1500 invested in 7 per cent. stock at 115?

What rate of interest is received on the money invested in each case?

4. Prove that a vulgar fraction may always be reduced to a terminated or to a repeating decimal.

Explain any short method of reducing $\frac{18}{19}$ to a repeating decimal.

5. Explain the method of contracted multiplication of decimals.

Employ this method to find the number of cubic yds. in a cubic metre correct to 4 decimal places, a metre being equal to 1·09363 yds., linear measure.

6. A rectangular solid is hammered until its length is increased 10 per cent., and its width 15 per cent.; by how much per cent., has its thickness been diminished?

7. In 1837 the U.S. half-dollar was changed in weight from 208 grains to $206\frac{1}{4}$ grs., and in fineness from ·8924 to ·900; find the least whole numbers which will show the relative values of the coins before and after the change.

8. The cost of manufacturing a certain article depends partly on the cost of labor and partly on the cost of the raw material. Wages rise 25 per cent., but a reduction of one-sixth in the cost of material enables the manufacturer to produce 16 of the articles for what 15 cost him before the change. How much does the raw material for $100 worth of the manufactured article now cost him?

9. The expense of constructing a railroad is $2000000, two-fifths of which was borrowed on mortgage at 5 per cent., and the remaining three-fifths was held in shares. What must be the average weekly receipts so as to pay the shareholders 4 per cent., the expenses of working the road being 55 per cent. of the gross receipts?

10. A person buys a house and lot—the lot being worth $\frac{2}{5}$ as much as the house—and lets to a tenant at a monthly rental of one per cent. on the cost of the property. He finds that the lot will rise 5 per cent. and the house depreciate 4 per cent. in value every year, that insurance (on $\frac{4}{5}$ of the value of the property insured) will cost him $1\frac{1}{4}$ per cent., every three years, that his taxes will be 18 mills on the dollar, and that the assessors have valued his property at 10 per cent. less than he gave for it. What rate per cent. will he receive on the money he has invested?

11. An invoice of British merchandise, amounting to £20,000 and subject to an ad valorem duty of 35 per cent., is received at New York and converted into U.S. money at the rate of $4·844 to the pound sterling instead of $4·8665, the true value: how much is gained or lost by the difference and by whom?

12. A leaves P for Q, 39 miles distant, at the same time that B leaves Q for P; they travel at uniform rates of speed till they meet. B then increases his speed one-eighth and reaches P in 5 hours from the time he met A; while A, after resting for an hour, proceeds at $\frac{9}{10}$ his former rate and reaches Q at the same time that B reaches P. Find the rate at which each person set out?

SECOND CLASS AND JUNIOR MATRICULATION, 1890.

1. Define a fraction and establish a series of propositions based upon your definition leading up to the rule for the division of fractions.

Examine how far each of the following statements is true:

(*a*) $\frac{2}{3}$ of $\frac{5}{7} = \frac{5}{7} \times \frac{2}{3}$.

(*b*) $\frac{1}{3}$ of $\frac{7}{8} = \frac{7}{8} \div 3$.

(*c*) A fraction represents the quotient of the numerator by the denominator.

2. A bought goods to the value of $5191.53, and gave in payment his note due in 3 months. What must be the face of the note so that when discounted at 7 per cent. it will realize the amount required?

3. A person having a quantity of gold bullion may either dispose of it at once at the rate of £3 17s 9d per oz., or take it to the mint and have it coined for him at £3 17s $10\frac{1}{2}$d per oz., waiting in the latter case 10 days for his money. Which plan had he better adopt, if money is worth 6 per cent. per annum to him?

4. Find the distance (in inches correct to 3 decimal places) between the opposite corners of a cube whose volume is 2 cubic yards.

5. In a certain municipality $33\frac{1}{3}$ per cent. of the taxes (at 18 mills on the dollar) go to pay interest on its indebtedness, the remainder being apportioned in the ratio 3:5 to school and to city purposes. In a subsequent year when 16 per cent. of the debt has been paid off and the interest on the remainder reduced to $\frac{5}{6}$ of the former rate, it was found that for city purposes there would be needed $13\frac{1}{3}$ per cent. more than before, and for school purposes $22\frac{2}{9}$ per cent. more. The value of the taxable property having increased by $3\frac{1}{3}$ per cent., what rate of assessment will now be required?

6. A train 110 yds. long overtakes A who is going at the rate of 4 miles an hour, and passes him in 9 seconds. Ten minutes after leaving A the train meets B and passes him in $7\frac{1}{5}$ seconds. In what time after meeting the train will B meet A?

7. A government which derives a revenue of 20 million dollars from the duty on imported goods finds it necessary to obtain an additional two millions from this source. Assuming that if the rate of duty be increased by any fraction, say, one-fifth of itself, the value of the goods imported will be diminished by one-tenth, and so on, find approximately by what per cent. of itself the rate of duty must be increased in order to produce the revenue required?

8. A cubic inch of water weighs 252·458 grains. Gold is 19·3 times and silver 10·5 times as heavy as water. Find the weight of a cubic inch of a mixture containing gold and silver in the ratio (by weight) of 11:1.

9. Find, correct to the nearest digit in the fourth decimal place,

(a) the value of

$$2\left(\frac{1}{5}+\frac{1}{3\times5^3}+\frac{1}{5\times5^5}+\frac{1}{7\times5^7}\right);$$

(b) the number of cubic centimetres in a cubic inch (1 metre = 1·09363 yds.).

10. A dealer buys a quantity of liquor at $\frac{4}{5}$ of its value, which he keeps for two years and then sells. The value increases 10 per cent. per annum by age, 1 per cent. is lost each year by evaporation, and there is a waste of 2 per cent. in handling while it is being sold. What rate per cent. per annum (compound) interest does he make on his money if he sells at the enhanced value?

11. A person invests money

(a) In bank stock at 128 paying half-yearly dividends of 4 per cent., subject to an income tax of 18 mills in the dollar; and

(b) In city property yielding a rental of 10 per cent., costing him one-fifth of the rent for insurance and repairs, and $18\frac{1}{2}$ mills on the assessed value (90 per cent. of the cost) for taxes.

If the whole amount invested is $4989 how shall he divide it so that the net income from the two investments may be the same?

12. "In dividing by 73000 it is advantageous to do so by the following method: Having written down the number to be divided we write under it one-third of itself, then one-tenth of this second number, neglecting remainders, and lastly one-tenth of this third number. The sum of these four numbers, with the last five figures reckoned as decimals, will be the quotient required."

Establish the correctness of this method.

To what extent can its accuracy be depended upon?

Indicate a slight extension of the method which will enable any required degree of accuracy to be obtained.

FIRST CLASS, 1883.

1. Define a recurring decimal, and classify the several kinds.

Prove, in any way, a rule for converting a mixed circulating decimal into its equivalent vulgar fraction, and apply your rule to convert $\cdot 1013\dot{2}5\dot{7}$.

2. Perform the operations here indicated, employing contracted multiplication and division, and retaining six decimals throughout:

$$\frac{\cdot 3472 + \cdot 03172}{6146 \cdot 38} \div \cdot 0004675.$$

3. In the expression "six per cents are at 103," explain fully what is meant.

A person sells a certain amount of 5 per cents for 86, and invests in the 6 per cents at 103, and by so doing changes his income by one dollar.

Is the change an increase or a decrease? How much stock did he sell?

4. A man buys a note, drawn for 2 years at 6 per cent. interest, and which is now 6 months old, at 15 per cent. true discount. After keeping it 9 months, and receiving one payment of interest, he sells it to a bank at 8 per cent. bank discount. What per cent. does he make upon his money while invested?

5. A, B and C, whose rates of walking are $3\frac{1}{2}$, 4 and 5 miles an hour respectively, walk on circular tracks whose circumferences are 8, 10 and 15 miles respectively, and whose centres are in the same straight line. At the same instant they start from points on this line, and on the same side of the centres. Find (1) when first they will be all on this line at the same time; (2) all at same time at the points from which they started; (3) whether they will ever be all at the same instant at points on opposite sides of the circles to the starting points.

6. Lead is 11·4 times, and zinc 7·2 times as heavy as water. If 3 pounds of lead and 2 pounds of zinc be melted together, compare the weight of the alloy with that of water.

7. A, B and C start at the same time, and from the same point, to travel around an island 26 miles in circuit. A goes 10 miles and B 4 miles per hour in the same direction, and C goes 5 miles per hour in the opposite direction. When and where will they first be all together again?

Algebraical symbols will be allowed in the three following questions:—

8. It is required to make a hollow leaden cylinder open at both ends, 10 inches long, with its wall one inch thick, and which is to weigh 25 pounds. Find its outside diameter.

9. A conical vessel 6 inches deep and 3 inches across the mouth is filled to 5 inches with water. Find the diameter of the sphere which, when dropped into the cylinder, will raise the water so as just to fill the vessel.

10. The diagonals of a quadrilateral plane figure are 10 and 12, and they intersect at an angle of 60°. Find the area of the figure.

FIRST CLASS, 1884.

1. A gallon of water weighs 10 lbs.; a litre is equal to 1·761 pints, and a kilogram to 2·205 lbs. How many kilograms would a litre of water weigh?

2. A tradesman marks all his goods at an advance of $22\frac{1}{2}$ per cent. on cost. In selling he uses a yard stick an inch too short, and a pound-weight a quarter of an ounce light. In 43 per cent. of his sales, reckoned on their total value, the goods are measured with the short yard stick, in 36 per cent. they are weighed with the light pound-weight, and in the rest they are sold at the marked price independently of weight or of measure. What is the actual rate of profit on the whole of the sales?

3. Skilled workmen and labourers are employed on a work, a skilled workman receiving $1.75 per day more than a labourer. The average of their daily wages is $12\frac{1}{2}$ cents more than what it would be if skilled workmen and labourers were employed in equal numbers. If 6 men of each kind were discharged, the average of the daily wages would be raised by 5 cents. Find the number of men of each kind employed?

4. The nearest approximate value in thousandths of a certain vulgar fraction is ·539, the numerator of the fraction is 187. What is its denominator?

5. Four men start together from the same point, and run in the same direction round a ring at different uniform speeds. The first runs at the rate of 10 miles, the second at the rate of $10\frac{3}{4}$ miles, the third at the rate of $11\frac{1}{8}$ miles, and the fourth at the rate of $12\frac{1}{4}$ miles, each per hour. At what part of the ring will they be first all together again after starting?

6. The discount off a note drawn at 4 months is $10.50; the interest on the proceeds reckoned for the same time, and at the same per cent. as the discount, would be $10.20. Find the amount for which the note is drawn, and the percentage taken off as discount.

7. Point off any number into periods of three figures each, beginning at the right. If the difference between the sum of the odd periods, numbering them from the right, and the sum of the even periods, be divisible by 7 or by 13, the number itself will be divisible by 7 or by 13, as the case may be. *Prove this.*

8. The perimeter of a semi-circle is three yards, find its area in square feet.

9. Find the surface and the volume of a right circular cone, given the diameter of the base 174 inches, and the slant height 145 inches.

10. The areas of two regular polygons of the same number of sides, the one inscribed in a circle, the other described about it, are to each other as 3 to 4. Find the number of sides of the polygon.

FIRST CLASS, 1885.

1. "Every operation of Division may be viewed as giving the answer to two different questions." Explain and illustrate this statement.

2. Show that if the greater of two integers be divided by the other, the greatest common measure of the two numbers is the same as the G.C.M. of *remainder* and *divisor*.

3. Divide the fraction $\frac{17}{18}$ into two such parts that 4 times one of them added to $5\frac{1}{2}$ times the other may make $1\frac{1}{3}$.

4. When a vulgar fraction is to be reduced to a decimal, show how to determine *(a)* whether the result will be a finite decimal or a pure circulating decimal, or a mixed circulating decimal; *(b)* the number of non-repeating digits in each case.

5. A man barters 120 yds. of silk, which cost $1.50 a yard and sells at $2.50, giving nine months' credit, for cloth which sells at $2 on six months' credit. How much cloth ought he to receive?

6. A, B, C and D together do a work for which A by himself would require two hours less than B. A and B together could do it in $\frac{50}{63}$ of the time C and D together would take, A and C in $\frac{56}{75}$ of the time B and D would take, and B and C in $\frac{65}{68}$ of the time A and D would take. Find the time each person singly would require to do the work.

7. Find the square root of ·00013 to within less than a millionth, and the value of $\sqrt{\{2+\sqrt{(2+\sqrt{2})}\}}$ to within less than a ten-thousandth.

8. Two trains, whose lengths are 120 feet and 160 feet respectively, pass each other in 30 seconds when moving in the same direction, and in $7\frac{1}{2}$ seconds when moving in opposite directions. Find the rate of each train in miles per hour.

9. The circumference of one circle is $27\frac{1}{2}$ feet longer than that of another, and 11 times the diameter of the first is equal to 5 times the circumference of the second. Find the diameter of each, π being assumed $= 3\frac{1}{7}$.

10. The length of an iron cylindrical vessel with closed ends is four feet, and its outside circumference is 40 inches, and the thickness of the metal one inch. Find the entire weight when the cylinder is filled with water, iron being $7\frac{1}{9}$ times heavier than water, and water weighing $62\frac{1}{2}$ lbs. per cubic foot.

11. I hold some three per cent. stock; on receiving my first half year's dividend I invest it in the same stock at $93\frac{3}{4}$, and my next half year's dividend is \$1905. What amount of stock had I at first?

12. The area of each of the longer walls of a room is 360 ft., that of each of the other walls is 192 ft., and that of the floor is 180 ft. How many yards (linear) of paper 18 inches wide, will be needed for the walls, deducting one twenty-fifth of the whole area for doors, etc.

13. Find the depth of a ditch, the transverse section of which is a trapezoid, of which the longer side is 20 ft., the slopes of the sides 2 in 1 and 3 in 1, respectively, and the area 146·25 square feet.

ANSWERS.

PAGE.

16.—(1) $3\frac{1}{4}$. (2) 9. (3) $5\frac{7}{8}$. (4) $8\frac{6}{7}$. (5) $27\frac{1}{4}$. (6) 9. (7) 32. (8) $80\frac{42}{43}$. (9). $45\frac{60}{107}$. (10) $11\frac{1}{909}$. (11) $100\frac{1}{101}$. (12) 1001. (13) $845\frac{748}{987}$. (14) $427\frac{375}{8076}$. (15) $1\frac{111111}{6552732}$.
(1) $\frac{7}{2}$. (2) $\frac{21}{4}$. (3) $\frac{55}{8}$. (4) $\frac{143}{12}$. (5) $\frac{124}{7}$. (6) $\frac{579}{10}$. (7) $\frac{211}{15}$. (8) $\frac{455}{19}$. (9) $\frac{31285}{9}$. (10) $\frac{22306}{23}$. (11) $\frac{159023}{6789}$. (12) $\frac{322928}{577}$. (13) $\frac{103121}{102}$. (14) $\frac{16000}{8357}$. (15) $\frac{10000000}{916524}$. (16) $\frac{50043583}{6541}$. (17) $\frac{72151070}{8008}$. (18) $\frac{1000199}{100}$. (19) $\frac{1000347}{1000}$. (20) $\frac{873087}{100}$. (21) $\frac{9342856}{1000}$.

17.—(1) $\frac{24}{7}$, $\frac{36}{7}$, $\frac{48}{7}$, $\frac{60}{7}$, $\frac{96}{7}$. (2) $\frac{153}{19}$, $\frac{221}{19}$, $\frac{391}{19}$, $\frac{799}{19}$. (3) $\frac{4900}{137}$, $\frac{9800}{137}$, $\frac{25600}{137}$.

18.—(1) $\frac{1}{6}$. (2) $\frac{2}{9}$. (3) $\frac{5}{48}$. (4) $\frac{18}{77}$. (5) $\frac{253}{288}$. (6) $\frac{799}{912}$.
(1) $\frac{1}{24}$. (2) $\frac{56}{135}$. (3) $\frac{1}{720}$. (4) $\frac{720}{1463}$. (5) $\frac{1683}{80000000}$. (6) $\frac{16000}{969}$. (7) $\frac{32}{9}$. (8) $\frac{2288}{81}$. (9) $\frac{2508374}{2040}$. (10) $\frac{21989000}{2997000}$.

19.—(1) $\frac{5}{14}$, $\frac{5}{21}$, $\frac{5}{28}$, $\frac{5}{35}$. (2) $\frac{7}{144}$, $\frac{7}{180}$. (3) $\frac{148}{48875}$.
(1) $\frac{9}{12}$, $\frac{15}{20}$, $\frac{36}{48}$, $\frac{75}{100}$. (2) $\frac{10}{14}$, $\frac{14}{18}$, $\frac{24}{28}$. (3) $\frac{20}{28}$, $\frac{21}{28}$. (4) $\frac{49}{63}$, $\frac{54}{63}$. (5) $\frac{40}{60}$, $\frac{45}{60}$, $\frac{48}{60}$. (6) $\frac{20}{24}$, $\frac{21}{24}$. (7) $\frac{10}{18}$, $\frac{3}{18}$, $\frac{12}{18}$. (8) $\frac{30}{36}$, $\frac{28}{36}$, $\frac{33}{36}$, $\frac{18}{36}$, $\frac{10}{36}$.

20.—(1) $\frac{1}{3}$. (2) $\frac{3}{4}$. (3) $\frac{3}{4}$. (4) $\frac{2}{3}$. (5) $\frac{4}{5}$. (6) $\frac{7}{8}$. (7) $\frac{4}{5}$. (8) $\frac{3}{5}$. (9) $\frac{3}{5}$. (10) $\frac{2}{7}$. (11) $\frac{2}{3}$. (12) $\frac{3}{4}$. (13) $\frac{1}{3}$. (14) $\frac{1}{3}$. (15). $\frac{5}{33}$. (16) $\frac{7}{9}$. (17) $\frac{5}{9}$. (18) $\frac{11}{13}$. (19) $\frac{13}{17}$. (20) $\frac{5}{7}$. (21) $\frac{1547}{1881}$. (22) $\frac{5}{7}$. (23) $\frac{1547}{1881}$. (24) $\frac{47957}{54549}$.

21.—(1) $\frac{1}{5}$. (2) $\frac{1}{2}$. (3) $\frac{3}{4}$. (4) 1. (5) $\frac{3}{20}$. (6) $\frac{33}{7}$. (7) $\frac{107}{7}$. (8) $\frac{915}{98}$. (9) $\frac{113}{123}$. (10) $\frac{1}{4}$.

22.—(1) $\frac{54}{6}$, $\frac{63}{7}$, $\frac{90}{10}$, $\frac{900}{100}$. (2) $\frac{546}{14}$, $\frac{741}{19}$, $\frac{897}{23}$.

23.—(1) $\frac{70}{90}$, $\frac{48}{90}$, $\frac{15}{90}$. (2) $\frac{40}{60}$, $\frac{45}{60}$, $\frac{48}{60}$, $\frac{50}{60}$. (3) $\frac{2352}{2520}$, $\frac{980}{2520}$, $\frac{2415}{2520}$, $\frac{2280}{2520}$. (4) $\frac{1000}{10000}$, $\frac{100}{10000}$, $\frac{10}{10000}$, $\frac{1}{10000}$. (5) $\frac{49}{273}$, $\frac{208}{273}$, $\frac{77}{273}$, (6) $\frac{255}{357}$, $\frac{90}{357}$, $\frac{7}{357}$, $\frac{238}{357}$, $\frac{17}{357}$. (7) $\frac{690}{2760}$, $\frac{300}{2760}$, $\frac{2185}{2760}$, $\frac{552}{2760}$, $\frac{2720}{2760}$. (8) $\frac{3}{8}$, $\frac{55}{8}$, $\frac{32}{8}$. (9) $\frac{28}{65}$, $\frac{720}{65}$, $\frac{325}{65}$. (10) $\frac{143}{15015}$, $\frac{39}{15015}$, $\frac{35}{15015}$.

24.—(1) $\frac{7}{8}$. (2) $\frac{7}{8}$. (3) $\frac{1}{2}$, $\frac{1}{3}$, $\frac{1}{4}$. (4) $\frac{17}{24}$, $\frac{2}{3}$, $\frac{7}{12}$. (5) $\frac{5}{7}$, $\frac{1}{12}$, $\frac{3}{40}$. (6) $\frac{7}{12}$, $\frac{11}{21}$, $\frac{29}{60}$, $\frac{5}{16}$. (7) $\frac{5}{11}$, $\frac{4}{9}$, $\frac{3}{8}$, $\frac{2}{7}$. (8) $\frac{15}{14}$, $\frac{6}{11}$, $\frac{13}{27}$, $\frac{26}{75}$. (9) $\frac{23}{24}$, $\frac{22}{23}$, $\frac{19}{20}$, $\frac{18}{19}$.

25.—(1) $\frac{7}{8}$. (2) $\frac{18}{19}$. (3) $\frac{12}{13}$. (4) $\frac{27}{45}$. (5) $\frac{13}{36}$. (6) $\frac{16}{25}$. (7) $\frac{6}{7}$. (8) $\frac{6}{7}$. (9) $\frac{29}{37}$. (10) $\frac{4}{5}$. (11) $\frac{8}{9}$. (12) $\frac{13}{14}$.

26.—(1) $\frac{3}{19}$, $\frac{5}{31}$, $\frac{6}{37}$. (2) $\frac{7}{92}$, $\frac{8}{105}$, $\frac{82}{111}$. (3) $\frac{3}{97}$, $\frac{5}{161}$, $\frac{7}{225}$, $\frac{11}{353}$.
(1) $1\frac{4}{5}$. (2) $\frac{12}{13}$. (3) $\frac{1}{3}$. (4) $\frac{5}{9}$. (5) 1. (6) $\frac{13}{15}$.

27.—(1) $1\frac{11}{35}$. (2) $1\frac{29}{33}$. (3) $1\frac{17}{24}$. (4) $1\frac{1}{3}$. (5) $1\frac{53}{72}$. (6) 1. (7) $2\frac{13}{60}$. (8) $2\frac{23}{60}$. (9) $3\frac{307}{840}$. (10) $2\frac{13}{18}$. (11) 1. (12) $\frac{31592}{118881}$. (13) $\frac{14432}{34467}$. (15) $13\frac{3}{4}$. (16) $10\frac{2}{35}$. (17) $20\frac{84}{187}$. (18) $19\frac{701}{1260}$. (19) $22\frac{2321}{2520}$.

PAGE.

28.—(20) $37\frac{713}{840}$. (23) $15\frac{17}{924}$. (24) 1.

29.—(1) $\frac{3}{5}$. (2) $\frac{1}{4}$. (3) $\frac{2}{17}$. (4) $\frac{16}{29}$. (5) $\frac{1}{2}$. (6) $\frac{2}{9}$. (7) $\frac{6}{29}$. (8) $\frac{38}{113}$. (9) $3\frac{1}{2}$. (10) $7\frac{1}{3}$. (11) $\frac{1}{8}$. (12) $\frac{1}{24}$.

30.—(1) $\frac{1}{6}$. (2) $\frac{1}{12}$. (3) $\frac{2}{143}$. (4) $\frac{3}{221}$. (5) $\frac{1}{90}$. (6) 3. (7) $\frac{1}{198}$. (8) $\frac{3}{2701}$. (9) $\frac{327}{6940}$. (10) $23\frac{38}{63}$. (11) $\frac{19}{420}$. (12) $1\frac{3}{7}$. (13) $\frac{241}{420}$. (14) $\frac{2}{19}$. (15) $\frac{19}{24}$. (16) 0. (17) $\frac{1}{36}$. (18) $\frac{1}{342}$. (19) $\frac{1}{812}$. (20) $\frac{1}{7}$. (21) $\frac{29}{140}$. (23) $1\frac{71}{99}$. (24) $1\frac{6}{117}$. (25) $15\frac{6949}{69160}$. (26) $\frac{11}{35}$.

31.—(29) $1\frac{29}{56}$. (30) $9\frac{85}{126}$. (31) $13\frac{7}{8}$. (32) $\frac{125}{144}$. (33) $\frac{59}{377580}$.

33.—(1) $\frac{8}{9}$. (2) $\frac{19}{21}$. (3) $\frac{10}{21}$. (4) $\frac{8}{105}$. (5) $\frac{63}{220}$. (6) $\frac{343}{720}$. (7) $\frac{3}{4}$. (8) $\frac{1}{17}$. (9) 1. (10) $2\frac{7}{31}$. (11) 2. (12) $\frac{9}{38}$. (13) $162\frac{1}{2}$. (14) $\frac{35}{72}$. (15) $\frac{9}{400}$. (16) $1\frac{5}{8}$. (17) $\frac{1}{10}$. (18) $\frac{2}{3}$. (19) $1\frac{5}{7}$. (20) $\frac{1}{7}$. (21) $\frac{13}{56}$.

34.—(24) $6\frac{32}{33}$. (25) $9\frac{85}{126}$. (26) $3\frac{111}{374}$. (27) $\frac{1}{3}$. (28) 1. (29) $\frac{301}{360}$. (30) $100\frac{1}{20}$. (31) $2\frac{49}{128}$.

35.—(1) $\frac{4}{5}$. (2) $\frac{2}{7}$. (3) 2. (4) 34. (5) $1\frac{5}{91}$. (6) $\frac{110}{111}$. (7) $\frac{427}{495}$. (8) 3. (9) $\frac{18}{19}$. (10) $10\frac{10}{111}$. (11) $\frac{19}{100}$. (12) $\frac{162}{423}$. (13) $\frac{1}{11}$. (14) 99. (15) $5\frac{125}{329}$. (16) $\frac{5}{18}$. (17) $\frac{25}{32}$. (18) 1. (19) $1\frac{1}{110}$. (20) 9. (21) $1\frac{19}{27}$.

37.—(24) $1\frac{1}{4}$. (25) $7\frac{1}{9}$. (26) $1\frac{40}{299}$. (27) $7\frac{35}{39}$. (28) $\frac{25}{144}$. (29) $4\frac{412}{2337}$.

38.—(1) $1\frac{11}{45}$. (2) $5\frac{7}{10}$. (3) $8\frac{2}{3}$. (4) $\frac{7}{30}$. (5) $\frac{1}{5}$. (6) $\frac{8}{21}$. (7) $1\frac{1}{7}$. (8) $\frac{45}{301}$. (9) $1\frac{109}{210}$. (10) $7\frac{7}{19}$. (11) $3\frac{475}{1008}$.

39.—(12) $\frac{3}{19}$. (13) $1\frac{41}{69}$. (14) 8. (15) $\frac{912247}{3757000}$. (16) $\frac{3057}{1081228}$. (17) $\frac{937}{960}$. (18) $\frac{113}{355}$. (19) $\frac{39}{37}$. (20) $1\frac{1}{35}$. (21) $1\frac{35}{1296}$. (22) $\frac{3}{4}$. (23) 18. (24) $\frac{75}{77}$.

44.—(1) Seven, and six-tenths. (2) Thirty-nine, and three-tenths. (3) Four, and eighty-nine hundredths. (4) Seven hundred and sixty-two units. (5) Seven hundred and sixty-two thousandths. (6) Seven hundred and sixty-two, and seven hundred and sixty-two thousandths (7) One thousand two hundred and thirty-four, and five thousand six hundred and seventy-eight ten-thousandths. (8) One hundred and twenty-three, and forty-five thousand six hundred and seventy-eight hundred-thousandths. (9) Two thousand four hundred, and thirty-six ten-thousandths. (10) Two thousand four hundred and thirty-six ten-thousandths. (11) Six ten-thousandths. (12) Six millionths. (13) Six billionths. (14) 76·89. (15) 14·003. (16) 100·0003. (17) ·0103. (18) 30070·001083.

46.—(1) 3525·9774. (2) 61·23737. (3) 2503·61876243. (4) 14654·5429118. (5) 1·0011101.

49.—(1) 2·4493944. (2) 5·477082. (3) 44·395031413. (4) 57·8687277. (5) 7·7123875248. (6) ·26484624. (7) 12025·21200096. (8) ·00000072. (9) ·0000001. (10) 1.

50.—(1) 1·2, ·12, ·012. (2) ·4, ·04, ·16, 12·8, 64. (3) 8·1, ·81, ·9, 2·7, 2 18·7, ·0003. (4) ·015625, 78·125, 625, ·000025, ·625. (5) ·07, ·00049, ·00343, 2401.

PAGE.

51.—(7) ·00121, ·161051, ·014641, ·01331. (8) 1·368, ·05472, 2·736, ·1368, ·0684. (9) 1210. (10) 2799360, 77760, 129600, 216000. (11) 360, 216000, 600. (12) 10, 100, 1000, 10000. (13) ·5, ·2, ·125, ·04, ·015625, ·008, ·00390625, ·0016.

52.—(14) ·054. (15) 33·080. (16) 1·732. (17) ·47712. (18) ·43241. (19) ·69315. (20) ·31830.

53.—(1) $\frac{31}{32}$, $\frac{13}{16}$, $\frac{4}{625}$, $\frac{1}{3125}$.
(1) ·25, ·5, ·75, ·125, ·375, ·625, ·875. (2) ·1875, ·3125, ·4375, ·5625, ·6875, ·8125, ·9375.

54.—(3) ·2, ·4, ·6, ·8. (4) ·96, ·992, ·9984. (5) ·96875, ·984375, ·9921875, ·00390625. (6) ·142857, ·285714, ·428571, ·571428, ·714285, ·857142. (7) ·918918, ·675675, ·531531, ·438438, ·386386. (8) ·333333, ·666666, ·166666, ·833333, ·777777, ·727272, ·692307.

59.—(1) $\cdot\dot{9}4736842105263157\dot{8}$, $\cdot\dot{6}3157894736842105\dot{2}$, $\cdot\dot{5}88235294117647\dot{0}$, $\cdot\dot{3}52941176470588\dot{2}$, $\cdot\dot{6}1538\dot{4}$, $\cdot\dot{3}91304347826086956521\dot{7}$, $\cdot\dot{3}8095\dot{2}$.
(2) $\cdot\dot{6}08695652173913043478\dot{2}$, $\cdot\dot{4}48275862068965517241379310\dot{3}$, $\cdot\dot{9}6774193548387\dot{0}$, $\cdot\dot{9}7\dot{2}$, $\cdot\dot{0}243\dot{9}$, $\cdot\dot{3}58490566037\dot{7}$.
(3) $\cdot\dot{8}7912\dot{0}$, $\cdot\dot{2}6984\dot{1}$, $\cdot\dot{9}8701\dot{2}$, $\cdot\dot{3}8610\dot{0}$, $\cdot\dot{5}1\dot{8}$, $\cdot\dot{0}0035\dot{1}$, $\cdot\dot{9}9900\dot{0}$.

62.—(1) $\frac{8}{11}$, $\frac{6}{11}$, $\frac{9}{11}$, $\frac{10}{99}$, $\frac{1}{99}$, $\frac{1}{11}$. (2) $\frac{43}{111}$, $\frac{669}{1111}$, $\frac{5}{7}$, $\frac{1}{13}$, $\frac{1}{41}$.

63.—(1) $\frac{29}{101}$. (2) $\frac{1}{1001}$. (3) $\frac{6}{13}$. (4) $\frac{36}{185}$. (5) $\frac{103}{104}$.

64.—(1) $\frac{1691}{495}$. (2) $\frac{193649}{9990}$. (3) $\frac{2171}{10}$. (4) $\frac{9424801}{300}$. (5) $\frac{791}{7}$.
(1) 3·9283. (2) 1·6854. (3) 40·90586. (4) 2621·108687.

66.—(1) $\cdot6\dot{3}5972\dot{3}$. (2) $1\cdot\dot{7}1539946787\dot{4}$. (3) $1\cdot6766\dot{0}503$.
(4) $\cdot402\dot{3}9297\dot{0}$.

68.—(1) 148·862. (2) 1·3382. (3) 244·79. (4) 1·1918. (5) 781·86. (6) 4359·166. (7) 70·5839. (8) 56827005·9588. (9) 180 (10) 131·4285, 6689·6528, ·1985.

70.—(1) 29·995. (2) 856·967. (3) 1·413. (4) 29·956. (5) 13·593. (6) 1·407. (7) 262·856. (8) 30·940.

71.—(9) ·301030. (10) ·477121. (11) ·43429448.
(1) 2^2. (2) 3^2. (3) 10^2. (4) 17^2. (5) 20^2. (6) 100^2. (7) 300^2. (8) $(\cdot1)^2$. (9) $(\cdot4)^2$. (10) $(\cdot06)^2$. (11) $(\cdot235)^2$. (12) $(\frac{1}{2})^2$. (13) $(\frac{1}{7})^2$. (14) $(\frac{3}{11})^2$. (15) 5^3. (16) 10^3. (17) 6^4. (18) $(\cdot1)^3$. (19) $(\cdot02)^4$. (20) $(\frac{3}{2})^5$. (21) $(1\frac{2}{3})^4$.
(1) 16, 25, 529. (2) ·09, ·0016, ·000256, ·268324. (3) $\frac{1}{4}$, $\frac{9}{16}$, $\frac{121}{196}$. (4) .0470, ·0021, 21·3721. (5) 100, 10000, 1000000. (6) 343, 2744, 4330747. (7) ·001, ·008, ·000000000001. (8) $\frac{8}{27}$, $\frac{729}{2197}$, $\frac{2197}{1000}$. (9) ·33310, ·00001, 4437·21960. (10) 432, 36125, 144000. (11) 2401, 243, 128. (12) 1·4989, 151·2191, 2·3965.

PAGE.

72.—(1) 1, 2, 3, 4. 5, 6, 7, 8, 9, 10, 12, 14, 16, 18, 72, 60, 90, 140, 480, 800, 9000. (2) 1, 2, 3, 4, 5, 6, 7, 8, 9, 10, 14, 24, 45, 20, 50, 80, 126, 270, 200, 900. (3) 6, 8. 10, 24, 27, 32, 36, 42, 80, 33, 240, 800, 560, 280, 105. (4) 2, 3, 8, 10, 60, 30, 12, 21, 11, 110, 300, 35. (5) 2, 3, 5, 4, 6, 20, 21. (6) $\frac{1}{2}$, $\frac{1}{3}$, $\frac{3}{8}$, $\frac{7}{10}$, $\frac{9}{100}$, $\frac{8}{27}$, $\frac{10}{27}$, $\frac{8}{105}$, ·1, ·2, ·3, ·8, 2·4, 3·2, 4·2, 10·5, 1·05, $\frac{8}{27}$. (7) $\frac{1}{2}$, $\frac{1}{3}$, $\frac{4}{5}$, $\frac{7}{10}$, $\frac{10}{9}$, $\frac{4}{21}$, $\frac{35}{16}$, ·1, ·2, ·4, 1·2, 1·1, 2·1, 3·5, ·01, ·12, ·35.
(2) 25. (3) 150. (4) 900. (6) 1225. (7) 325.

73.—(11) 5. (12) 5. (13) 0. (15) 3, 5, 35. (19) 5. (21) 35. (22) 25, 0 ; 25, 6 ; 31, 0 ; 31, 9.

74.—(1) 9. (2) 10. (3) 100. (17) 754. (19) 917.

75.—(20) 829, 389, 213, 295.
(1) 1, ·01, ·0001, ·000001. (6) ·3, ·07, ·009, ·002, 1·1, 1·2, 7·1, 5·1, 82·2946, 2·13, 17·2522, 18·8281, 51·4732, ·6324.

76.—(12) 7, 4, 74. (17) 74. (18) 23, 1094 ; 23, 1102 ; 94, 0; 94, 13,

77.—(18) 917, 0 ; 917, 476 ; 251, 0 ; 251, 1446.

78.—(6) ·2, ·3, ·02, ·08, 1·1, 2·1, 7·3, 2·444, 3·086, ·464, ·430, ·064.
(2) 3. (3) 1. (6) 3. (7) 2. (9) 6. (11) 3. (12) 3. (13) 3 (14) 9. (15) 1. (16) 1. (18) 9.

79.—(3) 1 ft. (4) 4. (6) 40. (7) 10. (8) 10 sq. ft. (9) $38\frac{1}{2}$ sq. rods. (10) $8\frac{15}{16}$ sq. yds. (11) 6 sq. yds. 6 sq. ft. 40 sq. in. (12) 24·03662 sq. ft. (13) 1833·16 sq. per.

80.—(1) 13 ft. (2) 41 ft. (3) 85 ft. (4) 17 yds.

81.—(5) 27·784 in. (6) 4 ft. 11·933 in. (7) 18·870 ft. (8) 8·171 ft. (9) 8·452 yds. (10) 19·943 in. (11) 6 ft. (12) 40 ft. (13) 112 ft. (14) 12·649 yds. (15) 12·472 in. (16) 3 yds. 10·253 in. (17) 4·529 ft. (18) 83·288 in. (19) 2·682 yds.

82.—(1) $1\frac{12}{13}$, $11\frac{1}{13}$, $4\frac{8}{13}$. (2) $1\frac{10}{41}$, $39\frac{1}{41}$, $8\frac{32}{41}$. (3) $3\frac{83}{85}$, $166\frac{2}{85}$, $25\frac{52}{85}$. (4) $\frac{1}{16}$, $8\frac{1}{16}$, 3·999. (5) $3\frac{143}{456}$ ft., $4\frac{199}{456}$ ft., 5·002 ft. (6) $73\frac{8}{21}$ in., $50\frac{1}{21}$ in., 102·426 in. (7) 1·504, 6·134, 4·977. (8) 4·601, 3·853, 4·511.
(9) 6 sq. ft. (10) 3·72 sq. yds. (11) 72·727 sq. in. (12) 6 sq. ft. (13) 30·594 sq. yds. (14) 81·259 sq. in. (15) 295·076 sq. in.

83.—(1) $12\frac{4}{7}$ ft. (2) $18\frac{1}{14}$ yds. (3) 14 yds. 2 ft. $9\frac{2}{7}$ in. (4) 23·470 ft. (5) $2\frac{19}{22}$ ft. (6) 2 ft. $3\frac{13}{22}$ in. (7) 3·01 yds. (8) 5·546 yds. (9) 2·959 in. (10) 8·878 in. (11) $75\frac{3}{7}$ ft. (12) $27\frac{5}{21}$ yds. (13) 17·568 in. (14) 30 yds. 1 ft. 8 in. (15) 4·850 ft.

84.—(8) 20. (9) 20. (22) $38\frac{1}{2}$ sq. ft. (23) $50\frac{2}{7}$ sq. ft.

85.—(24) 100 sq. yds. $8\frac{2}{7}$ sq. ft. (25) 129·537 sq. ft. (26) 5·941 sq. in. (27) $9\frac{5}{8}$ sq. ft. (28) $6\cdot\dot{9}\dot{2}$ sq. ft.
(1) $\frac{1}{12}$. (2) $\frac{1}{12}$. (3) 28 ft. (4) $51\frac{6}{11}$ sq. ft. (5) 35 sq. ft. (6) $4\frac{1}{2}$ sq. ft. (5) 112 sq. in. (6) 84 sq. ft. (7) 66 sq. ft.

86.—(8) 6 ft. (9) $11\frac{5}{11}$ in. (10) 2 ft. $7\frac{2}{11}$ in.
(6) 21 sq. ft. (7) 21 sq. ft. (8) 54 sq. ft. (9) 242 sq. in. (10) 14 in. (11) $2\frac{5}{22}$ in. (12) $12\frac{1}{4}$ in. (13) 4 ft. (14) 5·744 ft. (15) 8·720 in.

PAGE.

87.—(16) 5 ft. (17) 11·496 in. (18) $204\frac{2}{7}$ sq. in. (19) 269·992 sq. in. (20) 128·192 sq. ft. (21) 117·5 sq. in.

88.—(14) 154 sq. ft. (15) 676.063 sq. ft., 1079·134 sq. in., 1169·825 sq. yds., 2665·062 sq. ft. (16) 1·103 ft., 7·937 in., 22·018 in., 6·281 in.

89.—(20) 60 cu. ft., 65 cu. yds., $29\frac{1}{4}$ cu. ft., 163·312 cu. ft., 37·360 cu. ft. (21) 144 sq. in. (22) 12 in. (23) 8 ft.

90.—(24) 4·773 yds. (25) 64·058 yds.
(12) 140 cu. ft. (13) 12 cu. in. (14) 140 cu. ft. (15) 36 cu. ft. (16) 295·160 cu. ft. (17) 111·811 cu. ft.

91.—(23) 108 cu. ft.
(2) 108 cu. ft. (3) 60 cu. ft. (4) 27 cu. ft. (6) 18 cu. ft. (7) 40 cu. ft. (8) 54 cu. ft.

92.—(11) 84 cu. ft. (12) 352 cu. in. (13) $124\frac{8}{11}$ cu. ft. (14) 16 sq. ft. (15) 2·256 ft. (16) $6\frac{3}{4}$ in. (17) 1·948 in. (18) 4·063 ft.

93.—(17) 60 cu. ft. (18) 32 cu. ft. (19) 18 sq. in. (20) 4·535 in. (21) $46\frac{2}{3}$ cu. ft. (22) 18 cu. ft. (23) 137·673 cu. ft.

94.—(6) 81 cu. ft. (7) $410\frac{2}{3}$ cu. ft. (8) $38\frac{1}{2}$ sq. in. (9) $3\frac{1}{2}$ in. (10) 3·349 yds.

95.—(17) $1437\frac{1}{3}$ cu. in. (18) 268·190 cu. in. (19) 21 in. (20) $179\frac{2}{3}$ cu. ft. (21) 296·508 sq. ft.
(1) 10 m., 100 dm., 1 Dm. (2) 1 Km., 100 Dm., 10000 dm. (3) 1200 Dm., 15000 m., 21000000 mm. (4) 1234567·89 dm., 123456·789 m., 123·456789 Km. (5) 856000 cm., 5632 mm., 12468000 μ.

96.—(6) 1635639·87 m. (7) 554151·1 cm. (8) 1152·48768 Km. (9) 91·441 cm. (10) $18\cdot\dot{3}$ m. (11) 50 m. (12) 179·07 cm. (13) 749·301 mm. (14) 4 m. 6 m. (15) 13 hr. 32 m. 30 sec. (16) 6250. (17) 900 mm. (18) 952, 400 mm. (19) 25·441 sq. m. (20) 123·4 m. (21) 12 Ha., 120000 ca., 120 Da. (22) 12·345 a., 5·678 Da., 10000 ca. (23) 1017·32 a.

97.—(24) 9·9 Da. (25) 20424·15 Ha. (26) 10·8 Ha. (27) 172 m. (28) ·1 ma. (29) 54 ca. (30) 1500000 sq. cm. (31) 13 ca. (32) 4·5 a. (33) 6000 cu. m. (34) \$547·50. (35) 1 a. (36) 721·1 m. (37) 2·471 acres. (38) 90 ca. (39) ·66 a. (40) 39·6 in. (41) 27 s. (42) 1000 cm.

98.—(43) 1000000 c. cm. (44) 1 s. (45) \$1080. (46) \$105. (47) 100 cu. dm. (48) 35·31 cu. ft. (49) ·2759 cord. (50) 24 s. (51) 3 m., 12 m., 15 m. (52) 45000. (53) 60 ma. (54) 2500 Kg. (55) $909\frac{1}{11}$ Kg. (56) 1·006 s. (57) 125 μ. (58) ·4 m. (59) 1 min. (60) 200 m.

99.—(61) 36 m. (62) 1000, 1000. (63) 1 Kg. (64) 2 m. (65) Equal. (66) 4·004004 s. (67) 1000. (68) 1 cu. dm. (69) ·03531 cu. ft. (70) 4·543. (71) 135·9 ml. (72) Equal. (73) 36 min. (74) 10 m. (75) ·157625 s. (76) 10 m.

100.—(77) 28·3 l. (78) 1·3 Kg. (79) 1·298 Kg. (80) 1000 c. cm. (81) 1000000 mg. (82) 1000000 g. (83) 10000 Kg.

PAGE.

100.—(84) 13·5 Kg. (85) 46:25. (86) 907·2 T. (87) 373·24 g. (88) 490 Kg. (89) 4·5 T. (90) 123456·789 Kg., 123·456789 T. (91) 2·204 lbs. (92) ·84 ca. (93) ·375 m. (94) 1·155 l. (95) 364·8 Kg.

101.—(96) 1311·566 Kg. (97) 5·76 min. (98) 138·8 cm. (99) 1·12 m. (100) 20·116 m. (101) 101·17 Ha. (102) 368·529 Km. (103) 1609·3 m. (104) ·984. (105) 14·21. (106) 39·6 in. (107) 4s. 8d. (108) 1000001. (109) 10001. (110) ·836 ca. ·404 Ha. (111) $509090909\frac{1}{11}$ Ha. (112) 1·539 s.

102.—(1) ·0202, ·2025, 202·522. (2) 2206·7. (3) 5·439. (4) ·033. (5) 13828·059. (6) ·83, 83, 830000. (7) ·03125, ·1764705882352941. (8) $\frac{4259}{19980}$, $\frac{30937}{99000}$, $\frac{8041}{9000}$. (9) 6·5504777, ·4. (10) $1\frac{12929}{874791}$. (11) ·000002. (12) ·0001 +. (13) $\frac{4}{21}$. (14) $\frac{317}{2035}$. (15) $\frac{8645}{21648}$.

103.—(16) 8934·084. (17) $4\frac{101261}{131340}$. (18) ·8398, ·8397. (19) ·3183098. (20) ·808. (21) 1. (22) £134 2s. 2d. (23) $\frac{10079}{35064}$. (24) $\frac{33}{182}$. (25) 16·8476190. (26) $110\frac{23}{215}$. (27) ·00015625. (28) $\frac{468}{2305}$, $1\frac{1}{3}$. (29) ·2. (30) $\frac{287}{291}$, $\frac{39}{37}$.

104.—(31) $\frac{25}{141}$. (32) $\frac{696}{1141}$. (33) $3\frac{29}{30}$, $1\frac{13}{30}$. (34) ·4047 or ·4048. (35) $\frac{455}{551}$, $\frac{3}{8}$. (36) ·102848, $28\frac{25}{38}$. (37) $\frac{251}{6750}$, $\frac{335}{9009}$. (38) ·30375. (39) $\frac{1}{47}$. (40) 7925·591 m. (41) 1. (42) $\frac{364834}{611187}$. (43) $47\frac{53}{209}$, $48\frac{129}{1232}$. (44) 1, $4\frac{1}{6}$. (45) $\frac{15}{26}$.

105.—(46) £2 0s. 6d. (48) £4 5s. 4d. (49) $38\frac{1601}{6150}$. (50) $\frac{479}{1170}$. (51) 1171 × fr. (52) $1\frac{3}{104}$. (53) $113\frac{84603}{240623}$. (54) 3·11. (55) $\frac{2721}{1001}$. (56) 3. (57) $\frac{1}{7}$.

106.—(58) ·52. (59) $\frac{71}{1008}$. (60) $\frac{16432}{22149}$. (61) 3·141592. (62) 7. (63) 6. (64) ·0000002. (65) $10\frac{4}{7}$. (66) $\frac{17}{18\frac{1}{10}}$. (67) $\frac{74}{77}$. (69) $\frac{1}{4}$, 18. (70) 66.

107.—(71) $\frac{9001}{1000}$. (72) $1\frac{1}{2}$, $1\frac{1}{4}$, $\frac{3}{4}$. (73) ·3183. (74) ·43429. (75) $7\frac{1}{9}$. (76) $\frac{90}{714}$, $\frac{90}{715}$. (77) ·131870. (78) ·00003. (79) 3·1415926. (80) ·0042. (81) ·410686. (82) 615·0703204.

108.—(84) 1·3999... (85) 7. (86) 2·7182818285. (87) $\frac{5}{8}$ m. (88) ·9984097. (89) ·1. (90) 3·14285, 3·14159. (91) $\frac{5}{13}$, $5\frac{1}{3}$, $4\frac{11}{38}$.

109.—(92) 409. (93) \$4866·66$\frac{2}{3}$, \$4866·56. (94) ·1715. (95) ·00004. (97) 1·6961. (99) 3·14159. (100) 26·2856. (102) 7·745966.

110.—(103) The metre would be nearly 39·41 in. (104) ·00001... (105) 20·06. (106) $5\frac{5}{14}$ m. (107) 704·037 lb. (108) $909\frac{1}{11}$ oz. (109) $2\frac{46}{77}$ oz. (110) 60. (111) $17015\frac{5}{8}$. (112) $\frac{7395}{38896}$. (113) \$2·20.

111.—(114) 6. (115) 100. (116) 286750 ft. (117) 70. (118) \$440. (119) \$23·25. (120) Merchant 50c. (121) 10·52 p.m. (122) £13 9f. 1c. $3\frac{13}{21}$m.

112.—(123) 6d. (124) ·00278... (125) \$36300. (126) $6\frac{6}{7}$ ft. (127) 30. (128) $3319\frac{3}{8}$ lb. (129) $11\frac{121}{864}$ oz. (130) $81\frac{9}{11}$ hr. (131) $\frac{15}{19}$d. (132) 21. (133) 84. (134) $13\frac{899}{1332}$. (135) 13.

PAGE.

113.—(136) 15·49. (137) $10\frac{20}{33}$. (138) 14. (139) 2240. (140) 7, 80. (141) $1365. (142) 37, 5162647, 14d. 7h. 11m. 17s., 2674d. 0h. 9m. 59s. (143) 60d. (144) $3645\frac{5}{6}$, $75834\frac{18}{35}$. (145) $83\frac{1}{3}$c. (146) $320. (147) £45.

114.—(148) $\frac{3}{25}$. (149) $19\frac{1}{2}$ yd. (150) 12. (151) $76\frac{8}{33}$c. (152) 2·2136. (153) 53. (154) 36·56. (155) 1 min. in 12 h. 1 min. (156) 11c. (157) 27 gal. (158) $\frac{1}{11}$. (159) House $1200.

115.—(160) $60. (161) $41·03. (162) $18·84. (163) $8\frac{2}{5}$ hr. (164) 1c. (165) $668\frac{4}{91}$, $785\frac{75}{91}$, $943\frac{11}{91}$, $1178\frac{82}{91}$. (166) 16s. $9\frac{1}{3}$d. (167) $\frac{55}{171}$ miles. (168) 2541 grs. (169) Nearly 8 gal. (170) 35.

116.—(171) $10\frac{10}{21}$ yd. (172) $1·31. (173) 113. (174) 5, $4\frac{1}{2}$, 4s. (175) $2290·41, $2495, $2295. (176) $\frac{9}{200}$. (177) $448·80. (178) $170\frac{1}{3}$ yd. nearly. (179) $349·72. (180) 346, 345. (181) 69·2 m.

117.—(182) $\frac{60}{79}$. (183) 9 yr. (184) 45. (185) $12937·50. (186) Every $65\frac{5}{11}$ min. (187) 50c. (188) 7·4. (189) 6 sec. (190) $\frac{8}{11}$. (191) $240. (192) 3627. (193) 7.

118.—(194) $5·40, $7·20, $3·60. (195) $3·57. (196) $10\frac{320}{353}$ hr. (197) 10 yr. (198) $\frac{99}{100}$. (199) 12. (200) 24. (201) $\frac{9}{10}$. (202) 2.

119.—(203) Dec. 3, 3 p.m. (204) £432 11s. 11d. (205) 3, 5, 7, 9, 13. (206) 7. (207) $1\frac{1}{34}$ hr. (208) 146097. (209) 146096d., 21 hr. 16 min. 40 sec. (210) $3526\frac{26}{49}$. (211) $3000 etc.

120.—(212) 12. (213) ·8 gr. (214) $168\frac{1}{18}$. (215) ·7433956. (216) $1\frac{1}{2}$ m. per hr. (217) Total $14·31. (218) $307\frac{5}{9}$.

121.—(219) $38·25, $38, $37·80. (220) $4076·163, $\frac{431}{910}$. (221) 73. (222) 15. (223) 2205. (224) 147. (225) 3456. (226) 64. (227) $26\frac{4}{21}$. (229) 91:81. (230) 3 ft. 2 yd. 7 in.

122.=(231) 2560. (232) $\frac{1769}{1306880}$. (233) $3\frac{2251}{3888}$. (234) $\frac{7}{2112}$. (235) $\frac{181}{480}$. (236) 1 yd. (237) 1 qt. (238) 972, $6\frac{3}{4}$, $\frac{3}{4}$. (239) $2\frac{1}{4}$, 10, 1. (240) 1 rod. (241) 1 link. (242) 1 yd., 1 chain, 100000, 6272640. (243) $4·03, 23·0384 ft. (244) 12 lb. av.

123.—(245) 16000. (246) 15 gal. (247) 60480. (248) $\frac{1}{16}$ sq. in. (249) 12·78. (250) 2 yd. (251) 33·84. (252) $\frac{125}{7989}$. (253) 883345·17. (254) 16·83. (255) 32 days.

124.—(256) 2160. (257) $1\frac{1}{4}$. (258) 30. (259) $\frac{27}{28}$ ft., $\frac{729}{784}$ sq. ft. (260) 1530. (261) Between 237626·5 and 237633. (262) 55. (263) Gold ·48 in., and 9·31. (264) 59·92 lb. (265) $103\frac{8}{19}$ lb.

125.—(266) 510 lb. (267) 27000. (268) ·00021 in. (269) 4th. (270) 87293·9. (271) 4290·8 lb. (272) $\frac{9}{44}$. (273) $11\frac{1}{12}$. (274) $2\frac{47}{54}$g. (275) $1323, $1260, $1200. (276) $28571\frac{3}{7}$ ft.

126.—(277) 46·5 gal. (278) 2·75. (279) 5. (280) $\frac{27}{35}$. (281) $18\frac{1}{3}$°, $39\frac{1}{5}$, $-17\frac{7}{9}$°. (282) 44802. (283) 880 ft., 54. (284) 3:2.

127.—(285) £45·55. (286) 415:408. (287) 8672 yd. (288) 20:7. 5s. $1\frac{1}{2}$d. (289) Men, $1251·76. (290) 3:7. (291) $433\frac{1}{3}$.

PAGE.

128.—(292) 6·66. (293) 2 hr. 40′. (294) 2 m. (295) 14d. 9p. 1c. $6\frac{2}{3}$ c. s. (296) 8 dr. (297) 76 grs. each. (298) 5:112:2800.

129.—(299) $\frac{7}{20}$. (300) $235·20. (301) 1·15$\frac{5}{10}$ p.m. (302) $62.50. (303) Better.

140.—(13) 20 , $16\frac{2}{3}$%, 5 %, $5\frac{5}{9}$%, $4\frac{16}{21}$%. (14) 216 men, 108 bush. (15) £5 16s. $4\frac{7}{39}$d., $4·7753. (16) $2915000; 11s. $6\frac{3}{5}$d. (17) 420 m., 57·60 fr. (18) $6\frac{1}{4}$%. (19) $2\frac{1}{2}$%. (20) $12\frac{1}{2}$%. (21) 130. (22) 96. (23) 200. (24) 250. (25) 800. (26) 4. (27) 500. (28) $3360. (29) 35·2. (30) $42\frac{6}{7}$%.

141.—(31) $8800. (32) $13\frac{1}{3}$. (33) 120, 240. (34) $153\frac{11}{13}$, $184\frac{8}{13}$. (35) ·3003001 %. (36) 75 %. (37) $9. (38) 30·7+ %. (39) 125. (40) $42\frac{1}{2}$%. (41) $120000. (42) $2000. (43) 600, 480, 360. (44) 40 %. (45) 76·8%.

142.—(46) $1\frac{19}{21}$%. (47) 125. (48) $21\frac{19}{21}$. (49) $16000. (50) 45·8 pts. (51) 600. (52) $1029\frac{7}{17}$.
(1) $540. (2) $150. (3) 15%. (4) $4·$28\frac{3}{4}$. (5) $16·$66\frac{2}{3}$. (6) $4·$57\frac{1}{7}$.

143.—(7) $3.75. (8) 81%. (9) $1\frac{1}{3}$c. (10) $6. (11) $10. (12) $28·$12\frac{1}{2}$. (13) 45c. (14) Loss $$333\frac{1}{3}$. (15) Loss 2 %. (16) $3\frac{1}{33}$%. (17) $1\frac{27}{73}$%. (18) $4. (19) $9\frac{1}{11}$%. (20) $11\frac{1}{9}$%.

144.—(21) 20%. (22) $16\frac{1}{9}$%. (23) $8\frac{8}{13}$%. (24) 28 %. (25) $5\frac{5}{7}$%. (26) 20%. (27) 10%. (28) 20%. (29) $\frac{1}{4}$%. (30) $12\frac{1}{2}$%. (31) $120. (32) $4·05. (33) $2·$33\frac{1}{3}$.

145.—(34) 50c. (35) 22·5+ %.
(1) $240. (2) $3000. (3) 50c. (4) $10200. (5) $2·10. (6) $2887·50. (7) $10·$10\frac{10}{13}$. (8) $41\frac{1}{3}$c. (9) $22\frac{2}{9}$c. (10) $500.

146.—(11) No. (12) $71\frac{3}{7}$%. (13) 50 . (14) 50 %. (15) $7\frac{1}{2}$%. (16) $4050. (17) 15 %. (18) $5·75. (19) $6\frac{1}{4}$%. (20) $5\frac{5}{64}$%. (21) $10. (22) $11\frac{1}{9}$%. (23) 1 %. (24) $37\frac{1}{16}$c.

147.—(25) $33\frac{1}{3}$%. (26) $12\frac{1}{2}$. (27) 12 . (28) $63\frac{7}{11}$%. (29) $\frac{4}{21}$ of amount of whisky. (30) $1\frac{37}{63}$. . (31) 50%. (32) $1\frac{9}{16}$%. (33) 25%. (34) $33\frac{1}{3}$%. (35) 220 %. (36) $235. (37) 116%. (38) 20%.

148. (39) $$1129\frac{16}{41}$. (40) 48c. (41) Lost $9\frac{73}{303}$ %. (42) $83\frac{1}{3}$. (43) $33\frac{1}{3}$. (44) $$71\frac{3}{7}$. (45) $\left(\frac{ac+ce-ab-cd}{ab+cd}\right)100$. (46) $42\frac{6}{7}$%. (47) 46·6+c. (48) $200. (49) 10 . (50) $11520.

149. -(1) $240. (2) $11·$62\frac{1}{2}$. (3) $5850. (4) $66\frac{2}{3}$c. (5) $10·$12\frac{1}{2}$. (6) 2%. (7) 4 %. (8) 2 . (9) $400. (10) $945. (11) 80. (12) $\frac{3}{4}$.

150.—(13) $16·25. (14) $102. (15) $25. (16) $$1715\frac{35}{51}$. (17) $70. (18) $3\frac{1}{3}$. (19) $21\frac{1}{19}$%. (20) $25263\frac{3}{17}$ lbs. (21) $9720. (22) $$43\frac{3}{20}$, $30172\frac{2}{3}\frac{2}{5}$ lbs. (23) 311963 ft. (24) $294117\frac{11}{17}$ lbs.

151.—(25) $103291\frac{151}{269}$ lbs. (26) $1\frac{1}{8}$. (27) 120000 pounds. (28) $7847\frac{131}{367}$ lbs. (29) $54. (30) $8000. (31) $$5333\frac{1}{3}$. (32) $4060. (33) $399\frac{7}{17}$ cwt.

152.—(34) $40058\frac{54}{137}$ lbs. (35) $5970. (36) $\frac{^{\prime\prime}a+b}{100+b}$. (37) $1\frac{1}{2}$%. (38) $1\frac{3}{4}$%. (40) $1\frac{1}{2}$, 1 .

PAGE.

153.—(41) 2%. (42) 2%. (43) 1%, $\frac{1}{2}$%. (44) 2%. (45) $1\frac{1}{4}$%. (46) $8016·63.

(1) $9. (2) $105. (3) $56·25.

154.—(4) $160. (5) $\frac{3}{4}$%. (6) $\frac{3}{4}$%. (7) $\frac{2}{3}$%. (8) $\frac{21}{40}$%. (9) $1\frac{1}{3}$%. (10) $$2133\frac{1}{3}$. (11) $\frac{5}{8}$. (12) $\frac{3}{4}$%. (13) $40. (15) $8000. (16) $4000.

155.—(17) $16000. (18) $6000. (19) $1\frac{3}{28}$%. (20) $1\frac{1}{4}$%. (21) $312·50, $400, $875, $$15151\frac{17}{33}$, $$24242\frac{14}{33}$, $$60606\frac{2}{33}$. (22) $$1901\frac{47}{53}$. (23) $25600. (24) $\frac{12}{27}$%. (25) $42256·62. (26) $10·15.

156.—(27) $$4666\frac{2}{3}$. (28) $759. (29) $3000. (30) $1\frac{1}{4}$%. (31) $1\frac{1}{2}$%. (32) 1%. (33) $51750. (34) $$6283\frac{7}{11}$, $$4189\frac{1}{11}$. (35) $11970. (36) $$18406\frac{11}{131}$. (37) $\frac{72}{97}$.

157.—(1) $100. (2) $37·50. (3) $240. (4) $30. (5) $9375 (6) $11\frac{1}{9}$ mills. (7) $14112. (8) $16\frac{2}{3}$ mills. (9) $14·40. (10) $16. (11) $1979. (12) 16 mills. (13) $$5714\frac{2}{7}$.

158.—(14) 15 mills. (15) $507500. (16) $4. (17) $30. (18) $$14\frac{16}{31}$. (19) $1900. (20) $32·$37\frac{1}{2}$. (21) $6000. (22) $23·75.

(1) $3·70.

159.—(2) $3·25, $3·34. (3) 75c. (4) $300. (5) $3·78. (6) $7\frac{1}{2}$c. (7) $2·35 nearly. (8) $2·88. (9) $20. (10) $162·73+. (11) $76·17+. (12) $1. (13) $43\frac{3}{4}$%.

160.—(14) $20. (15) 10%. (16) 20 doz. (17) 27 in. (18) Gain $$14\frac{2}{13}$, Loss $$5\frac{11}{13}$. (19) 3 yds. (20) $2\frac{1}{2}$ lbs.

(1) $3200.

161.—(2) $9375. (3) $2100. (4) $4050. (5) $2922·75. (6) $4795. (7) $7350. (8) $7200. (9) $8000. (10) $8000. (11) $4000. (12) $4000. (13) $320. (14) $591·50. (15) $87·50. (16) $118·25. (17) $5\frac{5}{19}$%. (18) $5\frac{5}{19}$%. (19) 8%. (20) $4000, $3010.

162.—(21) 140. (22) $110\frac{7}{13}$. (23) Loss $4. (24) Gain $352. (25) $100500. (26) $150000. (27) $78\frac{1}{8}$. (28) $480. (29) 84. (30) $$947\frac{7}{19}$. (31) $11868·13. (32) 3%.

163.—(33) 5 per cent. (34) $$636363\frac{7}{11}$. (35) $120000. (36) $$291\frac{1}{9}$. (37) $4\frac{1}{6}$%. (38) 20%. (39) 75. (40) $6500, $250. (41) 108. (42) $41400. (43) $1043·29. (44) $$2666\frac{2}{3}$. (45) $62\frac{1}{2}$.

164.—(46) $7200, $10800. (47) 6 per cent. (48) 60c. gain. (49) $93\frac{3}{4}$. (50) £$4733\frac{1}{3}$, £63 11s 8d.

(5) $45, $5·40, $146·88.

165.—(6) $30, $$84·58\frac{1}{3}$, $18·612. (7) *(a)* $144, *(b)* $131·25, *(c)* $129·71, *(d)* $255·864, *(e)* $163·487, *(f)* $108·904, *(g)* $404·416 *(h)* £216 1s 6d. (8) $106. (9) $1344. (10) *(a)* $1488, *(b)* $2100. *(c)* $1099·446, *(d)* $2168.752, *(e)* $1252·374, *(f)* $1174·264, *(g)* $1587·332, *(h)* £2146 6s. $10\frac{4}{5}$d. (11) $60·449. (12) $180·369. (13) $1431·493. (14) $5610. (15) $4. (16) $65. (17) 3%. (18) 4%. (19) 3%. (20) 5%. (21) 5 yrs.

PAGE.

166.—(22) $2\frac{22}{189}$ yrs. (23) $16\frac{2}{3}$ yrs., \$1159·78. (24) 25 yrs. (25) $33\frac{1}{3}$ yrs. (26) $\frac{3}{25}$. (27) $\frac{1}{5}$. (28) *(a)* $\frac{3}{10}$, *(b)* $\frac{27}{200}$, *(c)* $\frac{21}{100}$, *(d)* 1, *(e)* $\frac{1}{25}$, *(f)* $\frac{1}{50}$, *(g)* $\frac{2}{25}$. (29) 12 yrs. (30) 20 yrs. (31) 16 yrs., $18\frac{2}{11}$ yrs., 50 yrs. (32) 50 yrs., $33\frac{1}{3}$ yrs., 28 yrs. (33) \$3200. (34) \$60. (35) 4%. (36) 2%, 4%. (37) $2\frac{1}{2}$%, 5%.

167.—(38) 4%, 3%. (39) \$108. (40) \$500. (41) \$800. (42) \$1300. (43) \$723·938. (44) – Latter. (45) \$$261\frac{19}{21}$. (46) \$250. (47) \$$149\frac{61}{111}$. (49) A's offer. (50) \$1389·15. (51) \$1170.

168.—(52) \$297·804. (53) \$384 at end of 6 mos. (54) $4\frac{1}{6}$%. (55) $1\frac{1}{9}$ yrs. (56) 5 yrs.
(1) \$1181·65. (2) *(a)* Oct. 4, *(b)* 111 days, *(c)* \$490·88, *(d)* \$9·12. (3) *(a)* Sept. 4, *(b)* 92 days, *(c)* \$12·60, *(d)* \$987·40.

169.—(4) *(a)* Ap. 18th, *(b)* 63 days, *(c)* \$24·40, *(d)* \$2332·10.
(5) *(a)* Ap. 18th, *(b)* 76 days, *(c)* \$16·92, *(d)* \$1233·08.
(6) *(a)* Sept. 26, *(b)* 86 days, *(c)* \$106·32, *(d)* \$5534·43.
(7) *(a)* Mar. 4th, *(b)* 70 days, *(c)* \$31·86, *(d)* \$2737·14.
(8) *(a)* Aug. 4th, *(b)* 61 days, *(c)* \$3·68, *(d)* \$271·32.
(9) *(a)* Mar. 3rd, *(b)* 92 days, *(c)* \$80·66, *(d)* \$3919·34.
(10) *(a)* Nov. 8th, *(b)* 157 days, *(c)* \$31·86, *(d)* \$1202·70.

170.—(11) *(a)* Mar. 3rd, *(b)* 33 days, *(c)* \$2·89, *(d)* \$397·11.
(12) *(a)* June 6th, *(b)* 97 days, *(c)* \$12·26, *(d)* \$564·49.
(13) *(a)* May 9th, *(b)* 80 days, *(c)* \$6·39, *(d)* \$479·66.
(14) *(a)* May 6th, *(b)* 63 days, *(c)* \$27·90, *(d)* \$1992·81.
(15) *(a)* Nov. 26th, *(b)* 86 days, *(c)* \$80·58, *(d)* \$4194·33
(16) *(a)* Mar. 3rd, *(b)* Mar. 3rd, *(c)* Mar. 2nd, *(d)* Mar. 3rd. *(e)* Mar. 3rd (17) $\frac{1}{100}$. (18) $\frac{57}{3650}$.

171.—(19) $\frac{57}{3650}$. (20) $\frac{1}{50}$. (21) $\frac{1798}{1825}$. (22) \$365. (23) \$888·50. (24) \$365. (25) 6%. (26) 5%. (27) 7%. (28) $12\frac{1}{2}$%. (29) \$365. (30) July 7th. (31) Nov. 2nd (32) \$1199·87.

172.—(1) \$35. (2) \$465. (3) \$465. (4) \$48·23. (5) \$951·77. (6) \$2083·23. (7) \$2104·06. (8) \$60. (9) \$20. (10) \$121·64. (11) \$1118·36. (12) \$3942·26.

173. (13) \$244·51. (14) \$286·25. (15) \$469·71. (16) \$397·60. (17) \$1266·31. (18) \$335·80.

174. (19) \$296·07.
(1) \$500. (2) \$500 (3) \$500. (4) 10 days. (5) 15 days. (6) 15 days. (7) 15 days. (8) 5 mos. (9) 3 mos. (10) \$3600. (11) 7 mos (12) $3\frac{3}{13}$ mos.

175.—(13) 20 days (14) 48 days. (15) 15 days. (16) 16 days (17) 9 mos. (18) 24 days. (19) 120 days. (20) 32 days after debt was due. (21) 44 days, July 15th. (22) June 7th. (23) July 23rd. (24) 7 days.

176—(25) 4 days. (26) Sept 10th. (27) Sept. 2nd. (28) \$923·87. (29) Apr. 15th (30) \$1299·97.

177.—(1) \$104 (2) \$104 (3) \$108·16. (4) \$108·16. (5) \$112·4864. (6) \$112·4864. (7) \$1·124864. (8) \$121·55. (9) \$182·33.

PAGE.

177.—(10) \$32·33. (11) \$109·29. (12) \$197·03. (13) 1·157625. (14) 1·169858. (15) \$689·83. (16) \$1201·31.

178.—(17) \$662·45. (18) \$9·72. (19) \$4448·99+. (20) \$1025·76. (21) \$1350·61. (22) \$111·49. (23) \$1689·67. (24) 11·35¼c. (25) \$4328·25 (26) 43923.

179.—(33) \$1789·25 $\{(1\cdot04)^3 - 1\}$. (34) $\$(1\cdot06)^3 \times (1\cdot03)$. (37) \$15400. (38) \$10000. (39) \$729·53. (40) 7+years. (41) 5%. (42) 6·09%.

180.—(43) 2·95+%. (44) $17\frac{1}{3}\frac{3}{8}$%. (45) 5%, \$200. (46) 4%, \$1250. (47) 4%, \$1500. (48) 5%. (49) \$10000. (50) \$1967·15.

181.—(7) \$769·23. (8) 96·15c. (9) 90·71c (10) \$7396·45. (11) \$23·54. (12) \$48·80. (13) (*a*) \$14·14, (*b*) \$97·97, (*c*) \$93·14, (*d*) \$2852·98, (*e*) 78·35c. (14) \$1089·30. (15) \$1200·62. (16) \$384·82. (17) \$97·60.

182.—(18) \$3013·86, \$3322·78, \$3663·36. (19) $\$(\sqrt[3]{\frac{b}{a}})^2$c. (20) $\$a\{1 - (\sqrt[3]{\frac{a-b}{a}})^2\}$. (23) $\frac{3}{23}$. (24) $\frac{515201}{6765201}$. (27) \$860·48. (28) \$3818·85.

183.—(29) \$6000. (30) 5%. (31) 10%. (32) \$56⅔. (33) \$27. (34) \$1200, 4%. (35) \$1500, 4%. (36) \$13·902+. (37) \$11000. (38) \$1922·75. (40) \$3463·09.

184.—(41) 5%, \$1260. (42) 2%.
(1) \$7500 (2) \$7000. (3) \$6666⅔. (4) \$6250. (5) \$10000. (7) \$5137·04. (8) \$4274·02. (9) \$1421·36. (10) \$1727·68.

185.—(12) 3·377 times the annuity. (13) \$295·92. (14) \$3290·81. (15) \$6177·39. (16) \$2289·15. (17) \$63·60. (18) \$131·57. (19) \$310·26.

186.—(22) \$7122·92. (23) \$1876·52. (24) \$9803·92. (25) \$5386·08. (26) \$4476·90. (27) \$962·91. (28) \$11118·39. (29) \$5965·57. (30) \$5167·20.

187.—(31) \$2675·49 (32) \$4178·91. (33) \$404·21. (34) $(1\cdot04)^3$, $(1\cdot04)^2$, (1·04), 1. (35) \$3491·39. (36) \$220·67. (37) \$77·88. (38) \$5385·12. (39). \$105·24. (40) \$350+.

188.—(41) \$33·43. (42) \$463·93. (43) \$28·38. (44) \$22·22. (45) About 10 years. (46) Never. (47) \$18395·44.
(1) \$1250 each (2) Jones \$1120, Smith \$1680.

189.—(3) \$4000 (4) \$2000. (5) \$20000. (6) \$1200 (7) A \$1001·74, B \$459·13, C \$939·13. (8) \$5422$\frac{18}{31}$. (9) \$2618·18. (10) 8, 10 and 12 mos. (11) \$2000 each.

190—(12) \$3764·25 (13) Lock \$1000, Smith \$1012, Knight \$960. (14) B \$6315·79, C \$5684·21 (15) B \$1333⅓, C \$1666⅔. (16) Smith \$8000, Jones \$9000, Cook \$3000. (17) \$14472+. (18) 8 mos.

191.—(19) Terry \$273⅓, Tucker \$133⅓, Taylor \$73⅓. (20) \$722·06.
(1) \$4010. (2) \$2509·37½ (3) \$796. (4) \$12090. (5) 5764·50. (6) \$7481·29. (7) £281 5s. (8) \$290·69+. (9) 9½%. (10) \$14433⅓.

PAGE.

192.—(11) \$7275. (12) £500. (13) $9\frac{1}{2}$. (14) Direct \$4010; via Chicago \$4009·85. (15) 1% Disc. (16) \$2716·875. (17) 5·16%. (18) 40 lbs.

193.—(20) £802·67. (21) £2321 17s. 4d; £276 8s. 3d. (22) \$12279·60. (23) \$4744·186. (24) \$342·857. (25) 1 rouble = 4·012 fr. (26) 1 dollar = 2·487 fl.

195.—(Use $\pi = 3\frac{1}{7}$) (1) 18 ft. (2) 56 ft. (3) 48 in. (4) $100\frac{4}{7}$ sq. ft. (5) 5·744 ft. (6) 120 yds. (7) 6·2 ft. (8) 2·699 ft. (9) $28\cdot\dot{6}$ sq. ft. (10) ·0000918.

196.—(11) 30·805 ft. (12) \$266. (13) $10\frac{2}{7}$ ft. (14) $15\cdot\dot{5}$ in. (15) 346 106. (16) 60 ft. (17) 40 yds. 6 in (18) 357·071 ft. (19) 2·121 ft. (20) 1281·7 sq. yds. (21) $69\cdot\dot{4}$ miles. (22) 40 (23) 15186·292. (24) 114 363 in.

197.—(25) 30 ft., 42 ft. (26) 364·641 sq. ft. (27) 672. (28) $22\cdot\dot{2}\dot{7}$ ft. (29) 24·64 sq. ft. (30) 190 114 sq. yds. (31) 21. (32) $2005\cdot\dot{3}$ ac (33) 152·168 sq. yds. (34) 1·68 ft. (35) 2000. (36) 525 sq. ft (37) ·541 yds. (38) 23 ft. 4 in.

198.—(39) 2710·19 cu. ft. (40) 18·973 ft., 12·649 ft. (41) 30·588. (42) 18·978 ft. (43) 8 ft. (44) 2 ac. 46 rods. (45) Square, $8\frac{6}{7}$ rods. (46) $8\cdot\dot{9}\dot{0}$ ft. (47) 20. (48) 3 chains. (49) $431\frac{1}{5}$ sq. ft. (50) 5 291 in.

199.—(51) 1·118 in. (52) 76·157 ft. (53) 243·721. (54) 28·284 ft. (55) 6995 cu. ft. (56) $64\frac{1}{24}$ sq. ft. (57) ·577 in. (58) £7 3s. 4d. (59) 2·487 rods. (60) $346\cdot\dot{5}$ sq. yds. (61) 4500. (62) 134·114. (63) 216000. (64) $1257\frac{1}{7}$.

200.—(65) $1\frac{6}{7}$ ft. per hr. (66) $444444\cdot\dot{4}$ (67) $12\frac{1}{32}$. (68) 547·764 sq. yds. (69) 792·608 sq. yds. (70) $179\frac{1}{7}$ sq. in. (71) 1 to 3 (72) 48·055. (73) Drive. (74) 46·084, 10·415 rods. (75) 62·353 rods. (76) $32\frac{1}{12}$ lbs.

201.—(77) $2\sqrt[3]{5}$ in. (78) $\frac{269}{512}$. (79) 8 to 27. (80) 104 ft (81) $2\frac{1}{12}$ in (82) 13 ft. 6 in. (83) $51\frac{2}{3}$ sq. yds (84) 107·158 yds (85) \$1·64 (86) 1. (87) 25 sq. in. (88) 11845 5 . . . lbs. (89) \$323·326. (90) 12 ft. (91) 44 in

202.—(92) \$203·20. (93) 12. (94) 82·683 ft (95) 123·81 sq. ft. (96) $5391\frac{5}{7}$. (97) $97\cdot\dot{7}$. (98) Square, 225 sq. yds. (99) $\$466\cdot66\frac{2}{3}$ (100) 5 ch. (101) 22·488 ft (102) 10·2. (103) 4708·97.

203.—(104) 1325·481 sq. yds. (105) $\frac{1}{3}\sqrt{12}$. (106) $1123\frac{1}{3}$ yds (107) 100 (108) \$158·56. (109) $15\frac{5}{7}$ in. (110) 2475 (111) 640 tons. (112) \$504. (113) $30\frac{1}{3}$ days. (114) \$362·50. (115) 15·512 in.

204.—(116) 1152 sq. yds. (117) $6788\frac{4}{7}$ lbs. (118) 655·896. (119) 22·563 in. (120) 208 sq. ch. (121) 8400 sq. yds. (122) $1437\frac{1}{3}$ cu. in. (123) $235\frac{5}{7}$. (124) 20 rods. (125) $3\frac{3}{7}$ sq. ft., $12\frac{1}{7}$ ft. (126) $31\frac{3}{7}$ in. (127) $113\frac{1}{7}$ in. (128) $172\frac{6}{7}$ in.

PAGE.

205.—(129) 1·445 sq. ft. (130) 368 sq. in., 90 in. (131) $5028\frac{4}{7}$ sq. ft. (132) 6 (133) \$98. (134) 65·918 ft. (135) 8·808 sq. ft. (136) $92\frac{4}{7}$ sq. ft, $37\frac{5}{7}$ ft. (137) 126 in., 674 sq. in. (138) 84 in., 339·477 sq. in. (139) 189·52 sq. ft.

206.—(140) 676·859 sq. rods (141) 80000 miles. (142) $113\frac{1}{7}$ cu. in. (143) $29\frac{1}{3}$ cu. in. (144) $\frac{3}{5}$ hrs. (145) \$79·65. (146) 398·353 sq. ft. (147) 8·64 in. (148) $212\frac{4}{7}$ sq. in. (149) $148\frac{1}{2}$ sq. in. (150) 24·589 in. (151) 14·661 in. (152) 40·737 sq. in.

207.—(153) $10879\frac{1}{7}$ sq. yds (154) 68·938 mi. (155) 34·469 mi. (156) 48·849 mi. (157) 5·128 (158) 2836 (159) $117333333\frac{1}{3}$. (160) $212\frac{1}{7}$ (161) $366\frac{2}{3}$ sq. in. (162) $229\frac{1}{3}$ sq. in. (163) 12 ft. (164) 96 sq. ft. (165) 2·659 ft., 3·159 ft.

208.—(166) $7\sqrt[3]{4}$ in. (167) $\frac{24}{5}\sqrt[3]{\frac{105}{11}}$ in. (168) 1396·825. (169) 488. (170) \$25·099. (171) 16·165 ft., 113·16 sq. ft. (172) 63 ft., 45 ft. (173) $22\frac{1}{2}$. (174) $28\frac{4}{7}$. (175) $553\frac{1}{7}$. (176) $\frac{12}{5}\sqrt[3]{60}$ in. (177) $\frac{5}{4}\sqrt[3]{1540}$ in.

209.—(178) 3748·075. (179) 3466145·382. (180) 792·261 cu. in. (181) 12·728 lbs (182) $11407·\dot{4}0\dot{7}$ oz (183) 25 (184) 49 in. (185) 2799·107. (186) 44·9. (187) $3943\frac{1}{3}$ square feet. (188) 758·556 cu. in.

210.—(189) $\frac{176}{7}\sqrt[3]{18}$. (190) 28704·761 sq. ft. (191) 1390027. (192) 43·216. (193) 49. (194) \$59·15. (195) 48 sq. ft., 105 sq. in. (196) $278\frac{7}{15}$. (197) 75 ft. (198) 973·531

211.—(199) 1896250 (200) 59·37 in., etc. (201) 15 ft., 36·496 ft. (202) 6·928 in. (203) 5·55. (204) $\sqrt[3]{\frac{96}{5025}}$ ft. (206) 25.

212.—(1) A \$30, B \$29·70. (2) \$1648. (3) \$325 (4) \$144, \$180, \$202·50 (5) $48\frac{2}{3}$c. (6) 196. (7) $\$133·87\frac{1}{2}$. (8) \$4383. (9) \$3·92 (10) A \$75, B \$126 (11) $\frac{1}{3}$ distance from starting point (12) $69\frac{1}{3}$c. (13) $20\frac{20}{23}'$ past 4. (14) \$54·25 \$71·75. (15) 26 lbs. (16) \$180, \$240, \$300. (17) \$270 (18) \$500. (19) 25 (20) $20\frac{20}{23}'$ past 5. (21) $\$9·18\frac{3}{4}$. (22) A \$69, B \$249. (23) $\frac{11}{20}$. (24) $6\frac{15}{37}$ days (25) $6\frac{1}{4}$c. loss. (26) $5\frac{1}{2}$. (27) $\frac{5}{7}$. (28) A \$900, B \$1800. (29) \$78. (30) $4\frac{2}{5}$.

214.—(1) $49\frac{1}{3}$c. (2) $5\frac{5}{9}$ sq. yds. (3) A \$12, B \$10 (4) 18′ past 3 (5) B pay A \$16. (6) 35. (7) $2\frac{1}{4}$ yrs. (8) \$2576·25 (9) \$1015, \$1421, \$2233. (10) \$22·50 (11) $\$14·28\frac{4}{7}$ (12) A \$6, B \$8, C \$17·50. (13) \$40. (14) \$891. (15) \$2826, \$2355. (16) 32. (17) A \$8·40, B \$7, C \$6. (18) \$217·50. (19) \$3485. (20) $27\frac{3}{11}'$ past 5. (21) \$1250. (22) \$52·25 loss. (23) 5·25. (24) $41\frac{17}{24}$c. (25) \$4. (26) \$240, \$192, \$180. (27) A \$550, B \$371·25. (28) 12c.

216.—(1) $5\frac{1}{3}$ days. (2) $\$4·85\frac{9}{25}$. (3) \$180. (4) A \$240, B \$210. (5) $\$140·41\frac{2}{3}$. (6) \$2517·50. (7) \$348. (8) \$65·40. (9) A \$8, B \$14, C \$20. (10) A \$49, B \$320. (11) \$6·44. (12) 720. (13) B $2\frac{2}{9}$ yards. (14) \$29·16. (15) 84.

PAGE.

216.—(16) $588·24. (17) 1000000. (18) $1000c. (19) 75, 60. (20) $643·50. (21) 25. (22) 8. (23) 21, 28. 42. (24) $5\frac{3}{5}$. (25) 40.

218.—(1) $21\frac{12}{13}$, $17\frac{2}{7}$. (2) $90. (3) $1980. (4) A $9·10, B $9. (5) A $128, B $120. (6) $24\frac{4}{9}$. (7) $2\frac{1}{2}$ loss. (8) ·0000315. (9) $3432. (10) 40 rods. (11) $8874. (12) $\frac{399}{497}$′ past 7. (13) A $840, B $882. (14) $3000. (15) $9\frac{81}{191}$. (16) $80\sqrt{2}$ rods. (17) $4\frac{1}{6}$. (18) $2600. (19) $17\frac{1}{47}$. (20) $41\frac{2}{3}$. (21) $11\frac{1}{9}$. (22) $20. (23) $875. (24) $32\frac{8}{11}$′ past 6.

220—(1) $\frac{1}{2}$. (2) 160 rods. (3) $674 (4) $152. (5) $4\frac{1}{3}$%. (6) $18. (7) $3. (8) $18·57$\frac{1}{7}$. (9) 4. (10) B 10 yds. (11) $6·01$\frac{8}{25}$. (12) $28·80. (13) $670. (14) $8·50 (15) $52\frac{8}{21}$. (16) A $12, B $24, C $26. (17) $41\frac{2}{3}$. (18) $3128, $1552. (19) $42\frac{6}{7}$. (20) $3·33$\frac{1}{3}$. (21) $102. (22) $\frac{1}{4}$ loss. (23) $7\frac{1}{3}$ days. (24) 26. (25) $1·19$\frac{13}{25}$. (26) $51\frac{2}{3}$.

222.—(1) $23\frac{3}{7}$. (2) $69·33$\frac{1}{3}$. (3) $55. (4) $2733·60 (5) $3331·12$\frac{1}{2}$. (6) $145·71$\frac{3}{7}$. (7) A $418·96$\frac{1}{4}$. (8) $1·06$\frac{2}{3}$. (9) 6% to 8. (10) $5180 (11) $34·046. (12) 30, 20 rods. (13) $2283·75. (14) $100 loss. (15) $31\frac{1}{4}$. (16) $540. (17) $5600 (18) 7. (19) A $25·50. (20) $11887·50. (21) $12\frac{1}{2}$. (22) 75c. (23) $12400. (24) $1389·60.

224.—(1) A $900. (2) 100 rods. (3) 96. (4) $3·33$\frac{1}{3}$. (5) 40 @ $7. (6) 136 lbs. hind quarter. (7) $998·40. (8) $300 @ 5%. (9) $19023·21$\frac{3}{7}$. (10) $1. (11) $7\frac{1}{4}$ loss. (12) $6\frac{1}{4}$. (13) $3\frac{1}{3}$′, $\frac{1}{2}$ way round. (14) 12. (15) $320, 6. (16) 230 rods. (17) $18·42. (18) $\frac{1}{2}$ way round. (19) $9600. (20) $26\frac{1}{2}$. (21) 60 @ $6.

226.—(1) $3571·42$\frac{6}{7}$. (2) A $480. (3) 54. (4) A $30·80. (5) 6\frac{2}{3}$. (6) $28·50. (7) 50 lbs. 56c. (8) $80. (9) 12000. (10) 40. (11) $41·903. (12) $11069·767. (13) $574·22. (14) 71·8c. (15) $600000. (16) $7·02. (17) $5\frac{1865}{2187}$ gal. (18) $28\frac{2}{3}$. (19) 43c. (20) 4·30 p.m (21) $30000. (22) 4, $600.

228.—(1) $33\frac{1}{3}$. (2) $94\frac{2}{7}$. (3) $5·62$\frac{1}{2}$. (4) $4800. (5) $5\frac{5}{11}$′ past 1. (6) $300 @ 6%. (7) A $28. (8) $2 gain. (9) $55\frac{1}{2}$. (10) $119. (11) $6750, $7200, $8000. (12) $2·50. (13) 180, 135, 108 days. (14) $150. (15) $45\frac{9}{11}$ sq. ft. (16) $935. (17) 50. (18) $451·20, $12. (19) $3\frac{1}{3}$ miles per hr. (20) 225. (21) $\frac{59}{73}$′ past 5.

230.—(1) $1656. (2) $1270. (3) 79. (4) 28. (5) $2·674. (6) $33·44. (7) $262·50. (8) 18. (9) 112. (10) 209·592 sq. ft. (11) $5\frac{2}{7}$. (12) 5·61. (13) $338.89. (14) $400. (15) 4 loss. (16) 5 miles per hr. (17) 50. (18) $37\frac{13}{15}$. (19) 360 (20) $38\frac{1}{4}$″. (21) $24\frac{3}{5}$ loss. (22) $450. (23) A $20.

232.—(1) 60. (2) 25. (3) $14·88. (4) 150. (5) 320. (6) £8750. (7) $\frac{549}{697}$′ past 3. (8) $66\frac{1}{4}$. (9) $1666·66$\frac{2}{3}$. (10) 108 yds. (11) 5·86c. (12) A $504. (13) $3965. (14) 333\frac{1}{3}$. (15) 10 horses. (16) $1000. (17) A $70. (18) 6 @ 10c. (19) 8. (20) $240\frac{1}{4}$ sq. yds. (21) $11\frac{1}{9}$. (22) $21\frac{9}{11}$′ past 3. (23) 11·49 rods. (24) $9000.

PAGE.

234.—(1) $\frac{1}{2}$. (2) \$1542·85$\frac{5}{7}$. (3) A \$840. (4) 3, 2. (5) Aug. 16. (6) 50, 25. (7) 25. (8) \$548·457. (9) 16$\frac{2}{3}$. (10) \$2838·75. (11) $7\frac{20}{127}$′ past 12. (12) \$4000. (13) 2$\frac{2}{3}$c. (14) \$878·87$\frac{1}{2}$. (15) March 26. (16) \$294. (17) 17 mills. (18) 5$\frac{1}{2}$ yrs., \$420. (19) \$9. (20) 40. (21) \$3. (22) 1, 5. (23) 9.

236.—(1) 175. (2) \$124·368. (3) 4$\frac{1}{2}$'s. (4) 16$\frac{2}{3}$. (5) $7\frac{3}{5}$′ past 12. (6) 5″. (7) 33. (8) A \$345·60. (9) \$27924·48. (10) \$3000, 6$\frac{2}{3}$. (11) 30 geese. (12) \$500. (13) \$248·64. (14) 1. (15) \$9483·12$\frac{1}{2}$. (16) \$14960·62$\frac{1}{2}$. (17) \$15300. (18) 28 men. (19) 24 miles. (20) \$67·80. (21) 74$\frac{2}{9}$.

238.—(1) A \$13. (2) 20$\frac{4}{7}$. (3) 24$\frac{121}{126}$. (4) $\frac{2}{3}$. (5) B \$2250. (6) 50, 35. (7) \$3600. (8) \$463 @ 4 %. (9) \$4000. (10) \$9·61$\frac{7}{13}$. (11) A 84 days. (12) 450. (13) 28$\frac{4}{7}$. (14) 35c. (15) \$96·385. (16) 6$\frac{2}{3}$. (17) $\frac{29}{31}$. (18) 1$\frac{1}{4}$. (19) A \$1714·28$\frac{4}{7}$. (20) 79$\frac{5}{6}$c.

240.— 1) \$3487·22. (2) 1372. (3) 55902 to 24565. (4) \$279·75 @ 6. (5) \$4315·78$\frac{8}{15}$. (6) 6. (7) 9 men. (8) 33. (9) 45$\frac{9}{11}$ hrs. (10) \$43·65 (11) \$4200. (12) \$66935067. (13). 30$\frac{7}{9}$c. (14) \$1·33$\frac{1}{3}$. (15) \$75. (16) \$2000. (17) 16$\frac{2}{3}$ gal. (18) \$45·36. (19) 5, 7. (20) 7.

242.—(1) \$1457·50. (2) \$2000. (3) \$829·637. (4) 123·582 cu. in. (5) \$1973684·21. (6) \$340·55. (7) \$1. (8) 8$\frac{4}{7}$. (9) 11478·582, 32·402. (10) \$472·50. (11) \$3400. (12) \$14400. (13) $\frac{25}{146}$. (14) 2. (15) 9. (16) 11$\frac{5}{12}$ mo. (17) 15 bush. (18) \$900. (19) 31 men full time, 1 man half time. (20) 240. (21) \$1267·86.

244.--(1) \$918·021. (2) \$5·25. (3) $\frac{5}{6}$ gal. (4) 1$\frac{13}{32}$. (5) \$105·263, (6) 164$\frac{1}{16}$. (7) 20. (8) 41. (9) \$3125. (10 \$1232. &c. (11) \$25·92. (12) 18 mills. (13) 910 alloy, 80 tin, 102 lead. (14) Dec. 3, '90. (15) 20·247 sq. ft. (16) \$500000. 17. 2$\frac{1}{2}$. (18) 37 in. (19) \$2000, &c. (20) 24 lbs. @ 42, &c.

246.—(1) 8$\frac{1}{2}$ % prem. (2) \$250. (3) 69$\frac{7}{8}$. (4) 3. (5) \$160. (6) \$15000. (7) \$1008, &c. (8) \$7·50. (9) 6c. (10) \$250, 6. (11) 8$\frac{16}{25}$ mo. (12) 65. 13. 4. (14) \$20·80. (15) 90 alloy, &c. (16) 4·795. (17) 65. (18) \$1491·856. (19) 3$\frac{11}{63}$. (20). \$6000.

248.—(1) \$2400. (2) 75c. corn. (3) 49 days. (4) \$325, 4. (5) 60 %. (6) 13$\frac{1}{3}$ mo. (7) 2 %. (8) \$11087·52. (9) 1st by 11c. (10) 105. (11) \$19000. (12) \$1·19$\frac{49}{51}$. (13) \$4000. (14) A \$139·12, &c. (15) 12 days. (16) 44 nitre, &c. (17) \$230, 4$\frac{1}{2}$ yrs. (18) 11$\frac{63}{67}$c. (19) \$400000.

250.--(1) 4$\frac{9}{20}$ mo. (2) 5, 1, 1, 3. (3) \$7·20. (4 \$1039·54. (5) 2$\frac{8}{11}$ ac. (6) 106$\frac{1}{4}$. (7) \$9·50. (8) 432 yds. (9) Br. \$725, &c. (10) 7$\frac{3}{5}$c. (11) \$22·50. (12) 1$\frac{1}{2}$. (13 10. (14) 50 yrs. (15) 300. (16) \$12·60. (17) \$45062·50. 18) 5. (19) 2592$\frac{4}{7}$ sq. ft. (20) $\frac{3}{8}$ gal.

PAGE.

252.—(1) 2 hrs. (2) $9000. (3) $2501·24$\frac{1}{9}$. (4) 94. (5) 5$\frac{1}{4}$. (6) $1681·35$\frac{1}{2}$, (7) $133·33$\frac{1}{3}$. (8) 29·56. (9) $17. (10) 3$\frac{9}{17}$ days. (11) $168·80. (12) 78 w. (13) 2 for 1 penny. (14) 1$\frac{1}{2}$% premium. (15) $655·36. (16) A $600. (17) 240 m. (18) $472·50. (19) $6979·50.

254.—(1) 42, &c. (2) 2265·3. (3) 3600. (4) 9$\frac{3}{13}$′ past 4. (5) 21·213 in. (6) 858. (7) 6. (8) $116. (9) 4. (10) $40, &c. (11) 36$\frac{1}{3}$ in. (12) $6000. (13) $4900·50. (15) 60 w. (16) 132. (17) $2241·28. (18) $4600. (19) $173·85. (20) $227·50. (21) A $3381·75.

256.—(1) 2$\frac{11}{64}$ g. (2) 4$\frac{1}{2}$ miles, &c. (3) 236·16 sq. in. (4) $28938·24. (5) 25′, 12·022″ past 9. (6) $126·75. (7) $11·33. (8). 81. (9) $\frac{6}{25}$. (10) A $444·44$\frac{4}{9}$. (11) $2436·66$\frac{2}{3}$. (12) $10·50. (13) 49$\frac{1}{3}$. (14) 5·679. (15) 5 w. (17) $1856·72$\frac{1}{2}$. (18) 8.

258.—(1) 4. (2) 80. (3) 3$\frac{1}{2}$ mills. (4) $3587·50 @ 8, &c. (5) $10000. (6) 1 ac., 55 sq. rods, 14 sq. yds., 6 sq. ft., 132 sq. in. (7) 1334$\frac{3}{8}$ lbs. (8) $\frac{5}{17}$ mile. (9) 19′, 36″. (10) $3000. (11) 19$\frac{5}{7}$ min. (12) $10557·50. (13) 127.$\frac{3695}{4633}$. (14) A. $35·43$\frac{3}{4}$. (15) $720. (16) $33333·33$\frac{1}{3}$. (17) 40 miles, &c. (18) 3 to 2. (19) 4$\frac{10}{11}$ mo.

260.—(1) 26′,40″. (2) 1·232 ′. (3) 36. (4) 14$\frac{2}{3}$ yrs. (5) 12′ to 5. (6) 7$\frac{1}{2}$. (7) $4950. (8) $734·40. (9) 72′. (10) 2s. 11·51d. (11) A 6 mo., &c. (12) 2·16 p.m. July 6. (13) $\frac{1}{4}$. (14) $8·60. (15) $3·50. (16) 7$\frac{11}{13}$ miles. (17) $400, &c. (18) $888·88$\frac{8}{9}$. (19) 8$\frac{16}{123}$. (20) $\frac{500}{891}$ oz.

262.—(1) $\frac{32}{243}$. (2) 7·26 a.m., Oct. 8. (3) 13$\frac{1}{3}$ days. (4) $4112·24. (5) 46$\frac{2}{3}$. (6) $3, $2. (7) $1120, &c. (8) $360. (9) A. $3612·50. (10) $\frac{192}{160}$. (11) 21$\frac{7}{8}$ c. (12) 9·13 a.m. (13) 45. 30. (14) $6·19$\frac{1}{8}$. (15) 2$\frac{1}{3}$. (16) 60 to 31. (17) A. 7$\frac{1}{2}$ mo. (18) $3200. (19) $325.

264.—(1) 17$\frac{3}{16}$. (2) $144 (3) 5′, 36″ to 8 a.m. (4) 2255\frac{5}{9}$. (5) $7·46. (6) $7880·299. (7) A. $3666·66$\frac{2}{3}$. (8) A. 1 yd. (9) $1·34. (10) $699·85$\frac{3}{5}$. (11) $1720. (12) $273·40. (13) 17$\frac{1}{7}$. (14) 7$\frac{1}{20}$. (15) 22·18$\frac{1}{3}$ (16) 2 p.m. (17) $340 nearly. (18) $1·64$\frac{1}{2}$. (19) 4000. (20) 273.

266.—(1) A. 4$\frac{2}{3}$ hrs., dist. 25$\frac{1}{3}$ miles. (2) 25·13c. (3) $570. (4) $175. (5) 12·35 a.m. (6) 327. 109. (7) $2·46. (8) 64870 lbs., 2 oz (9) 40$\frac{7}{11}$ days. (10) $350. (11) 4. (12) $13800. (13) $2482·165. (14) $81·60, 9$\frac{1}{2}$ mo. (15) $855. &c. (16) 384 sq. in. (17) 4 mo. (18) 12$\frac{1}{2}$ mills. (19) 10·37 p.m. (20) 14$\frac{7}{12}$.

268.—(1) 30. (2) 34$\frac{1}{2}$c. (3) $11·20. (4) $2250. (5) $300, &c. (6) 3$\frac{22}{32}$ (7) 36, 45. (8) $\frac{100}{481}$ (9) 9·02 p.m. (10) 50. (11) $6168·36$\frac{36}{49}$. (12) $150. (13) 50. (14) 6$\frac{2}{3}$. (15) $1787·50. (16) $4000. (17) 26$\frac{1}{3}$. (18) 28$\frac{4}{5}$. (19) $756.

270.—(1) 10·63 in. (2) 375$\frac{3}{5}$″. (3) $6·06$\frac{1}{4}$. (4) $18. (5) A $1443·75. (6) 50. (7) $730. (8) $10489·45. (9) 53·4375c. (10) 78$\frac{1}{8}$c.

PAGE.

270.—(11) \$9·75. (12) \$9107. (13) 4·375 in. (14) \$11200. (15) 4. (16) $35\frac{24}{31}$. (17) \$1404·90, &c. (18) \$26666·66$\frac{2}{3}$. (19) 38. (20) 2 miles. (21) 12.

272.—(1) \$1000, &c. (2) 20. (3) 40. (4) $\frac{1}{160}$. (5) 21, &c. (6) \$450. (7) \$3. (8) $8\frac{13}{22}$ cu. ft. (9) $16\frac{2}{3}$. (10) 45. (11) 40″, 20 yds. from starting point. (12) $\frac{240}{1427}$′ past 4. (13) 17·46c. (14) 15, &c. (15) 15 miles from Toronto. (16) \$30. (17) 92c. (18) 110. (19) $83\frac{1}{3}$c. (20) $1\frac{23}{57}$′ past 12. (21) $\frac{1}{3}$ dist. from starting point. (22) 10. (23) $2\frac{1}{16}$, &c. (24) $14\frac{2}{5}$.

274.—(1) B $\frac{5}{19}$ yd. (2) \$300. (3) \$750. (4) \$333·33$\frac{1}{3}$. (5) $40\frac{229}{480}$ miles. (6) 7, 50's, &c. (7) $8\frac{41}{48}$. (8) A \$7·05. (9) $17\frac{11}{17}$. (10) \$875. (11) \$5810. (12) $81\frac{2}{5}$. (13) $4\frac{4}{15}$ miles. (14) \$1179. (15) $5\frac{1}{2}$ mo. (16) 42c., $4\frac{1}{2}$ lbs. black. (17) \$70·50. (18) \$500·48. (19) $129\frac{1}{4}$. (20) 9·54. (21) $3\frac{1}{2}$ hrs. (22) $6\frac{1}{4}$.

276.—(1) 40 mi. (2) $22\frac{2}{13}$ mi. (3) A \$7, &c. (4) $228\frac{4}{7}$. (5) \$44·27. (6) 7600 π sq. ft. (7) 14. (8) \$40. (9) 3 to 2. (10) $37\frac{11}{12}$. (11) $7\frac{1}{7}$. (12) 7 +. (13) $17\frac{6}{7}$. (14) \$25. (15) 5. (16) 540 mi. (17) A \$225. (18) \$350., 5. (19) $7\frac{11}{23}$. (20) \$11·06. (21) $3\frac{1}{3}$ hrs. (22) 1600.

278.—(1) A 23. (2) $8\frac{1}{3}$. (3) $5802\frac{6}{7}$. (4) 5's. (5) \$10·73. (6) $5\frac{3}{5}$ hrs. (7) 1650 lbs. (8) 60c. (9) $4\frac{46}{71}$, $\frac{55}{71}$. (10) $19\frac{41}{241}$′ to 8. (11) \$9333. (12) 8·93. (13) \$16000, &c. (14) 10·261. (15) $61\frac{2}{3}$ rods clay. (16) $\$1\frac{12}{13}$. (17) 61 @ \$10. (18) $17\frac{1}{7}$′ past 12. (19) $28\frac{1}{3}$. (20) \$1000. (21) \$190, &c. (22) \$1240. (23) $5\frac{19}{20}$ days.

280.—(1) \$180, &c. (2) 80c. tea, &c. (3) 17, 25's, &c. (4) 48′. (5) \$135·60. (6) \$56·25. (7) \$50. (8) \$73·91$\frac{1}{4}$. (9) 1 gal. 1 qt. $\frac{4}{7}$ pt. (10) \$1804·275. (11) \$151·87$\frac{1}{2}$. (12) 2. (13) $7\frac{32}{37}$ days. (14) $4\frac{32}{157}$′ past 11 a.m. (15) 1st. (16) A. \$61·25. (17) \$5859·37$\frac{1}{2}$. (18) $8\frac{1}{3}$c. (19) 6. (20) \$4800.

282.—(1) Inc. \$15·60. (2) \$390. (3) 33, 33, 7, 7. (4) $7\frac{1}{5}$. (5) \$1·55. (6) \$6·72. (7) $29\frac{11}{20}$ yds. (8) £4500. (9) \$49. (10) 68. (11) $4\frac{1}{2}$. (12) $55\frac{5}{9}$. (13) \$1320. (14) $73\frac{1}{3}$. (15) 72, 48, 24. (16) \$3950, \$50. (17) 165, 162. (18) 12 mi. (19) \$3·45.

284.—(1) $1\frac{1}{20}$ yds. (2) £3005. 4s. 3·6d. (3) $94\frac{377}{1417}$. (4) A. \$175. (5) \$547·50. (6) \$8166$\frac{2}{3}$. (7) \$2·50. (8) B pays A \$18·75. (9) 9. (10) \$43. (11) \$4700. (12) 9 to 10. (13) \$1. (14) 6 lbs. tea. (15) \$160. (16) $135\frac{5}{13}$ yds. (17) 10 mo. (18) A \$21·309. (19) \$120, &c. (20) man \$2, &c. (21) \$250.

286—(1) \$1005. (2) $8\frac{24}{125}$, \$1000. (3) 2, 2, 2, 11. (4) \$441·66$\frac{2}{3}$. (5) \$57·72$\frac{1}{2}$. (6) $10\frac{5}{12}$. (7) \$4048·40. (8) $34\frac{11}{21}$. (9) \$1·68$\frac{8}{19}$. (10) 2, 3, 5. (11) A \$23, &c. (12) \$5000. (13) 45. (14) \$37·50. (15) A \$300. (16) \$900. (17) \$2250. (18) A \$646·434. (19) \$7760. (20) A \$4·78$\frac{1}{2}$.

288.—(1) \$100. (2) $7\frac{41}{107}$. (3) 1200. (4) $11\frac{2}{3}$c. (5) $44\frac{1}{4}$ days. (6) A \$234. (7) 2700 sq. yds. (8) \$8842·68. (9) $4\frac{296}{2401}$.

PAGE.

288.—(10) 30 men. (11) $83\frac{3}{4}$. (12) \$5000. (13) \$2·20. (14) 6. (15) \$1600. (16) A 44 yrs. (17) $2\frac{5}{8}$. (18) 20. (19) $18\frac{6}{65}$ days. (20) 2 to 5. (21) A \$$209·30\frac{10}{43}$. (22) 136.

290.—(1) $15\frac{1}{5}$. (2) \$9. (3) \$1462·50. (4) 7 oxen. (5) \$356. (6) $113636\frac{4}{11}$ oz. (7) 8. (8) \$1000. (9) 25. (10) $20\frac{24}{37}$. (11) 6 centres. (12) A \$480. (13) man 90 days, etc. (14) \$9·59. (15) \$27·50. (16) \$$67·32\frac{2}{3}$. (17) \$17·50. (18) 27 bush. (19) Draft on Toronto, 10c.

292.—(1) \$4·50. (2) $3\frac{33}{37}$ days. (3) $18\frac{3}{4}$. (4) $540\frac{10}{11}$ tons. (5) \$$333·33\frac{1}{3}$. (6) \$375·50, &c. (7) 16000. (8) A \$25. (9) \$560. (10) \$2940. (11) \$500. (12) $12\frac{1}{2}$. (13) $8\frac{3}{4}$ gal (14) \$625. (15) 11 oxen. (16) 8. (17) 40. (18) \$689·92. (19) \$1725·22. (20) $42\frac{80}{101}$. (21) $11\frac{1}{5}$%. (22) A \$956 97.

294.—(1) \$17080. (2) $5\frac{3}{5}$ (3) B. $8\frac{8}{99}$ yds (4) \$17550. (5) \$15·65. (6) $\frac{9}{28}$. (7) $10\frac{10}{13}$ hrs. (8) 10 (9) 40. (10) Midway between them, $4\frac{1}{2}$. (11) \$1256·25. (12) \$8983·50. (13) \$6250. (14) \$142180. (15) 7. (16) \$862·50. (17) $5\frac{1}{12}$. (18) $64\frac{27}{32}$c. (19) $4\frac{156}{161}$ black.

296.—(1) \$250. (2) \$10·89. (3) $83\frac{1}{3}$c. (4) $2\frac{11}{12}$. (5) \$694·72 (6) 6 days. (7) \$1000. (8) \$900 (9) \$840. (10) 58 gal. (11) \$142·57. (12) 50c, &c. (13) $22\frac{1}{12}$ miles. (14) \$18·18. (15) $7\frac{31}{35}$. (16) $42\frac{1}{22}$. (17) \$1402·12. (18) \$7. (19) Midway between them 5 miles, (20) A. \$55.

298.—(1) \$830·22 loss. (2) $107\frac{9}{25}$ (3) \$$4444\frac{4}{9}$, &c. (4) 5 (5) 18000. (6) $1\frac{3}{7}$ gal. A., &c, (7) 22·4 in. (8) 1st. (9) \$10 80. (10) £21. 16s 1d. (11) 40c. (12) 5. (13) 60c. (14) $7\frac{7}{17}$. (15) \$1134. (16) \$795·45. (17) 384.

300. (1) $·08\dot{3}$. (2) $7\frac{1}{2}$c. (3) \$10. (4) \$117. (5) \$1773·23. (6) \$10000. (7) 90. (8) 25. (9) 4. (10) $12\frac{28}{31}$. (11) \$22400 (12) \$9600. (13) 40·114 oz. (14) 315 (15) 23 gal., 1 qt. (16) \$39 92. (17) 6·539 (18) Paris. (19) \$2519·38.

302.—(2) \$$3·43\frac{1}{3}$. (3) \$4000. (4) \$9800. (5) 8. (6) \$18·40. (7) £11. 5s. (8) $43\frac{3}{4}$. (9) 275. (10) Child \$1920·60. (11) 20. (12) 45', $38\frac{68}{89}$". (13) 6·647. (14) latter. (15) \$3200. 15 mo. (16) 5. (17) \$2000000. (18) \$$133·33\frac{1}{3}$. (19) $3\frac{1}{4}$c.

308.—(64) 6. (65) 12.

311.—(100) 19. (101) 27.

312.—(102) 651. (103) 1865. (105) 1331. (106) 343340. (107) 26375. (108) 1534. (109) 18537. (110) 2086. (111) 5662074. (112) 14643006. (113) 320001. (114) 12. (115) 3030: rem. 102, (117) 1, 10, 11, 100, 101, 110, 111, 1000, 1001, 1010. (120) $32+16+2+1=51$.
(121) $3^6-3^5-3^4+3^3-3^2+3+1=427$.

315. (3) $10\frac{12}{13}$ (5) Nearly 83.

316 (7) A \$$36\frac{2}{3}$, B \$$18\frac{2}{3}$, 4 : 3. (8) $7315\frac{1}{3}$ ft. (9) $1\frac{1}{19}$ yr.

317. (12) \$189·93. (13) 7·52, 7·38. (14) \$973·78.

318.—(16) \$20·88. (17) 50c. Gold basis, 30c. currency.
(18) 22·23 lb. Av.

PAGE.

319.—(19) 36·2. (20) A £126$\frac{3}{41}$. (22) £20. (24) 97$\frac{1}{2}$c.

320.—(25) 1·7320508075 68. (26) £168100. (27) 895·7. (28) $650, $647·91. (29) $52·67 gain. (30) 6$\frac{1}{4}$ loss. (31) 4, 6$\frac{2}{5}$. (32) $11000, $5500.

321.—(33) 6 in. (34) 1321·4530, 7733·7204, 29910·7176. (35) $212·97. (36) 1·326, 2·652, 3·979 ft. (37) 9·2 per cent. (39) $131·04 inc.

322.—(41) N. S. Cy., $14·48. (42) Oats $2400, &c. (43) Aug. 28. (44) $26·58. (45) $25000, $50000. (46) 2 : 15$\frac{135}{143}$.

323.—(47) 7$\frac{58}{107}$, $952·04. (48) 7$\frac{3}{56}$. (49) 7 yr. (50) 4. (51) 16·28864. (52) 8 m., 6 m.

324—(53) 445$\frac{1}{2}$ min. (54) $300. (55) 3 m. (56) $583·62. (57) 22·7 m. (58) 8$\frac{1}{3}$.

325.—(59) 3$\frac{2}{9}$ hr. (60) 8 ft. (61) 20 lb. 1 oz. 5 dwt. 6·5 gr. (62) 35 m. (63) 20·45 m.

326.—(64) $24·20. (65) $800, $11987·50. (66) 13$\frac{8}{9}$. (67) 5 m., 70 min. (68) 19 ft. 6 in. (69) $207. (70) 117$\frac{2931}{5332}$.

327.—(72) $32. (73) $17530·56. (74) $346·64. (75) $603. (76) 76$\frac{12}{13}$. (77) 30, 36, 45. (79) £5 inc. (80) 9 m. (81) 30 loss.

328.—(82) $13077·67. (83) £7 10s. (84) £7 10s. 7$\frac{1}{2}$d. (85) 7$\frac{17}{81}$ c. (86) 400. (87) A 40, B 60 d. (88) 339·7 grs. (89) 12·61. (90) 6$\frac{1}{2}$ %, 3$\frac{1}{13}$.

329.—(91) 5·17 %. (92) 314050, 9064800. (93) $948·015. (94) $1104·08. (95) 12·37. (96) 8·9. (97) $\frac{bd}{c+d}$ hr., $\frac{bc}{c+d}$ hr.; $\frac{1}{2}\frac{a}{b-d}(c^2-d^2)$ m. per. hr. (98) 2079 min. (99) $317·86.

330.—(100) 50 m., 30 m. per hr. (101) 80 lit., 80 kil. (102) 2·86$\dot{3}$ oz., 2·304 c. in., 2·1475. (104) $31.25. (106) $91·80, 8$\frac{7}{31}$ %. (107) 4 min. (108) 6$\frac{1}{4}$ nearly, $2796·82. (109) 1090 ft. per sec.

331.—(110) £39 12s. 1d., $198·02, 999·21 fr. (111) 4:5$\frac{5}{11}$, 4:38$\frac{2}{11}$. (112) Lead 36·35, cork 43·64 lbs. (113) 2·004353564672 gal (114) $2046·99. (115) 25c., 12$\frac{1}{2}$c. (117) 101$\frac{5}{8}$. (118) 70. (119) 2:21$\frac{9}{11}$.

332.—(120) 480000199999. (121) 4·005.... (122) 5 m per hr. (124) 5063·65. (125) £19$\frac{229}{673}$. (126) 10. (127) 65$\frac{55}{63}$. (128) 16$\frac{7}{8}$ ft

333.—(129) 428·787. (130) 11$\frac{5}{17}$. (131) $212·71. (132) 4 in. nearly. (133) 8 ft., 7 ft. (134) 4:3:1:5 by wt. (135) 720.

334.—(137) 113$\frac{1}{9}$, 48. (138) 47. (143) 18$\frac{31}{33}$. (144) 11·632 ft. (145) £500000, £54000. (146) 476. (147) $10000.

335.—(148) $2250, $2000, $1750. (149) 2400, 80c. (150) 35$\frac{35}{47}$. (151) $22047187·50. (152) $87·53. (153) 65. (154) 12000 oz., 99680 oz.

336.—(156) 8 hr. 10′ 10·4″. (157) 2057, 833. (158) 46·5 per cent. (159) $29·56. (160) 10·23 loss. (161) 1·942. (162) 9 m. 660 yd., 9 m. 1035 yd.

PAGE.

337.—(163) $4\frac{2}{3}$. (164) \$5333$\frac{1}{3}$, $9\frac{3}{5}$ m. (165) \$511. (166) 18, 24 ft. (167) 6 m per hr. (168) \$7995.06. (169) \$900.

338.—(170) \$30·91$\frac{2}{3}$. (171) 14. (172) $\frac{12}{11}$(5m + n) or $\frac{12}{11}$(5m − n) min. past m o'c. (173) 30 ft. (174) 18096$\frac{1}{4}$ ft. (176) $39\frac{79}{84}$. (177) \$1038·87.

339.—(178) R shares. (179) 13·05. (180) 80. (183) £1 = \$4·8665. (184) $3\frac{3}{7}$. (185) \$16·20.

340.—(186) \$1058·63. (187) U. S. 4's. (188) \$1625 per m. (189) 155785·9. (190) − 2,2. (191) 7,6 per cent. (192) 192000. (193) 7·92 in., 1414·2 links.

341.—(194) \$2400. (195) \$37·55. (196) \$545·75. (198) \$12166$\frac{2}{3}$. (199) \$17·49 inc. (200) $5\frac{5}{8}$ min. (201) \$2·09, \$2·20, \$2·31.

342.—(202) $\frac{23}{17}$ yr., $\frac{7}{26}$ yr. (203) \$120.

343.—(1) \$28512. (2) $7\frac{191}{550}$. (3) A \$2·10, B \$2·73, C \$·63, D \$7. (4) 800. (5) \$1827. (6) \$23·55. (7) \$2442$\frac{6}{7}$, 7 . (8) 36 yd. (9) 182 yd. (10) 225 lb.

344.—(2) $\frac{1}{5}$, 1·4142, $\frac{13}{150}$. (3) Even. (4) 20c. (5) \$242 ; A \$67$\frac{1}{3}$, B \$64$\frac{8}{9}$, C \$109$\frac{7}{9}$. (6) \$7·50 less. (7) Across, $6\frac{19}{27}$ yd. (8) $37\frac{1}{2}$.

345.—(9) $\$2\frac{18}{35}$.
(1) 1. (2) 1001 yd. (3) \$405. (4) 20 yd. long. (5) Each child \$1. (6) 35 25-cent pieces. (7) $82\frac{7}{24}$. (8) 1313, 2121. (9) 7, \$1128.

346.—(10) 3:7.
(1) £5 15s. 8$\frac{1}{4}$d., ·68589743. (2) £9 7s. 11·1606d. (3) 10·2. (4) $1\frac{1}{209}$. (5) 11:2:1. (6) 100, 300, 600. (7) 2:21$\frac{9}{11}$ p.m. (8) 800.

347.—(9) 365·2425d. (10) $\sqrt[3]{319}$: $\sqrt[3]{350}$.
(1) $\frac{17}{1650}$. (2) 8s. 9d. (3) \$22.50, \$37.50. (4) 4s. 1$\frac{1}{2}$d. (5) $131\frac{5}{18}$ m. (6) $\frac{199}{27}$. (7) 4·86 . . . (8) $\frac{1999}{6323}$ oz.

348.—(9) 516·2 oz. (10) 8 yd. (11) $17\frac{1}{9}$ in.
(2) \$4·60 nearly. (3) £2 17s. 4$\frac{5}{18}$d. (4) 63, 47. (5) \$2000.

349. (6) 480. (7) \$420. (8) A \$1785$\frac{75}{122}$, &c. (9) \$883·77 . . . (10) 40 lb. (11) $\sqrt{725}$.
(1) 46. (2) 249 m. (3) 4:24 p.m.

350.—(4) 21 hr. (5) 2 gal. (6) 8th, 15; 9th, 25; 10th, 35. (7) 22$\frac{2}{9}$c. (8) \$247·65. (9) 490·49.
(1) $\frac{1605}{3311}$. (2) \$12, \$22·50, \$67.50.

351.—(3) $5\frac{111}{253}$d (4) ·007916. (5) 80d. (6) £16251 7s. 9$\frac{3}{32}$d. (7) 12 yrs. (8) 1 to 92 of mix. (9) \$34000. (10) $110\sqrt{22}$. (11) ·9027027. (12) 14·3 min.
(1) $\frac{133}{514}$, $\frac{69126697}{725582180}$.

352. (2) 33, 66, 99, d. (3) \$67·60. (4) 7780·44. (5) 20 m., 32 w. (6) \$842·30, \$918·87, \$1598·83. (7) Tin 128·39. (8) 65. (9) $\frac{16}{29}$, $\frac{9}{29}$, $\frac{4}{29}$. (10) \$117·85.

353.—(11) 7c. (12) 4528.
(1) $\frac{11}{80}$, 10·1873. (2) 265d., 10. (3) 8 %. (4) 29 rods. (5) \$2472·94. (6) \$3045, \$812. (7) \$7·20.

PAGE.

354.—(8) A $1514·13. (9) $77000. (10) 6. (11) $2·20$\frac{2}{3}$. (12) 20.
(1) 10080, 110250. (2) £6 16s. 5d. (3) 3$\frac{1}{2}$. (4) 30720.

355.—(5) $\frac{1}{30}$ less. (6) 818 yd. (7) 234$\frac{2}{3}$. (8) 6 in. (9) $101·85.
(10) 6 m. (11) Aug. 24.

356.—(12) $112·09·
(1) 3$\frac{247}{252}$. (2) 43$\frac{1}{3}$ sec. (3) 21$\frac{3}{17}$ hrs. (4) $493·30. (5) $5600.
(6) 7. (7) 40. (8) 30$\frac{399}{1127}$ sec.

357.—(9) 23.
(1) 10·8. (2) 60 cts., 40 cts. (3) ·9027027. (4) 51$\frac{21}{29}$ cts.
(5) 15, 12, m. per hr. (6) 2·23608:1. (7) 20:21. (8) $2828·07.

358.—(9) 2205. (10) $2·69.
(1) $1000. (2) 336 ft. (3) 6·03. (4) $382·12. (5) 17·57 ft.

359.—(6) $3. (7) $5086·25. (8) 1350, 1200. (9) $46·76.
(10) 21 m. (11) 9, 6$\frac{1}{3}$, 3 ft.
(1) $60. (2) 3 in.

360.—(3) 39. (4) $445. (5) $\frac{5}{8}$ fr. starting pt. (6) 4096.
(7) 22500. (8) 4·6c. (9) 440.

361.—(1) ·$\dot{0}1098\dot{9}$. (2) $1255·37. (3) 16·635. (4) ·16025, 6 %.
(5) $1192·47. (6) $11420.54. (7) $15, $1·87$\frac{1}{2}$. (8) 8$\frac{18}{19}$%.
(9) $1400, $4100.

362.—(10) $121·05. (11) 818 yd. (12) 38$\frac{1}{2}$ sq. ft. (1) $9450.
(2) $51834·92. (3) A $15315·42. (4) ·447213, ·809016,
3·236067. (5) 7:2$\frac{8}{11}$ p.m.

363.—(7) 3$\frac{2}{3}$. (9) 56. (10) $3·86. (11) 75 at 34c., 69 at 37$\frac{1}{2}$c.
(12) 1884 $31250, &c.

364.—(13) 9. (1) $887·10. (3) $2110·14, 5$\frac{10}{13}$, 6$\frac{2}{23}$. (5) 1·3080.
(6) 20$\frac{349}{253}$. (7) 928096, 928125.

365.—(8) 66\frac{2}{3}$. (9) $3750·38. (10) 8$\frac{107}{156}$. (11) $157·50.
(12) A 4$\frac{22}{27}$, B 3$\frac{23}{27}$ m. per hr.

366.—(2) $5284. (3) The former. (4) 78·561. (5) 16·8 mills.
(6) 40′ 36″.

367.—(7) 27$\frac{2}{3}$. (8) 4554·36 grs. (9) ·4055, 16·3862. (10) 20$\frac{1}{2}$.
(11) $2534, $2455.

368.—(2) ·131870. (3) Inc., $10300. (4) 14·7 nearly.

369.—(5) 120 hrs., 240 hrs., no. (6) 9$\frac{9}{37}$. (7) 8$\frac{1}{3}$ hrs., $\frac{1}{3}$ round
in A's direction. (8) 2$\frac{971}{1045}$. (9) $\frac{3}{2}\sqrt[3]{4}$ in. (10) $30\sqrt{3}$.
(1) ·998.

370.—(2) 24·69 %. (3) 24 w., 18 l. (4) 347. (5) $\frac{1}{3}$ round.
(6) $357, 2$\frac{16}{17}$%. (8) 4$\frac{13}{16}$.

371.—(9) 39647$\frac{1}{7}$ sq. in., 919813$\frac{5}{7}$ c. in. (10) 6″.
(3) 2$\frac{31}{37}$, −1$\frac{17}{27}$. (5) 130 yd. (6) A 8 hr., B 10, etc.
(7) ·011401, 1·9615.

372.—(8) 50, 30. (9) 29$\frac{1}{8}$, 20$\frac{5}{12}$ ft. (10) 651·96 lb. (11) $125000.
(12) 235·52. (13) 9.

www.ingramcontent.com/pod-product-compliance
Lightning Source LLC
Chambersburg PA
CBHW030555240426
43664CB00048B/293

* 9 7 8 3 3 3 7 1 5 8 8 7 3 *